Springer Series in Synergetics

Editor: Hermann Haken

Synergetics, an interdisciplinary field of research, is concerned with the cooperation of individual parts of a system that produces macroscopic spatial, temporal or functional structures. It deals with deterministic as well as stochastic processes.

Volume 40 **Information and Self-Organization** A Macroscopic Approach to Complex Systems By H. Haken

Volume 41 **Propagation in Systems Far from Equilibrium** Editors: J.E. Wesfreid, H.R. Brand, P. Manneville, G. Albinet, N. Boccara

Volume 42 **Neural and Synergetic Computers** Editor: H. Haken

Volume 43 **Cooperative Dynamics in Complex Physical Systems** Editor: H. Takayama

Volume 44 **Optimal Structures in Heterogeneous Reaction Systems** Editor: P.J. Plath

Volume 45 **Synergetics of Cognition** Editors: H. Haken, M. Stadler

Volume 46 **Theories of Immune Networks** Editors: H. Atlan, I.R. Cohen

Volume 47 **Relative Information** Theories and Applications By G. Jumarie

Volume 48 **Dissipative Structures in Transport Processes and Combustion** Editor: D. Meinköhn

Volume 49 **Neuronal Cooperativity** Editor: J. Krüger

Volume 50 **Synergetic Computers and Cognition** A Top-Down Approach to Neural Nets By H. Haken

Volume 51 **Foundations of Synergetics I** Distributed Active Systems By A.S. Mikhailov

Volume 52 **Foundations of Synergetics II** Complex Patterns By A. Yu. Loskutov, A.S. Mikhailov

Volume 53 **Synergetic Economics** Time and Change in Nonlinear Economics By W.-B. Zhang

Volume 54 **Quantum Signatures of Chaos** By F. Haake

Volume 55 **Synergetics of Rythms** Editors: H. Haken, P. Koepchen

Volumes 1–39 are listed on the back inside cover

Neuronal Cooperativity

Editor: Jürgen Krüger

With 125 Figures

Springer-Verlag
Berlin Heidelberg New York
London Paris Tokyo
Hong Kong Barcelona
Budapest

Professor Dr. Jürgen Krüger

Abteilung Neurologie und Neurophysiologie
Klinikum der Albert-Ludwigs-Universität
Hansastraße 9
7800 Freiburg, Fed. Rep. of Germany

Series Editor:

Professor Dr. Dr. h. c. Hermann Haken

Institut für Theoretische Physik und Synergetik der Universität Stuttgart, Pfaffenwaldring 57/IV, 7000 Stuttgart 80, Fed. Rep. of Germany, and Center for Complex Systems, Florida Atlantic University, Boca Raton, FL 33431, USA

ISBN 3-540-53155-6 Springer-Verlag Berlin Heidelberg New York
ISBN 0-387-53155-6 Springer-Verlag New York Berlin Heidelberg

Library of Congress Cataloging-in-Publication Data

Neuronal cooperativity/Jürgen Krüger (ed.). p.cm. – (Springer series in synergetics; v. 49)
Includes bibliographical references.
ISBN 3-540-53155-6 (Berlin: acid-free paper).
ISBN 0-387-53155-6 (New York: acid-free paper)
1. Cerebral cortex. 2. Visual cortex. 3. Neurons. 4. Neural circuitry. I. Krüger, Jürgen. II. Series.

Typesetting: ASCO Trade Typesetting Limited, Hong Kong; Printing: Mercedes-Druck, Berlin; Bookbinding: Lüderitz & Bauer, Berlin
54/3020-543210 – Printed on acid-free paper

Preface

Soon after the technique of microelectrode recording had become a routine method in the 1950s, many researchers felt that an obvious next step was to investigate the behaviour of larger networks of interacting cells. However, despite many early efforts, very little progress was made in that direction: a conceptual basis had not yet evolved, and the computer as an analytical tool was not available. This has now changed, with the result that the cooperation of neurons is again becoming a focus of interest.

In 1987, with the support of the Volkswagen Foundation, I was able to organize a conference on the subject of neuronal cooperativity near Freiburg, W. Germany, bringing together theoreticians and experimentalists working in this field. It became clear that many new theoretical concepts had evolved since the early days but that experiments on the subject were still difficult.

Later, with the encouragement of Professor H. Haken, editor of the Springer Series in Synergetics, I decided to ask a number of experimentalists, many of whom had participated in the conference, to contribute to a book. The aim was not to focus exclusively on the newest results but rather to give an overall idea of how the difficult question of cooperativity can be approached experimentally. The main emphasis is on the physiological relevance. Thus the methodological details are for the most part considered briefly, although some of the experimental procedures are unusual and complicated. To my knowledge, there is no comparable collection of reports in the literature. This is in contrast to theoretical activity in the field, on which compact information is already available. I felt that this imbalance should be reduced by inviting contributions from researchers who investigate neuronal cooperativity mainly by experiment.

The volume is intended for all brain researchers interested in the levels intermediate between the cellular and the behavioural. Experimentalists not familiar with cooperative phenomena, and also theoreticians, are offered an overview of the still somewhat fragmentary state of the art. I hope that this volume will help to stimulate new research efforts in the fascinating field of neuronal cooperativity.

I express my gratitude to the Volkswagen Foundation for their generous support, which was more than a mere sum of money. I also thank H. Haken for his encouragement, and the staff of Springer-Verlag for their care in the preparation of this book.

Freiburg, February 1991 J. Krüger

Contents

1. Introduction
By J. Krüger 1

2. The Structure-Function Problem in Visual Cortical Circuitry Studied by Cross-Correlation Techniques and Multi-Channel Recordings
By K. Toyama (With 13 Figures) 5

2.1 An Approach to the Structure-Function Problem by Cross-Correlation Technique 5
2.2 Experimental Techniques for Cross-Correlation Analysis 6
2.3 Basic Neuronal Interaction Demonstrated by Cross-Correlation Analysis 7
2.4 Signal Transmission Demonstrated by Cross-Correlation Analysis 9
2.4.1 Geniculo-Cortical Transmission 9
2.4.2 Cortico-Geniculate Transmission 11
2.5 Intracortical Transmission Along a Single Functional Column or to Adjacent Functional Columns 12
2.5.1 Nonspecific Transmission of Geniculo-Cortical Excitation 12
2.5.2 Specific Transmission of Cortical Excitation 14
2.5.3 Specific Transmission of Cortical Inhibition 15
2.5.4 Contribution of Interactions to Visual Responses 16
2.6 Transmission to Distant Functional Columns 17
2.7 Functional Scheme of Visual Cortical Circuitry 18
2.8 Synaptic Interaction Studied by Spike-Triggered Averaging . . . 21
2.9 Multi-Channel Optical Recording of Neural Activity 24
2.10 Concluding Remarks 27
References 28

3. Optical Imaging of Neuronal Activity in the Living Brain
By R. D. Frostig, E. E. Lieke, A. Arieli, D. Y. Ts'o, R. Hildesheim, and A. Grinvald (With 12 Figures) 30

3.1 Why Optical Imaging? 30
3.2 Optical Imaging of Neuronal Activity with Voltage-Sensitive Dyes 31

3.3 Optical Imaging Based on Intrinsic Signals 33
3.4 Studies of the Organization and Interactions of Neuronal Populations in Vivo Utilizing Voltage-Sensitive Dyes 33
3.4.1 Imaging Evoked Responses 34
3.4.2 Surround Inhibition Revealed by Optical Imaging . . . 37
3.4.3 Selective Visualization of Neuronal Assemblies in Cat Visual Cortex 40
3.5 Optical Imaging of Activity Using Intrinsic Signals 42
3.5.1 Determining the Source and Nature of the Intrinsic Signals 42
3.5.2 Mapping of Functional Organization in Visual Cortex . . 43
3.5.3 Increasing the Spatial Resolution of Imaging with Intrinsic Signals 43
3.5.4 Imaging of Functional Organization through the Intact Dura and Thinned Skull Preparation 47
3.5.5 Revealing Inhibitory Interactions Between Adjacent Cortical Regions 48
3.6 Comparison of Optical Imaging With and Without Voltage-Sensitive Dyes 48
3.7 Conclusions 50
References 50

4. Dynamic Aspects of Neuronal Cooperativity: Fast Stimulus-Locked Modulations of Effective Connectivity
By Ad. M. H. J. Aertsen and G. L Gerstein (With 5 Figures) 52

4.1 Introduction 52
4.2 Computational Methods 54
4.2.1 Cross-Correlation Analysis 54
4.2.2 Joint Peri Stimulus Time Histogram 54
4.2.3 Gravitational Clustering 55
4.3 Results 56
4.3.1 General Results 56
4.3.2 Cat Area 17 57
4.3.3 Gravitational Clustering 57
4.3.4 Joint Peri Stimulus Time Histogram 59
4.3.5 Stimulus Coding by Correlated Firing 64
4.4 Discussion 65
References 66

5. Temporal Structure of Cortical Information Processing: Cortical Architecture, Oscillations and Non-Separability of Spatio-Temporal Receptive Field Organization
By H. R. O. Dinse, K. Krüger, and J. Best (With 15 Figures) 68

5.1 Introduction 68
5.1.1 Receptive Fields as Spatial Filters 68
5.1.2 Functional Impact of Cortical Architecture 69

5.1.3 A Network Model of Cortical Information Processing Incorporating Some Basic Aspects of Cortical Architecture 69
5.1.4 Nonlinear Behavior vs. a Complex Temporal Response Pattern 70
5.1.5 Hierarchical-Parallel Processing in the Cat Visual System 70
5.1.6 Synopsis 71
5.2 Temporal-Spatial Aspects of Receptive Field Organization . . . 71
5.2.1 Low-Frequency Oscillation in Neuronal Response Following Photic Stimulation 71
5.2.2 Dynamics of Receptive Field Organization 77
5.2.3 Impacts of Non-Separability for RF Organization 82
5.2.4 Evidence for the Use of a Time Code 86
5.2.5 Evidence for Positive Feedback 89
5.2.6 General Outline of a Theory of RF Organization 90
5.2.7 Implications of RFs with Non-Separable Spatio-Temporal Behavior for Processing of Moving Stimuli 92
5.2.8 Evidence For Parallel Processing Across Different Hierarchical Levels 96
5.3 Summary 97
5.3.1 Low-Frequency Oscillation of Neuron Response 97
5.3.2 Non-Separability of Spatial and Temporal Behavior . . . 98
5.3.3 Further Implications of the Oscillatory Response Pattern 98
5.3.4 The Origin of the Oscillatory Pattern 99
5.4 Perspectives – Anything is Temporal 99
References 101

6. Spike Train Correlations on Slow Time Scales in Monkey Visual Cortex
By J. Krüger (With 12 Figures) 105

6.1 Spike Train Correlations 107
6.2 Correlations in the 50 ms Range 108
6.3 Correlations in the Range Fractions of Seconds to Minutes ("Covariability") 115
6.4 Operation of the Visual System in the Presence of Correlated Signals 120
6.5 Identification of Stimuli from Neuronal Responses 123
6.6 Signal Contents and Signal Importance 129
References 131

7. Connectivity and Functional Organization in the Mammalian Visual Cortex
By D. Y. Ts'o (With 18 Figures) 133

7.1 Introduction 134
7.2 Horizontal Interactions Among Oriented Cells 139

7.3 Functional Organization and Connectivity in the Blob and Interblob Regions of Primate Striate Cortex 148
7.3.1 Receptive Field Properties of Blob Cells 151
7.3.2 Functional Organization of Color Opponency 152
7.3.3 Connectivity of Blob and Non-Blob Cells 156
7.4 Conclusions 162
References 163

8. The Formation of Cooperative Cells Assemblies in the Visual Cortex
By W. Singer (With 5 Figures) 165

8.1 Coherence of Feature as Criterion for Scene Segmentation . . . 165
8.1.1 Representation of Coherent Features by Assemblies of Coherently Active Neurones 166
8.1.2 Constraints on Connectivity 167
8.2 The Development of Binocular Correspondence as an Example for Experience Dependent Selection of Neuronal Connections . . 169
8.3 The Control of Local Circuit Modifications by Global Gating Systems 171
8.4 Involvement of a Molecular Coincidence Detector in Use-Dependent Plasticity 172
8.5 Cooperativity Between Distributed Neuronal Responses as Criterion for Local Circuit Modifications 173
8.6 Activity-Dependent Self-Organization at Higher Levels of Cortical Processing 175
8.7 Detection and Representation of Coherent Features by Synchronously Oscillating Cell Assemblies 177
References 181

9. Stimulus-Specific Synchronizations in the Visual Cortex: Linking of Local Features into Global Figures?
By R. Eckhorn (With 16 Figures) 184

9.1 Search for Concepts of Global Cortical Processing 185
9.1.1 Processing of Visual Information in a Distributed System 185
9.1.2 Pattern Recognition Requires "Early" Feature Linking Independent of Cognitive Aspects 186
9.1.3 Figure Segmentation Defined by Correlated Activities . . 186
9.1.4 Invariant Local and Flexible Global Pre-Processing . . . 187
9.2 Development of Experimental Techniques for the Analysis of Global Cortical Processing 188
9.2.1 Search for Automatic Grouping Mechanisms 188
9.2.2 Recording of Different Signal Types with Multiple Electrodes 188
9.2.3 Methods for Multiple-Channel Correlation Analysis Require Improvements 189
9.2.4 Difficulties with Single Cell Correlations in the Visual Cortex 190

9.2.5 Local Mass Signals Provide Relevant Information about Cortical Grouping Mechanisms 190
9.3 Local Properties of Stimulus-Induced Oscillations in the Visual Cortex 191
9.3.1 Stimulus-Specific Oscillations 191
9.3.2 Internal Temporal Structure of Stimulus-Induced Oscillations is not Stimulus-Correlated 192
9.3.3 Correlations Between LFPs and Single-Cell Spikes 194
9.3.4 Stimulus Dependencies of Oscillations: Local-Global Coding Relations 195
9.4 Stimulus-Induced Synchronizations in Spatially Separated Assemblies 197
9.4.1 General Results 197
9.4.2 Details About Stimulus-Induced Cortical Synchronizations 198
9.4.3 Synchronizations Between Two Cortical Areas 201
9.4.4 Other Stimulus-Specific Variations of Interareal Synchronizations 207
9.5 Mechanisms for Stimulus-Induced Synchronizations Between Remote Cortical Assemblies 210
9.5.1 Neurophysiological Arguments for Synchronized Local Oscillators 210
9.5.2 Arguments from Network Models for Synchronized Local Oscillators 212
9.5.3 Breaking a Link 214
9.5.4 Stimulus-Forced and Stimulus-Induced Synchronization 216
9.5.5 Linking Fields 217
9.6 Data from Awake Cats 218
References 219

10. Cross Correlation Analysis of Motoneuron Inputs in a Coordinated Motor Act
By P.A. Kirkwood and T A. Sears (With 7 Figures) 225

10.1 Patterns of Activity and of Motoneuron Synchronization . . . 226
10.2 The Primary Correlation Operator 232
10.3 Patterns of Input Connections to Intercostal Motoneurons . . 238
10.4 Direct Measurement of the Respiratory Input to Intercostal Motoneurons 240
10.5 Further Analysis of Motoneuron Synchronization 244
10.6 Conclusions 246
References 247

11. Correlated Activity of Neurons: A Neural Code for Higher Brain Functions?
By E. Vaadia, E. Ahissar, H. Bergman, and Y. Lavner (With 13 Figures) 249

11.1 Introduction 249
11.2 Functional Properties of Single Neurons 251
11.3 Firing Rates and Response Properties of Neighboring Neurons . 254
11.4 Cross-Correlation Analysis: Significance and Limitations . . . 258
11.5 Estimating the Efficacy of Neuronal Interactions 261
11.6 Cross-Correlation Analysis: Experimental Results 266
11.7 Dynamics of Neuronal Interactions 270
11.8 Concluding Discussion 275
References 276

12. Systems Neuronal Activity in Learning
By U. G. Gassanov (With 9 Figures) 280

12.1 Cortical Network Activity 281
12.2 Two Views on the Function of Cortical Neurons in Learning . . 283
12.3 Two Types of Cortical Neuronal Network in Learning 284
12.4 Specificity of Cortical Networks 292
12.5 Conclusion 295
References 296

Subject Index 299

1. Introduction

J. Krüger

Neurologische Universitätsklinik, Hansastr. 9, D-7800 Freiburg

Human beings consider themselves as "individuals". However, they relate their individuality not so much to their physical uniqueness as to the unique character of the thoughts and accumulated experiences in their heads. They proudly defend the inaccessibility of these, and derive from them the freedom of thought. On the other hand, they are intrigued by the functioning of the material substrate of their individuality. This question provides the major source of motivation for investigations of the nervous system. Thus it is not surprising that many more scientists are attracted by research into the brain and its functioning, than by studies of any other organ of the body. Man wants to inquire into his nature.

In his book "Man on his Nature" Sir Charles Sherrington (1941) gave a famous description of the subject of inquiry:

> "A scheme of lines and nodal points, gathered together at one end into a great ravelled knot, the brain, and at the other trailing off to a sort of stalk, the spinal cord. Imagine activity in this shown by little points of light. Of these some stationary flash rhythmically, faster or slower. Others are travelling points, streaming in serial trains at various speeds. [. . .] Suppose we choose the hour of deep sleep. Then only in some sparse and out of the way places are nodes flashing and trains of light-points running. Should we continue to watch the scheme we should observe after a time an impressive change which suddenly accrues. [. . .] The great topmost sheet of the mass [. . .] becomes now a sparkling field of rhythmic flashing points with trains of travelling sparks hurrying hither and thither. The brain is waking and with it the mind is returning. [. . .] Swiftly the head-mass becomes an enchanted loom where millions of flashing shuttles weave a dissolving pattern, always a meaningful pattern though never an abiding one; a shifting harmony of subpatterns. Now as the waking body rouses, subpatterns of this great harmony of activity stretch down into the unlit tracks of the stalk-piece ot the scheme. [. . .] This means that the body is up and rises to meet its waking day."

This description, fantastic though it may be, is remarkable in two respects: it does not focus on reactions of the brain to inputs but rather on its intrinsic signal content, and it stresses the role of the "harmony" of cellular activity as a determinant of behaviour. Sherrington saw very clearly what needed to be understood, and what would have to be observed in order to approach an understanding: it is

Springer Series in Synergetics
J. Krüger (Ed.)
Neuronal Cooperativity

the neuronal "harmony". Less poetically, today we call this neuronal cooperativity.

Thus, neuronal cooperativity is not just one of the many topics within the field of brain research, but it is the missing link between the events at the cellular level and the behaviour of the entire organism. To state this differently, the brain can be considered to be understood if the incoming sensory signals as well as the outgoing motor signals can be explained entirely in terms of internal signal patterns: A stimulus is not recognized because some neurones fire 100 ms after stimulus onset. Rather, since the instant of stimulus onset is not known to the brain, the recognition is due to a particular configuration of excitations that appears in the brain. Similarly, each movement is defined by a corresponding spatio-temporal excitation pattern. In large brains, it is often impossible to detect any reproducible relationship at all between internal patterns and peripheral events, and yet well-defined processing goes on. These processing events are what we refer to as higher brain functions. They are causally related to a host of earlier events in the outer world, some of which occurred in the very remote past. The link to the present is memory, and of course genetic information.

It is a huge task to try to understand higher brain functions but it is also the principal source of our inquisitiveness. First, it is necessary to specify how this goal can be approached experimentally. In the past, unfortunately this has often been avoided; Sherrington's description, written when he was 79, was interpreted as poetry presented by a grand old man, rather than a delineation of the problem. The general response (still frequent today) was a discouragement that gave rise to a host of subordinate research topics which were often declared to be prerequisites for a later, true understanding ("bottom-up" analysis). While subjecting a seemingly endless "bottom" to a close scrutiny one was hoping that somehow the way "up" might emerge on its own.

The present volume contains a series of chapters written by authors engaged in an active search for that way "up". It is the path that leads directly to the functional complexity of the nervous system.

The general scientific procedure for understanding a function is to identify material substrates, constituents, and interactions, and to define *types* of these. The subject of investigation are the signal contents (to which "meaning" is coupled) of a neural network and their evolution in time. Since many causal determinants are experimentally inaccessible because they lie in the past, and therefore cannot be applied repeatedly, the signal flow is essentially unique at any instant in time. Nonetheless, it has to be broken down into (and is likely to consist of) types of other entities ("subpatterns" in Sherrington's terminology) that can be observed regularly and repeatedly. In other words, sequences of patterns have to be translated into correlations, i.e. into neural and/or behavioural events that regularly occur together. The sequence of patterns depends on the way in which the neurons communicate with each other, and on the parameters that determine this cooperation, memory being the most important factor.

Our most elementary knowledge of neuronal cooperativity stems from neuroanatomy. This wide field is not directly treated in the present volume but important concepts, such as the hypercolumn and the cortical area, which appear in a number of chapters, are derived from it. Even temporal aspects of cortical processing can be understood by comparison with anatomy-based models, as exemplified in Chap. 5 by Dinse et al.

The next step is to observe directly how the activities of pairs of neurons are related to each other, or, in other words, how one neuron "responds" to activity received from another one. This is the domain of cross-correlating spike trains, i.e. of counting how often a spike due to one neuron is followed or accompanied by a spike in another neuron. Most chapters of the present book are concerned with this method, or with important variants in which one member of the correlated pair is an intracellular potential (Chap. 2) or a multineuronal potential (Chaps. 8–10). In the cortex, the first problems encountered are that the spike activity is low, and that the average neuron receives synaptic input from 1000–10 000 other neurons. Each of these inputs has a correspondingly weak effect on the firing of the receptor neuron so that spike pairs giving evidence of such connections may be extremely rare. Another difficulty is that all types of direct and indirect interactions between two neurons manifest themselves in a superimposed manner in the same correlogram. However, this can also be an advantage because otherwise no information at all could be gained on the importance of the many indirect connections existing in a network. Chapters 2, 6, 7, and 10 are based on essentially this type of reasoning. Beyond the individual findings reported here, lies the general message that under some conditions the correlational approach is a very sensitive tool: correlations occurring over rarely considered temporal and spatial scales have been found; but these have yet to be incorporated into existing theories.

The authors of other chapters go another step further: They investigate the dependences of interactions on stimulation and on time. Here the notorious problem of normalization becomes a crucial one. Related problems are treated by Aertsen et al. (Chap. 4) and their implications are demonstrated. The results of Singer and Eckhorn, presented in Chaps. 8 and 9 are highly intriguing: the authors independently claim that oscillatory correlations between different stimulus-evoked activities imply that the stimuli "belong together". Their findings go far beyond the classical stimulus–response relationships based on the reactions of single cells.

The sensory side of the brain attracts a good deal more research interest than the motor side. Thus there are fewer established facts concerning the functional requirements of a motor system. In chap. 10 some problems arising in the respiratory system are highlighted. It is the task of a neural network not only to *detect* input correlations but also to *produce* output correlations, which in turn are the basis of coordinated muscle activation.

Brain functions are classified as "higher" if no rigid temporal relationship to external events can be established. In these cases, memory is the main causal determinant. The corresponding internal processes are certainly no less well-defined than responses to stimuli, but the identity of a given process is based only on neuronal correlations. In this domain, there is no alternative to the correlational approach. Chapters 6, 11 and 12 deal with aspects of higher brain functions. It is astonishing that relatively little effort is at present devoted to the direct observation of the changes in the nervous system corresponding to learning and memory, although these are clearly the clues to higher brain functions.

The intention of focussing mainly on results rather than methods has not always been strictly observed in this volume because unfortunately it is exceptional in this domain of the neurosciences that an established and standard methodology exists. The aim of Frostig et al. (Chap. 3) is an ambitious one, namely to visualize the activity of many neurons by optical means. This would fulfil Sherrington's dream of each active neuron being represented by a flashing light. Once this is achieved (and progress towards this goal is already evident in their contribution), a new era will begin in our understanding of the behaviour of large numbers of cells.

At present, research efforts devoted directly to cooperative phenomena are relatively rare. I hope that the present volume will help to attract other brain researchers into this exciting area, and to push the "bottom-up" analysis of the brain in the upwards direction.

Reference

Sherrington, C. (1941): Man on his Nature. The Gifford Lectures Edinburgh 1937–38 (Cambridge University Press, Cambridge)

2. The Structure–Function Problem in Visual Cortical Circuitry Studied by Cross-Correlation Techniques and Multi-Channel Recordings

Keisuke Toyama

Department of Physiology, Kyoto Prefectural University of Medicine, Kawaramachi, Kyoto, 602, Japan.

The anatomical structure of a neural net and its physiological function are usually investigated by independent experimental techniques. This means that only average relationships between the two features can be deduced. A major problem of modern neuroscience is to more directly link together these two types of knowledge.

2.1 An Approach to the Structure–Function Problem by the Cross-Correlation Technique

Analysis of the receptive field structures and photic responsiveness in the visual cortical cells produced by visual stimulation demonstrated the existence of various response selectivities, such as orientation, width, length of light slits and the direction of motion. Two basic response types have been identified:

1) the simple cells whose receptive area consisted of ON and OFF subareas and response selectivities predictable by linear integration of those evoked through ON and OFF areas, and
2) complex cells with superimposed ON and OFF subareas and with the selectivities explained by nonlinear interaction of the ON and OFF responses (Hubel and Wiesel 1962, 1965).

Analysis of the postsynaptic responses evoked by electrical stimulation of the visual pathway (Toyama et al. 1974; Stone and Dreher 1973; Mitzdorf and Singer 1978), and analysis of dendritic and axonal morphology of the cortical cells (Gilbert 1983) have elucidated the existence of four basic structures in cat visual cortical circuitry (area 17).

Firstly, the geniculate (both X- and Y-) axons terminate at layers IV and VI, and feed monosynaptic excitation to the somas or dendrites of layer III–VI cells. Secondly, some of the layer III and IV cells convey excitation to layer II cells through their main axonal trunks or axon collaterals, while others transmit inhibition to layer III–VI cells. Thirdly, some of the layer II cells are inhibitory, and transmit trisynaptic inhibition to other layer II cells or to themselves.

Springer Series in Synergetics
J. Krüger (Ed.)
Neuronal Cooperativity

Fourthly, in addition to these basic cortical pathways there are abundant pathways interconnecting the cells of layers II–VI. These pathways probably transfer excitation or inhibition in a rather diffuse fashion.

Previous studies of receptive field organization and those of neuronal circuitry have been conducted rather independently, and few studies have been conducted to combine these two approaches and to elucidate how the cortical circuitry yields neuronal responsiveness. The cross-correlation technique developed by Perkel et al. (1967a,b) gave a new approach to the problem of the structure–function relationship.

This technique reveals interactions between two simultaneously recorded cells by cross-correlating their impulse discharges. The neuronal connectivity yielding the interactions can be deduced from the correlations. Although the cross-correlation analysis is a more indirect way of demonstrating neuronal connectivity than the analysis of postsynaptic potentials evoked by electrical stimulation, there are some advantages over the latter method.

Firstly, cross-correlation analysis shows cell-to-cell interaction that is usually not demonstrable by electrical stimulation. Secondly, it is free of the problems of current spread, activation of passing axons and stimulus artifacts, which are regularly associated with electrical stimulation. Thirdly, it reveals the ongoing interaction between the cells during their visual responses and determines their contribution to the responses, if the analysis is made during visual stimulation. Given these advantages, cross-correlation analysis provides a powerful tool for studying the structure–function problem in the visual cortex (Toyama and Tanaka 1984).

2.2 Experimental Techniques for Cross-Correlation Analysis

To date, there have been three ways to simultaneously record extracellular impulse discharges from at least two cells, which is required for cross-correlation analysis. The first is two-channel recording with a pair of microelectrodes (Lee et al. 1977; Tsumoto et al. 1978; Toyama et al. 1981a,b; Tanaka 1983; Ts'o et al. 1986). This method has the advantage of enabling sampling from neuronal pairs at any relative location, such as those at some distance in the same (Toyama et al. 1981a,b) or different functional columns of the visual cortex (Ts'o et al. 1986), or a neuronal pair in different brain structures, e.g. those in the geniculate nucleus or the retina vs. the visual cortex (Lee et al. 1977; Tsumoto et al. 1978; Tanaka 1983).

The second method is multi-neuronal recording using a multi-wire electrode assembly (Kruger and Bach 1981; Michalski et al. 1983; Kruger and Aiple 1988; Bach and Kruger 1986), which has the advantage of enabling sampling of more than two neurons but at restricted locations. The last method is multi-neuronal recording with a single microelectrode combined with a window discriminator which separates impulse discharges of two or more cells according to the difference in their shape parameters, such as the amplitude and time course. This method is of limited use since it provides recordings only from cells in very close proximity.

Sampling of a large number of impulses is essential for the determination of reliable cross-correlograms (CCGs). Roughly a few thousand impulses are required in each cell in the neuronal pair, even for detection of relatively strong interaction. Since the visual cortical cells rarely discharge spontaneously, visual stimulation has been a common way to enhance discharges and collect the required number of impulses. Naturally, the neuronal pair tend to be excited synchronously by visual stimulation, and this will produce pseudo-correlation due to coactivation of the neuronal pair, superimposed on the correlation representing the true neuronal interaction.

Therefore, the CCG determined for the impulse trains which were obtained in the neuronal pairs under visual stimulation should represent a sum of the pseudo-correlation due to coactivation by visual stimulation and the net correlation representing the true neuronal interaction. In cases where visual stimulation is repeated at regular intervals, the contribution of the pseudo-correlation can easily be assessed as the shift predictor, which is CCG-determined for the impulse trains where the time of impulse occurrence in one cell of the neuronal pair is shifted by one or more stimulus intervals, and the net correlation representing the true neuronal interaction is determined by subtracting the shift predictor from the CCG determined for the unshifted impulse trains (Perkel et al. 1967b).

The second method to enhance neuronal discharges is chemical stimulation by the electrophoretic ejection of a chemical stimulant such as glutamate from the microelectrode (Tsumoto et al. 1978; Toyama et al. 1981a). Double-barrel microelectrodes are used for this purpose: one barrel for recording and the other for chemical stimulation. Both methods have advantages over each other and are selected according to the experimental purposes of investigation (Toyama and Tanaka 1984). The first method demonstrates the neuronal interaction due to the common inputs shared by the neurons and the interaction exerted between them which, however, are frequently difficult to discriminate, while the second method demonstrates only the latter interaction. On the other hand, the contribution of the interaction to the visual responses in the recipient cell can only be determined by the first method.

2.3 Basic Neuronal Interaction Demonstrated by Cross-Correlation Analysis

Three basic types of neuronal interaction have been demonstrated by cross-correlation analysis of neuronal discharges in the geniculate nucleus and visual cortex: common excitation, serial excitation and serial inhibition.

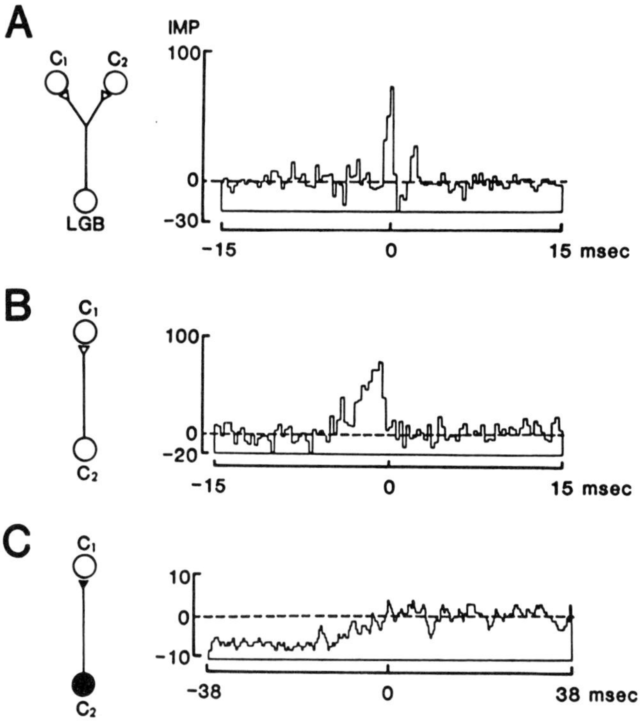

Fig. 2.1. Basic patterns of interneuronal interactions demonstrated by cross-correlation analysis (From Toyama and Tanaka 1984)

Common Excitation. The first type of interaction represents synchronized excitation of the two neurons. The CCG of Fig. 2.1A, determined for impulse discharges in two cortical cells (C_1 and C_2), represents a sharp positivity extending only for a millisecond around zero time, indicating that the C_1 and C_2 are excited in precise synchronism. Common excitation may occur if the two cells receive a common input, either excitatory or inhibitory, and tend to discharge conjointly (Perkel et al. 1967b). However, the precise synchronization of their activation indicates that they share potent common excitatory inputs.

Serial Excitation. The second type of interaction represents excitation of one cell by another. The CCG of Fig. 2.1B represents serial excitation exerted by C_2 onto C_1. The positivity is asymmetric at about zero time. It occurs with a negative delay (−0.6 ms), attains a maximum in a millisecond and declines gradually for a few milliseconds, indicating serial transfer of the excitation from C_2 to C_1.

Serial Inhibition. The third type of interaction is inhibition of one cell by another. The CCG of Fig. 1C represents a negativity occurring with a delay of −1 ms and continuing for a period of several tens of milliseconds, indicating serial transfer of inhibition from C_2 to C_1.

2.4 Signal Transmission Demonstrated by Cross-Correlation Analysis

2.4.1 Geniculo-Cortical Transmission

Parallel geniculo-cortical transmission. Geniculo-cortical and retino-cortical transmission have been investigated by cross-correlation analysis of impulse discharges in simultaneously recorded geniculo-cortical (Tanaka 1983) or retino-cortical neuronal pairs (Lee et al. 1977).

Cross-correlation analysis under visual stimulation demonstrated serial excitation only in about one-third of the geniculo-cortical pairs with overlapped receptive fields (82/243), suggesting the existence of factors other than the retinotopy which determine the geniculo-cortical connectivity. The connected pairs consisted of X- or Y-geniculate cells and cortical cells including Eon (subtype of simple cells with exclusively ON response area), Eoff, simple, standard and special complex cells (Table 2.1), indicating that the geniculate excitation was transferred onto practically all types of cortical cells: pure Y-geniculate inputs to the special complex cells, mixed X- and Y-inputs (but predominantly X-inputs) to the simple cells and their subtypes and mixed X- and Y-inputs (but predominantly Y-inputs) to the standard complex cells (Table 2.1).

Double-line model. Another important finding is the correspondence between the response polarity in the geniculate vs. Eon/Eoff cells and simple pairs that exhibited excitatory connection: the ON-center area of the geniculate cells overlapped with the ON-response area of the cortical cells (Fig. 2.2 and Fig 2.3A, B, D), and the geniculate OFF-center area with the cortical OFF-response area (Fig. 2.2 and Fig. 2.3B, C).

These findings are consistent with the double-line model, assuming that the ON- and OFF-response areas of the simple and Eon/Eoff cells are constructed by convergence of ON- and OFF-center geniculate cells, respectively (Fig. 2.2). Since the diameter of the ON- or OFF-center area of the geniculate cells is approximately the same as the width of the ON- or OFF-response area of the cortical cells (cf. Fig. 2.3A–D), the elongated response areas of the cortical cells may be constructed by the convergence of excitations from several or more

Table 2.1 Frequency of occurrence of geniculocortical connections. Denominators represent the number of the geniculate-striate pairs studied and numerators represent the number of pairs exhibiting the serial correlation (Modified from Table 1 in Tanaka 1983)

Geniculate cells	Cortical cells				
	Eon/Eoff	Simple	Standard complex	Special	Total
X	8/22	18/49	17/45	0/16	43/132
Y	1/6	4/14	20/33	11/29	36/82
Total	9/28	22/63	37/78	11/45	79/214

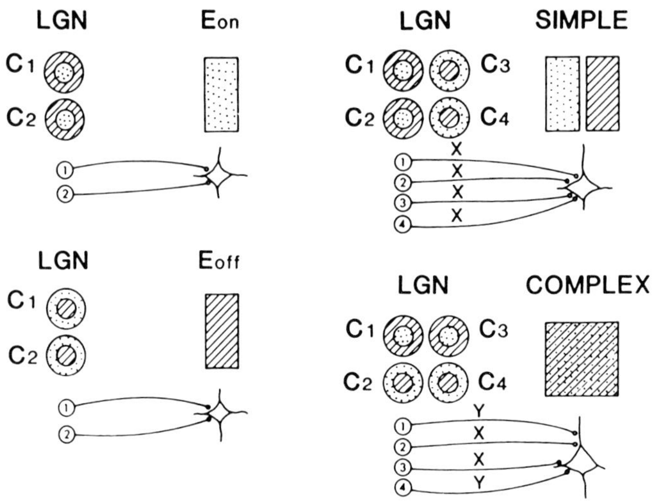

Fig. 2.2. Diagram of geniculo-cortical connections. The figure illustrates the convergence of X- and Y- and ON- and OFF-center geniculate inputs to simple, the subtype of simple cells (Eon and Eoff) and complex cells. LGN, lateral geniculate cells; ON response area, stippled; OFF area, hatched (Toyama and Tanaka 1984)

geniculate cells, whose receptive areas are lined up along the axial orientation in the receptive area of the cortical cells.

The following evidence supports this idea. Firstly, in the simple cells and their subtypes (Table 2.2), the contribution of the serial excitation to the response of the cortical cells, determined by the ratio of the number of impulses contained in the serial excitation to that in the response to visual stimulation (cf. Fig. 2.7I), was roughly 0.1, which indicates that approximately 10 geniculate axons converge onto these cells (Fig. 2.2). The contribution was much smaller (about 0.03) in the standard and special complex cells, suggesting greater geniculate convergence (about 30) onto the latter types of cells.

Secondly, cross-correlation analysis made in multiple geniculate vs. single cortical cell pairs (Tanaka 1983) demonstrated, in fact, convergence of excitation to a single cortical cell from multiple geniculate cells, whose receptive areas were lined up along the axial orientation of the receptive area of the cortical cell (Fig. 2.3).

A similar serial excitation to that demonstrated in the geniculo-cortical pairs was found in the retino-geniculate pairs (Lee et al. 1977). Cross-correlating the neuronal pairs with overlapped receptive areas during visual stimulation, a positive CCG was found in about one-fifth of the pairs (6/29) including both X- and Y-retinal cells. The CCGs closely resembled those found by Tanaka in the geniculo-cortical pairs except for the longer delay (4–5 ms) and longer duration. This difference may be attributed to the longer conduction distance and an additional synaptic relay at the geniculate nucleus.

Interestingly, the contribution of the CCG to the visual responses (determined by the author from their figures) was roughly the same (0.2–0.3) as that

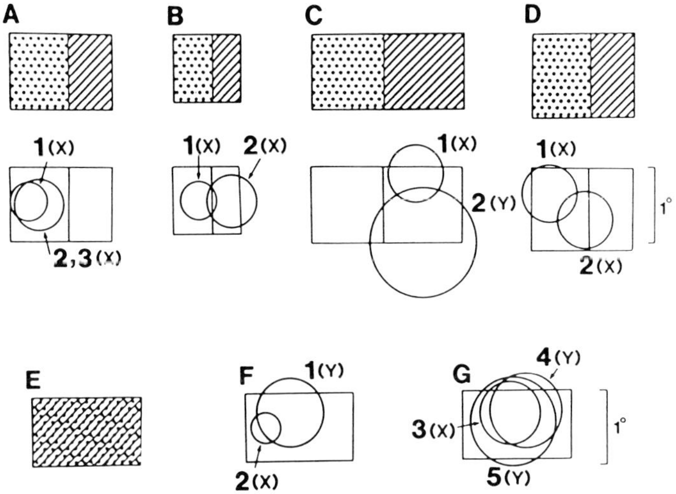

Fig. 2.3. Convergence of geniculate inputs to single cortical cells. **A–D** Four examples of the convergence demonstrated by cross-correlation analysis in multi-geniculate vs single cortical cell pairs. There was agreement in all neuronal pairs on the polarity of response subfields. The center subfields of all ON-center geniculate cells fell on the ON-response area of simple cells, as did those of OFF-center cells, except for the one ambiguous case of the ON-center X-cell in D; **E–G** the convergence from five geniculate cells to a single complex cell (Toyama and Tanaka 1984)

Table 2.2. Geniculo-cortical contribution. Values are means, and ranges of contribution are determined for responses to stimulation with a moving stimulus (From Tanaka 1983)

From	To			
	Eon/Eoff	Simple	Standard complex	Special complex
X	0.10 (0.05–0.17)	0.11 (0.04–0.25)	0.03 (0.01–0.07	
Y	0.03	0.11 (0.06–0.13)	0.04 (0.01–0.08)	0.03 (0.02–0.06)

reported by Tanaka for the geniculate-simple pairs, which is consistent with the finding in the cross-correlation study by Levick et al. (1972) that an almost one-to-one connection is established between the retinal ganglion and the geniculate cells representing the central visual field.

2.4.2 Cortico-Geniculate Transmission

Although Tanaka (1983) found no cross-correlation representing cortico-geniculate excitation, Tsumoto et al. (1978) reported serial excitation from cortical cells to the geniculate cells with overlapping receptive areas (18/114), and serial inhibition to those with slightly displaced receptive fields (8/114).

2.5 Intracortical Transmission Along a Single Functional Column or to Adjacent Functional Columns

The intracolumnar transmission in a single orientation column and intercolumnar transmission between the adjacent orientation columns were studied using simultaneous recordings from two microelectrodes (Toyama et al. 1981a,b) or a multi-electrode assembly (Michalski et al. 1983). The results obtained with the two different methods were in general agreement.

2.5.1 Nonspecific Transmission of Geniculo-Cortical Excitation

In the study where activities in a number of cells were sampled with a vertically advanced microelectrode, and cross-correlated with those in a single reference cell in the same or an adjacent column recorded with another microelectrode advanced obliquely (Fig. 2.4A), the neuronal interaction most frequently found (51%, 95/185) was common excitation (Toyama et al. 1981a,b), characterized by synchronized activation of neuronal pairs within a few milliseconds (Fig. 2.4C–E). The space where the target cells demonstrated common excitation extended over a vertical distance of several hundred microns around layer IV and horizontal distances of many orientation columns (Fig. 2.4A). The sharp synchronization and spatial distribution around layer IV indicate that common excitation represents excitation by the shared geniculate inputs.

Common excitation was not strictly specific to either response type, orientation or direction preference, but was specific to ocular preference. Although common excitation was most frequently found between neuronal pairs with the same response types, it could occur in neuronal pairs with different response types (i.e.; Eon/Eoff vs. simple, simple vs. complex, etc., Table 2.3) and in those with different orientation and directional preference, but never in those with different ocular preference.

Similarly in the study using a multi-electrode assembly (Michalski et al. 1983), common excitation was found in about two-thirds (61%) of the 286 neuronal pairs tested including all combinations of simple and complex cells, even in those separated by a horizontal distance of 1 mm and probably located in different functional columns.

All of the findings obtained in the studies using either two microelectrodes or a multi-electrode assembly indicate that single geniculate axons terminate in a rather nonspecific fashion onto cortical cells of diverse response types and different selectivities. The cortical space innervated by a single geniculate axon may extend horizontally at least across many orientation columns, and possibly throughout the entire extent of a single ocular dominance column and vertically through layers III–V. This view is consistent with Tanaka's finding (1983) that either Eon/Eoff, simple or complex cells received excitation from the X-geniculate cells.

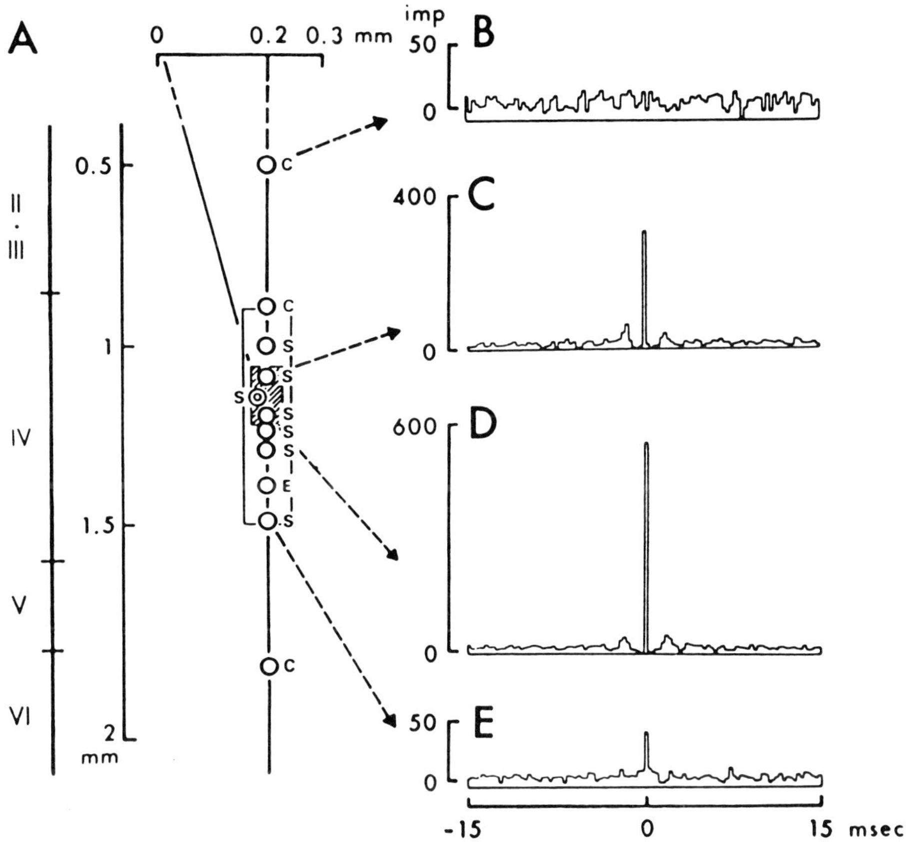

Fig. 2.4. Distribution of common excitation. **A** Diagram illustrating the locations of a reference cell (double circle) recorded with an oblique electrode, and target cells (single circles) sampled with a vertical electrode; **B–E** cross-correlograms (CCGs) of activities in four representative cells determined in reference to those in the reference cell. The hatched and blank rectangles indicate areas where strong and weak common excitation was demonstrated (Toyama et al. 1981b)

Table 2.3 Frequency of occurrence of common excitation. Figures in the table represent the frequency of occurrence of common excitation using the same conventions as in Table 2.1 (From Toyama et al. 1981b)

To cell	From cell			
	Eon/Eoff	Simple	Complex	Hypercomplex
Eon/Eoff	8/12	6/16	8/17	0/2
Simple		21/35	26/36	0/6
Complex			23/45	2/13
Hypercomplex				1/3

2.5.2 Specific Transmission of Cortical Excitation

In contrast to the frequent occurrence and broad distribution of common excitation, serial excitation was rarely encountered and confined to neuronal pairs of a specific combination of response types and also in a limited cortical space. As studied by the two-electrode method (Toyama et al. 1981b), serial excitation was found only in a small fraction of the neuronal pairs (10%), and was transferred from complex cells located at the border area between layers III and IV to complex or hypercomplex cells located at distances of a few hundred microns in a more superficial zone of the same orientation column (Fig. 2.5). Therefore, the transmission of cortical excitation is specific to the response type, the orientation and also to the laminar location of the target cells.

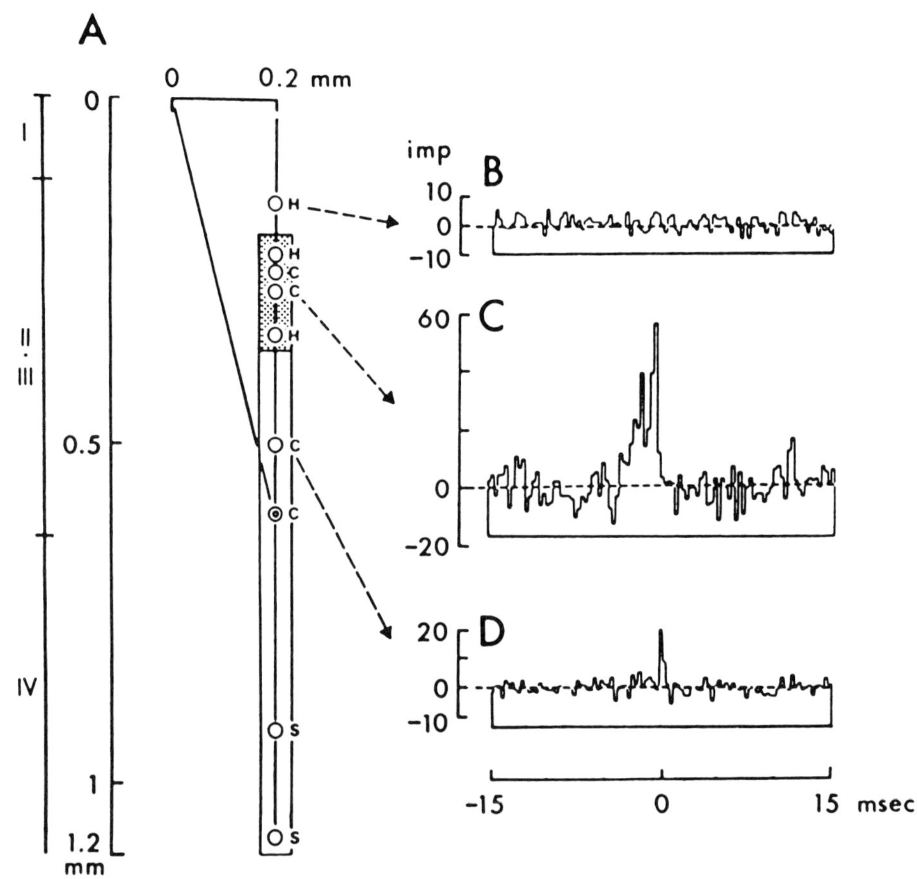

Fig. 2.5. Distribution of cortical excitation. **A** similar diagram to Fig. 4, illustrating the location of the source and target cells; **B–D** CCGs

Essentially similar findings were also obtained by the multi-electrode method (Michalski et al. 1983), except for a more frequent occurrence of serial excitation (20%), which was specific to the orientation preference.

2.5.3 Specific Transmission of Cortical Inhibition

Cortical inhibition was found by the two-electrode method at roughly the same incidence of occurrence (10%) as for serial excitation in neuronal pairs. The source cells of cortical inhibition were found in layers III–IV and were transferred to adjacent target cells within a vertical distance of a few hundred microns (Fig. 2.6). The inhibition was demonstrated between neuronal pairs of different

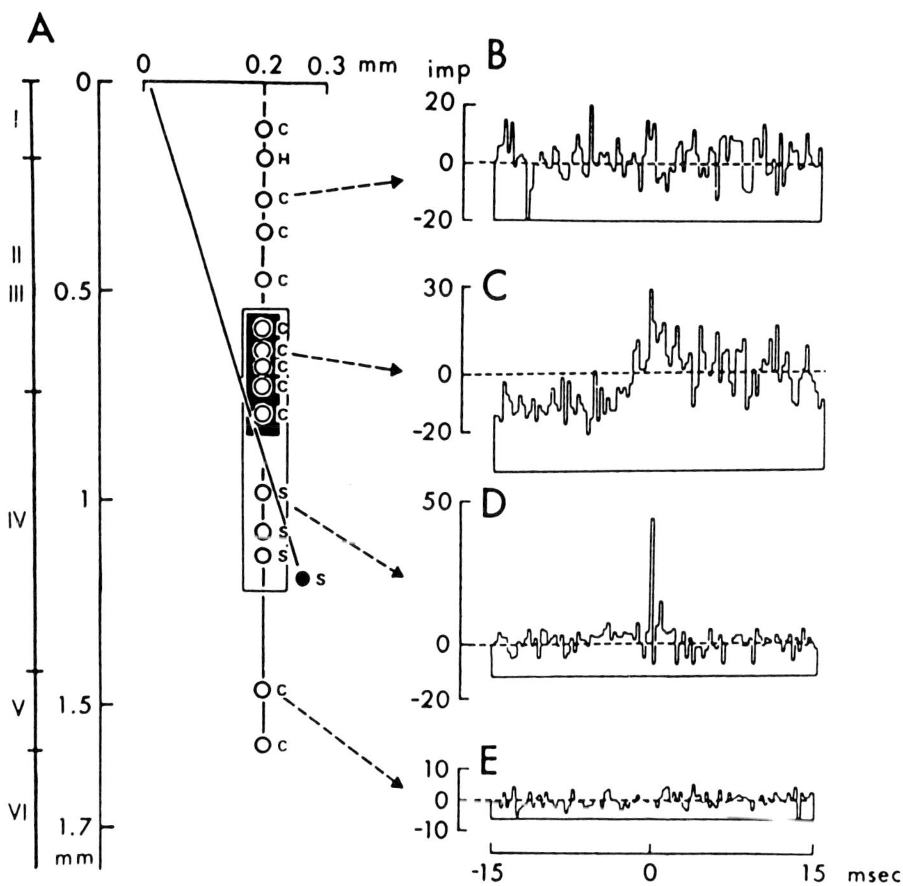

Fig. 2.6. Distribution of cortical inhibition. The conventions are the same as in Fig. 2.5 (Toyama et al. 1981b)

response types and with different (up to a range of degrees of a few tens) or the same orientation preference, suggesting that cortical inhibition is transferred across different orientation columns to cells with different response types. Serial inhibition was found in four combinations: mutually between Eon and Eoff, from Eon/Eoff to simple, from Eon/Eoff to complex and from simple to complex cells (cf. Fig 2.8). Therefore, it is likely that cortical inhibition is sequentially transferred from cells with a simple receptive field organization to those with a more complex organization.

Serial inhibition was detected by the multi-electrode method but was not related to the neuronal response types. Serial inhibition was slightly less frequent (7%) and much weaker than that demonstrated by the two-electrode method. This difference may probably be accounted for by differences in the experimental methods: the former study was based on cross-correlation analysis during visual stimulation, while the latter study was based on the analysis during chemical stimulation. The latter method demonstrates serial inhibition in isolation from common excitation, while the former method almost always yields common excitation and serial inhibition, which makes detection of serial inhibition difficult.

2.5.4 Contribution of Interactions to Visual Responses

The contribution of neuronal interactions to visual responses was also studied by the two-electrode and multi-electrode methods. By the two-electrode method, the contribution of common excitation varied according to the response types. In general, it was larger between the same response types than between different response types. In cell pairs with the same response types, common excitation was largest for Eon/Eoff, average for the simple and smallest for the complex cells (Table 2.4).

A contribution of a similar order of magnitude was found for serial excitation (0.1–0.3) and inhibition (0.1–0.2). Contribution of the same magnitudes (0.1–0.6) was also found for common excitation by the multi-electrode method.

Table 2.4 Contribution of common excitation. The conventions are the same as in Table 2.2 (Toyama et al. 1981b)

To cell	From cell		
	Eon/Eoff	Simple	Complex
Eon/Eoff	0.76 (0.38–0.92)	0.11 (0.01–0.29)	0.07 (0.03–0.09)
Simple		0.3 (0.02–0.81)	0.05 (0.01–0.17)
			0.15 (0.01–0.58)

2.6 Transmission to Distant Functional Columns

The transmission of cortical excitation to distant functional columns was studied by sampling a number of cells across many columns with one microelectrode advanced horizontally, while holding one reference cell with another electrode (Ts'o et al. 1986). Cross-correlation analysis under visual stimulation frequently demonstrated a mixture of common and serial excitation, but in some cases only common or serial excitation. The magnitude of correlation depended on multiple factors, including the horizontal separation and ocular and orientation preference of the neuronal pairs, but was relatively independent of the directional preference (Table 2.5). Excitatory interaction was detectable between pairs at distances less than 2 mm, and with orientation preference differing as much as 90°, but it was much stronger between pairs where the difference was smaller than 30° (Table 2.6).

The source of common excitation was unclear. It could partly be accounted for by the shared inputs transferred through the geniculate axons extending horizontally over a few millimeters (Humphery et al. 1985), and partly by the cortical excitation transferred by axon collaterals of the supra- and infragranular

Table 2.5. Relationship of ocular dominance and directionality with correlation strength. The figures represent the number of neuronal pairs included in the correlation strength groups. The analysis was made for 45 pairs which were distanced at least 50 μm apart and exhibited an orientation-preference difference of less than 30° (Modified from Ts'o et al. 1986)

	Correlation strength groups			
	0	1	2	3
Ocular dominance				
opposite	2	1	0	0
same	5	0	6	2
Directionality				
opposite	3	1	5	2
same	4	1	8	5

Table 2.6 Relationship of orientation preference to correlation strength. The figures represent the number of neuronal pairs for the different combinations of correlation strength and the orientation difference groups (Modified from Ts'o et al. 1986)

Orientation	Correlation strength groups			
difference	0	1	2	3
0–10°	0	2	10	9
20–30°	4	1	7	11
40–50°	10	6	4	3
60–70°	17	7	0	0
80–90°	18	6	0	0

cells. However, serial excitation may probably be ascribed to the cortical excitation transferred through those axon collaterals. Interestingly, the excitation transferred to remote cortical columns demonstrated a similar dependence (on vertical distance, ocular dominance and orientation preference) to that for the geniculate excitation.

It would be interesting to determine the relative strength between the transcolumnar and intracolumnar interaction. Direct comparison of this strength is difficult, because different measures were used to express these interactions. The transcolumnar interaction, however, appears to be at least one order of magnitude smaller than the intra-columnar interaction, as judged from the number of impulses used to determine the interactions. Roughly five times as many impulses were used for the transcolumnar than for the intracolumnar interaction (about 15000 in Ts'o et al. 1986 vs. 3000 in Toyama et al. 1981a,b), and the strength of the cross-correlation increases as a function of the product between the numbers of impulses used for the analysis.

2.7 Functional Scheme of Visual Cortical Circuitry

Cross-correlation studies have provided important information about structure-function relations in the visual cortex. Analysis of geniculo-cortical transmission revealed the existence of parallel geniculate channels (Xon/Xoff and Yon/Yoff) connected to a variety of cortical cells (Fig. 2.8). A convergence of X- and Y-inputs to single cortical cells was another feature (but X-inputs are dominant in the simple cell and Eon/Eoff, while Y-inputs are dominant in the standard complex cells), with the exception of the special complex cells receiving pure Y-inputs. Therefore, X- and Y-geniculate inputs are not completely specific to the response types, although they may contribute to response type differentiation.

In support of the double-line hypothesis, multi-geniculate vs. single cortical cell recording demonstrated that the ON and OFF subfields of the simple cells and their subtypes are constructed by convergence of excitation from several ON- and OFF-geniculate cells.

These findings indicate that geniculate excitation is probably important in establishing the receptive fields (ON and OFF subfields in simple cells and their subtypes, etc.) in each type of cortical cell, but probably not very relevant to the functional differentiation of the response type and establishment of the response selectivity. The evidence also indicates that a single geniculate axon extends across many orientation columns to the entire extent of an ocular dominance column, and feeds signals to cortical cells of diverse response types with the same or opposite directional preference. Therefore the geniculate inputs are neither specific to orientation nor directional selectivity.

Evidence supporting this view is provided by simultaneous recording from geniculo-cortical pairs where excitatory connections are demonstrated by cross-correlation analysis. In the geniculo (X-ON)–cortical (simple) pair illustrated in

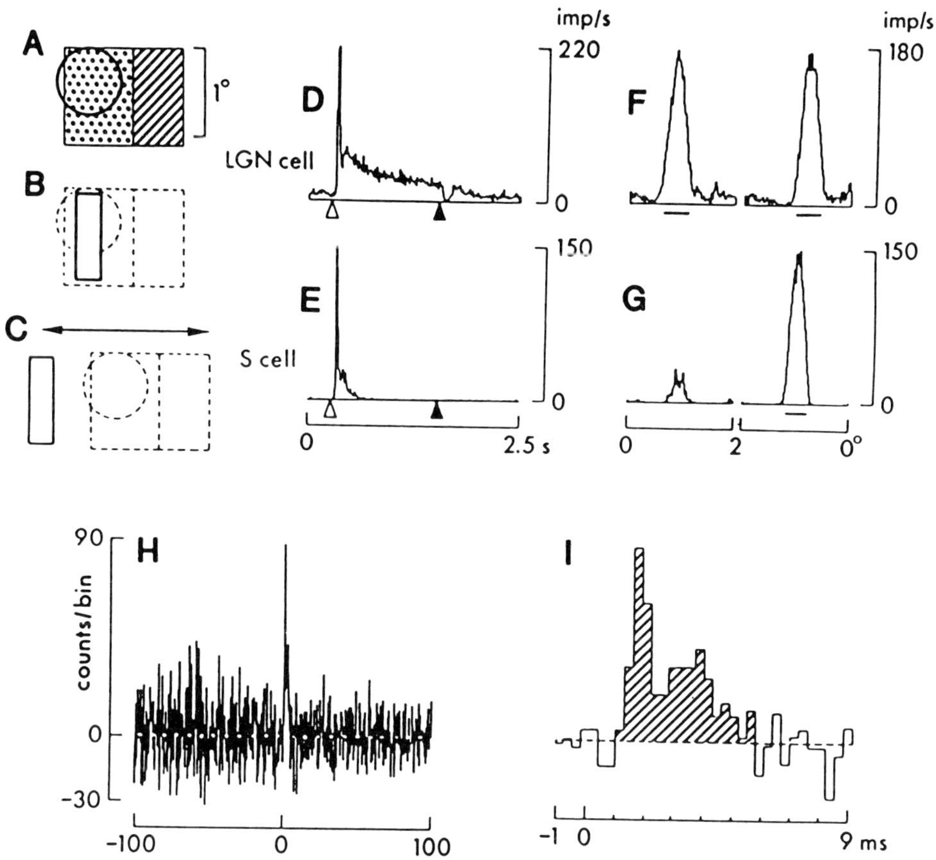

Fig. 2.7. Simultaneous recording from a correlated geniculo-cortical pair. **A–C** Diagrams illustrating the receptive fields and visual stimuli. A rectangle in **A** represents the receptive field of a simple cell, and a circle represents the ON-center field of a geniculate cell. The conventions to represent ON- and OFF-response areas are the same as in the previous figures; **B** and **C** stationary and moving light slits; **D** and **E** responses of the geniculate (**D**) and simple cell (**E**) to the stationary stimulus; **F** and **G** responses of the geniculate (**F**) and simple cell to the moving stimulus (**G**); **H** and **I** CCGs determined for responses in F and G (Tanaka 1983)

Fig. 2.7, the contribution of serial excitation (Fig. 2.7H, I) to the cortical response to a stationary stimulus (Fig. 2.7E) was 0.09, indicating that roughly 10% of the simple cell response was accounted for by excitation from the geniculate cell. Interestingly, the simple cell response is more phasic than the geniculate response (Fig. 2.7D), suggesting the involvement of cortical inhibition in transformation of the response pattern.

Comparison of the responses to a moving stimulus also suggests the contribution of cortical inhibition in the emergence of directional selectivity in the simple cell. The geniculate response (Fig. 2.7F) is exactly bidirectional, while the simple cell response is strongly unidirectional, which suggests that direc-

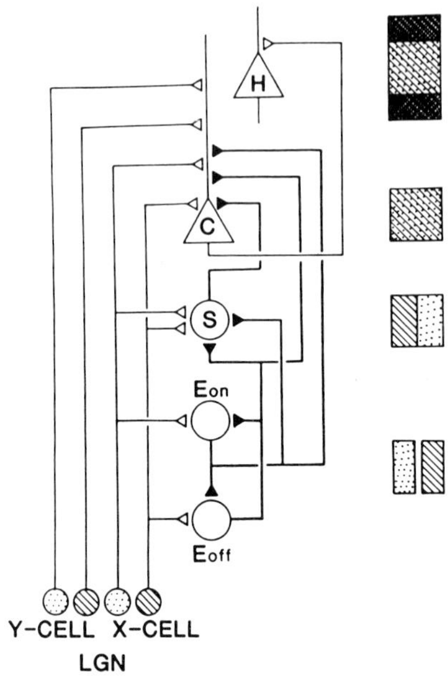

Fig. 2.8. A model of the visual cortical circuitry, representing geniculate connectivity to four different classes of striate cells and intrinsic excitatory and inhibitory connectivity. The pathways and synapses presumed to be inhibitory are indicated by a thick solid line and filled in black, respectively. Eon and Eoff are subtypes of simple cells with exclusively ON- or OFF-response field. *S* simple cell. *C* complex cell. *H* hypercomplex cell. Conventions to indicate ON- and OFF-response fields are the same as those in Fig. 2.2 (Toyama and Tanaka 1984)

tional selectivity in the simple cell is produced by direction-selective cortical inhibition.

Analysis of the cortical transmission also provided important information concerning the emergence of the response type and response selectivity. The evidence indicates the existence of a serial pathway of cortical inhibition (mutually between Eon and Eoff, from Eon and Eoff to simple and from simple to complex) (Fig. 2.8). This suggests the possibility that the response type and response selectivity are established in a sequential fashion along a serial neuronal chain of cortical inhibition.

Firstly, a weak direction and orientation selectivity may emerge in Eon and Eoff due to mutual inhibition (Fig. 2.8). The orientation and direction selectivity is established in simple cells due to antagonistic inhibition between ON and OFF subfields, which is mediated by Eoff and Eon cells with the receptive fields overlapped with ON and OFF subfields of the simple cells, respectively. The orientation and direction selectivity are further transferred to complex cells by inhibition exerted by simple cells. The evidence shows that the inhibition is transcolumnar and serves as the lateral inhibition across different orientation columns to sharpen the orientation and direction selectivity.

The cortical excitation does not seem to be very important in establishing the response selectivity, since it is transferred to cells of the same or a similar response type and selectivity. It probably provides a device to simply convey the response selectivity established by inhibition to higher order cells without much transformation.

2.8 Synaptic Interaction Studied by Spike-Triggered Averaging

Cross-correlation analysis of impulse discharges in neuronal pairs has provided a great deal of important information concerning interactions in the visual cortical circuitry. However, the application of the impulse cross-correlation has been restricted to the basic circuitry of the visual cortex which is probably constructed of focal and dense axonal projection, and many neural connectivities suggested by morphological studies (Gilbert 1983) still remain undetected by the impulse cross-correlation technique. These pathways probably provide diffuse and weak axonal projection, the effects of which would probably be difficult to detect using the present technique of impulse cross-correlation.

The study of single-fiber-evoked excitatory postsynaptic potentials (EPSPs) in the spinal motoneurons indicates that the EPSPs below 80 μV produced no significant cross-correlation peaks (Cope et al. 1987). A slow rise time of EPSPs could also reduce detectability by the impulse cross-correlation technique, particularly if synaptic noise is high (Fetz and Gustafsson 1983). Although the effect of a single axonal projection may be small, the convergence of many axons could eventually have a significant effect on the response selectivity of the cortical cells and information processing in the visual cortex.

Another problem might be the relatively low sensitivity of impulse cross-correlation to detect inhibition, as suggested by a simulation study of interneuronal interaction (Aersten and Gerstein 1985). However, this conclusion should be reviewed, in consideration of the strong shunting action of cortical inhibition demonstrated by spike triggered averaging (see below).

We attempted to demonstrate the interaction which might have remained undetected by the cross-correlation technique by applying spike-triggered averaging, which is supposed to be a more sensitive method than cross-correlation analysis (Kirkwood 1979), to slice preparations of visual cortex (Komatsu et al. 1988). A target cell was recorded intracellularly with one microelectrode, while extracellular recording was conducted from the other cell with another electrode under chemical stimulation with glutamate (Fig. 2.9).

Figure 2.10 illustrates an example of inhibitory transmission from the juxtagranular to the supragranular cell revealed by spike-triggered averaging. Simultaneous extracellular (upper trace of Fig. 2.10A) and intracellular recordings (lower trace) indicated that IPSPs (inhibitory post synaptic potentials) occurred in synchronization with the largest impulses in the extracellular trace. It was further confirmed, by averaging intracellular responses triggered by the largest impulses (Fig. 2.10B–E), that the impulses produced unitary IPSPs of roughly 2 mV with a delay of 0.6 ms, which is an order of monosynaptic delay. There-

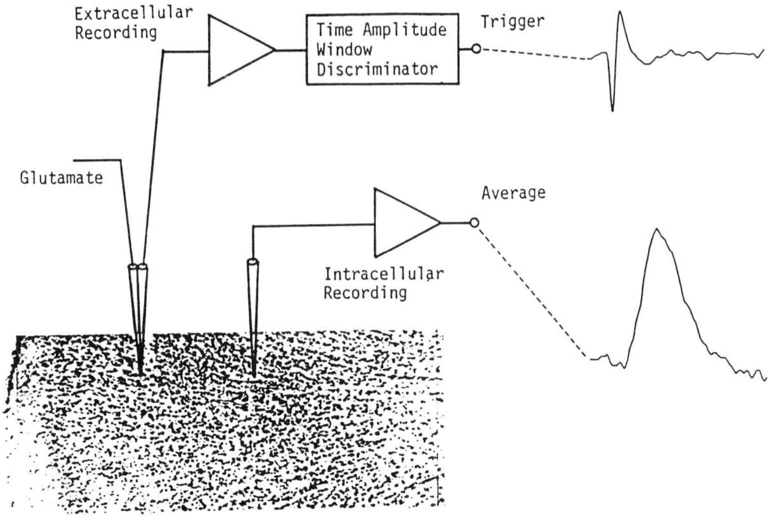

Fig. 2.9. A diagram illustrating the experimental arrangement for spike-triggered averaging in slice preparations of the visual cortex

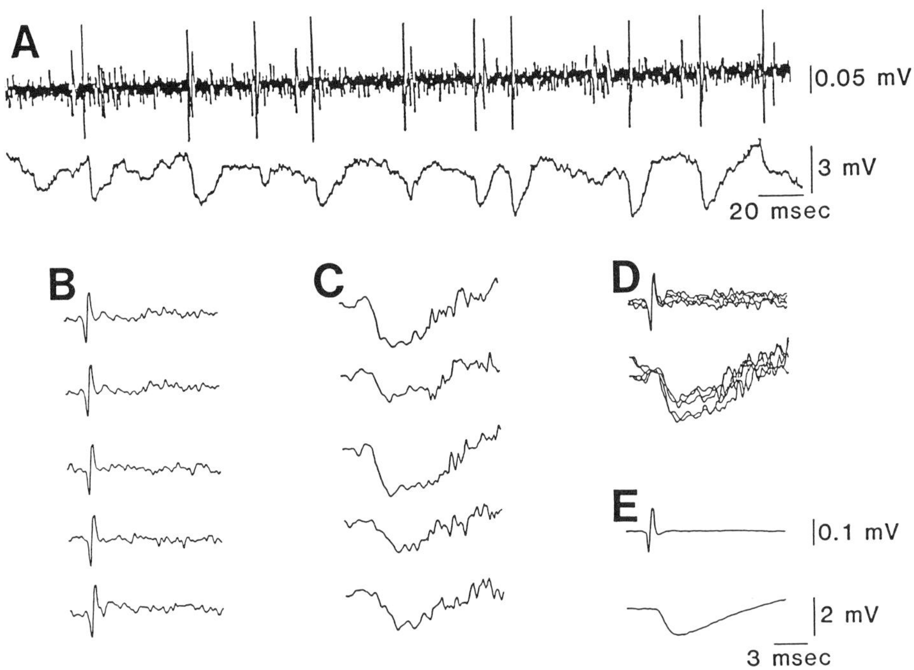

Fig. 2.10. Simultaneous recording from two visual cortical cells and spike-triggered averaging. **A** extracellular (upper trace) and intracellular recordings from cortical cells (lower trace). Extracellular recording was made from one cortical cell in the juxta-granular (JG) layer with one barrel of a double-barrel microelectrode during chemical stimulation by electrophoretic application of glutamate through the other barrel, while postsynaptic potentials were recorded intracellularly from the other cell in the supragranular (SG) layer; **B** extracellular impulses in a JG cell isolated by a time-amplitude window discriminator; **C** intracellular traces in a SG cell triggered by the JG impulses; **D** and **E** superimposed and averaged traces in JG and SG cells (Komatsu et al. 1988)

fore the cell generating the largest impulses is the source cell producing the IPSPs in the target cell.

Although the IPSPs were relatively large in the case of Fig. 2.10A–C, enhanced by slight (10 mV) depolarization of the membrane potential, they were much smaller (100 μV) in the resting membrane potentials. The strong enhancement effect of the membrane hyperpolarization in spite of the relatively small potential effect, indicates that the IPSPs are probably produced at the soma or proximal dendrites and have a large shunting conductance, probably more than 10 times that of the unitary excitatory postsynaptic potentials (EPSPs) of the juxtagranular-supragranular pathway (see below), as estimated by the effect of membrane depolarization. This inhibition would probably correspond to the serial inhibition shown to be transferred from layer IV to the neighboring cortical region by the impulse cross-correlation.

Likewise, spike-triggered averaging demonstrated unitary EPSPs transmitted from juxtagranular cells to supragranular cells (Fig. 2.11C), which probably corresponds to the serial excitation shown by impulse cross-correlation. The amplitude and time course of the unitary EPSPs were similar to those in the soma of the spinal motoneurons (Jankowska and Roberts 1972; Cope et al. 1987). The slightly longer delay (0.7 ms) of the juxtagranular-supragranular EPSPs compared to that of the juxtagranular-supragranular IPSPs is probably due to the larger distances (230 μm) between the source and target cells (cf. Fig. 2.11A).

In addition to the synaptic interaction already identified by the impulse cross-correlation technique, spike-triggered averaging demonstrated two additional types of excitatory transmission, transferred in dual ways between supragranular and layer V cells. The supragranular-layer V EPSPs were not much different from the juxtagranular-supragranular EPSPs (compare Fig. 2.11C and D)

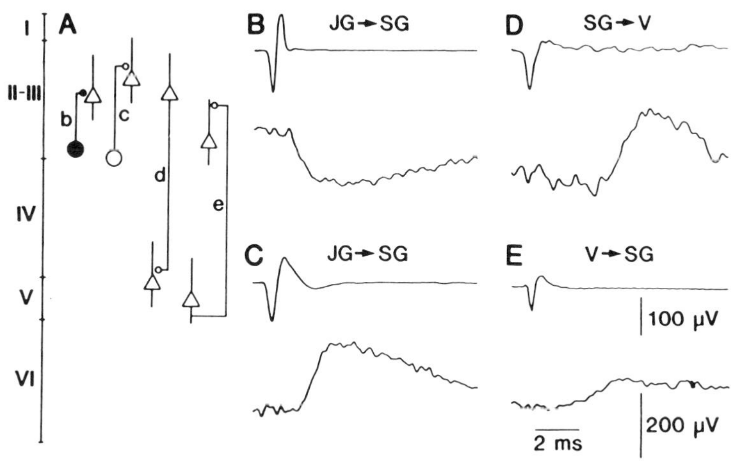

Fig. 2.11. Intrinsic connectivity in visual cortex and unitary postsynaptic potentials demonstrated by spike-triggered averaging. **A** Wiring diagram. Conventions to represent excitatory and inhibitory circuitry is the same as those in Fig. 2.8; **B–E** unitary EPSPs and IPSPs produced in visual cortical cells through the indicated pathways (Komatsu et al. 1988)

except for the longer latency (2.8 ms), which is probably ascribable to longer distances (700 μm) between supragranular and layer V cells. On the contrary, the layer V-supragranular EPSPs were smaller in amplitude and slow in time course (compare Fig. 2.11E and D), suggesting that the EPSPs are produced at remote dendrites of the supragranular cells (Fig. 2.11A).

An interesting finding made from the known pathways was that target cells commonly received excitation or inhibition from a single source cell. They tended to appear in a cluster, as do the source cells, suggesting the existence of dense convergence and divergence in the known pathways for synaptic interaction. On the contrary, neither the source nor target cell was ever shared between the adjacent neuronal pairs in the newly-found pathways. These results are consistent with the view that the old pathways probably establish focal and dense connections, while the new pathways are comprised of diffuse and weak connections.

The focal and strong pathway may be involved in classical neuronal responsiveness, i.e. orientation, direction, end-stop, disparity and color selectivity, which is probably produced by interaction in a small number of cells (the order of tens). In contrast, the weak and diffuse pathway may control responsiveness to random-dot stereograms (Poggio et al. 1985) which probably requires cooperative and antagonistic interaction in a much larger (hundreds or more) neuronal population (Poggio and Poggio 1984).

Altogether, the spike-triggered averaging technique was able to elucidate the neural interaction that remained undetected by the impulse cross-correlation technique, and is superior to the latter technique in detectability of weak and diffuse neuronal interaction.

2.9 Multi-Channel Optical Recording of Neural Activity

Although the cross-correlation and spike-triggered averaging techniques are useful in detecting cell-to-cell interaction, both of them are associated with the technical difficulty in multi-neuronal recording over the long period of time required for the detection of the interaction.

We further attempted to overcome this difficulty by applying optical recording of neural activity in a visual cortical slice (Grinvald et al. 1982, 1986; Blasdel and Salama 1986). A slice (500 μm thick) was dissected from rat visual cortex (area 17) and incubated in oxygenated Krebs-Ringer solution containing a voltage-sensitive dye (RH 155, 0.2 mg/ml) for 30 min. The cortical slice was placed in a recording chamber of an interface type, and the image of the slice was focused through a microscope (Nikon LABOPHOTO-YF) onto a modified version of a multi-channel photodiode array (Hamamatus Photonics S2307-128Q) which affords multi(128)-channel recording at a time resolution of 0.64 ms. The optical signals of the photodiode array were fed into a computer (IBM PC/AT) through a high-speed AD converter (Data Translation DT2821-G-8DI). The

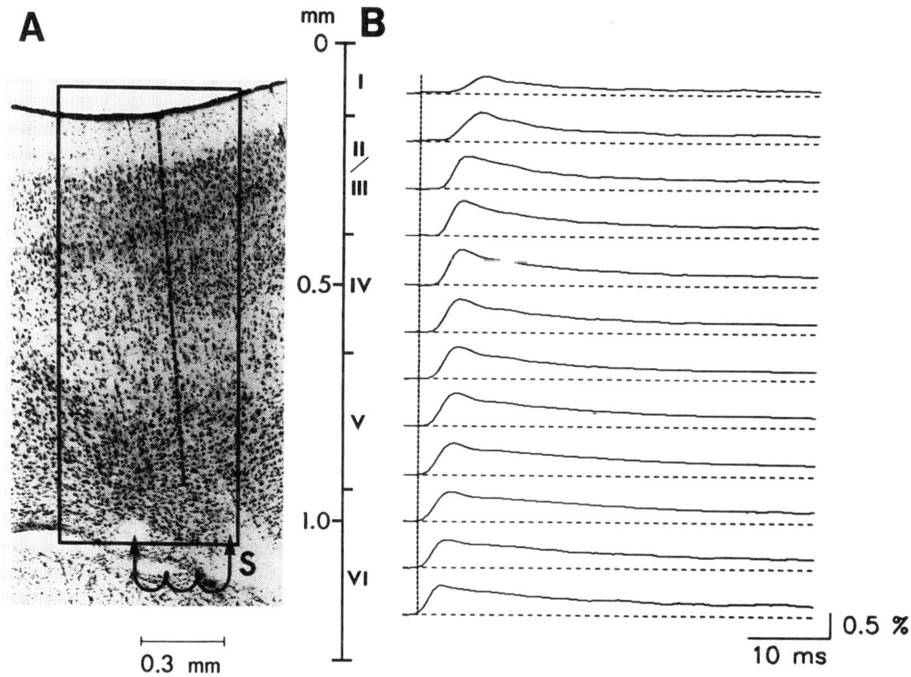

Fig. 2.12. A schematic illustration of multi-channel recording with a photodiode array and optical responses. **A** location of the photodiode array (solid rectangle) shown in the histological section of a cortical slice. *S* stimulating electrode placed in the white matter; **B** optical responses recorded with the photodiode array and averaged for 1000 trials of white matter stimulation at 0.5 Hz. Only 12 representative traces of 128 responses are illustrated (Tanifuji, Yamanaka and Toyama, unpublished)

diode array was usually aligned along the cortical column (Fig. 2.12A) so that the recording sites covered the entire depth from the cortical surface to the white matter (WM) at a precise spatial resolution (12.5 μm).

The optical responses (ORs) evoked by WM stimulation (at 0.5 Hz) were averaged for about 100–1000 trials, and were studied in comparison with the field potentials (FPs) which were recorded, before or after the optical recording, every 50 μm along the same cortical column as that for the optical recording.

The time course of the ORs (Fig. 2.12B) roughly resembled those of the intracellular responses reported in visual cortical cells (Fig. 2.12B), supporting the view that they represent the intracellular responses in the cortical cells (Grinvald et al. 1982, 1986; Ross and Krauthamer 1984; Saggau et al. 1986). However, a difference was found in that the ORs consisted only of a positive component corresponding to the EPSPs, and lacked a negative component corresponding to the IPSPs which usually follow the EPSPs (Toyama et al. 1974).

The correspondence of the neural responses was further studied by comparing ORs (Fig. 2.13A, and dotted trace in Fig. 2.13C) with FPs (solid trace in Fig. 2.13C). There was no exact match in the shape of the responses. The ORs were

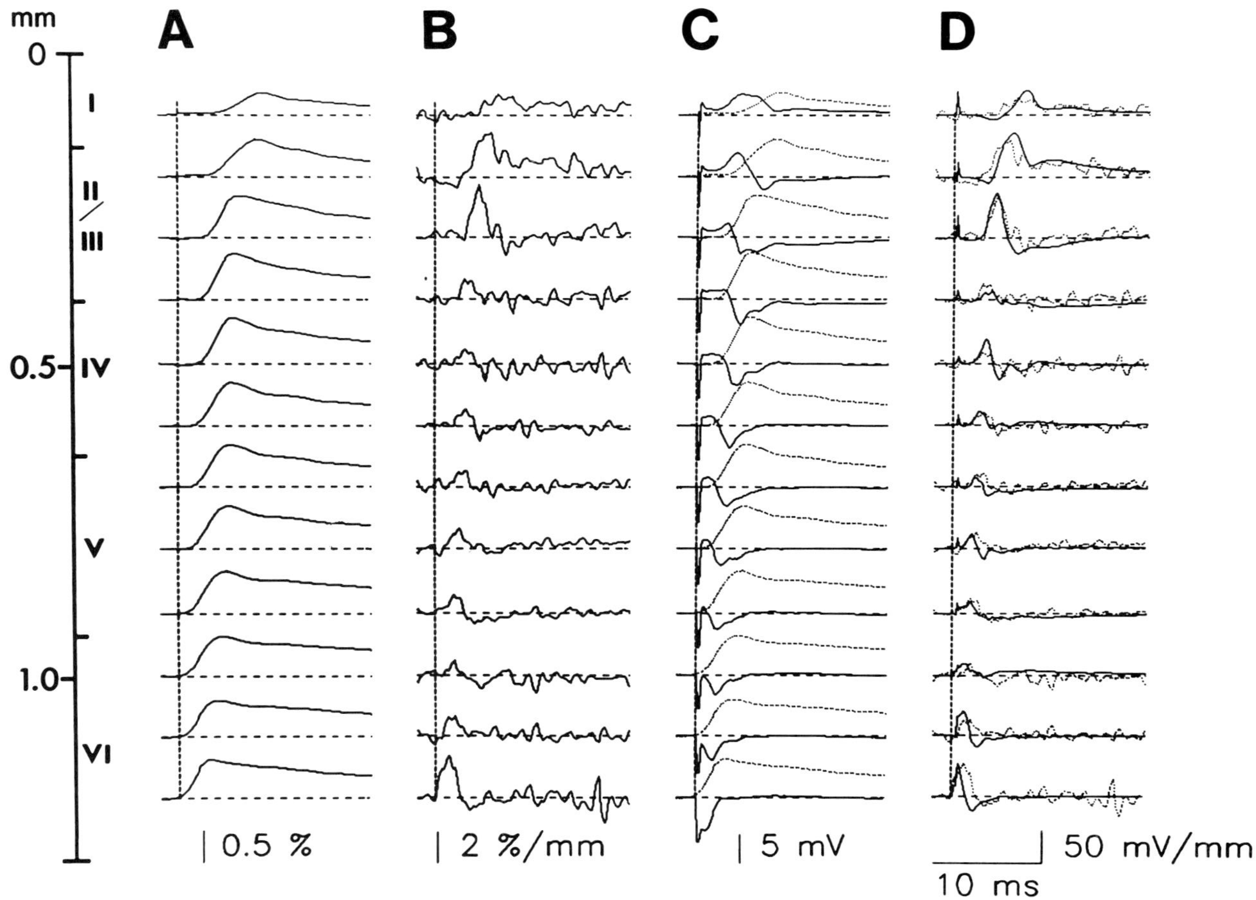

Fig. 2.13. Optical and field responses and their first-order differentials. **A** and **B** Optical responses and the first-order differentials; **C** and **D** similar to **A** and **B**, except for the field potentials. Traces in **A** and **B** are indicated in **C** and **D** by dotted lines (Tanifuji, Yamanaka and Toyama, unpublished)

positive all the way through the entire cortical layers, while FPs were positive in the superficial layer and became negative in the middle and deep layers.

Provided that the ORs represent the sum of the intracellular responses in the local cortical cells produced by WM stimulation, as has been suggested in hippocampal slices (Grinvald et al. 1986; Saggau et al. 1986), and also that the apical dendrites of cortical cells are oriented along the cortical column, the first-order differentials of ORs would give the sum of intracellular currents conveyed along dendrites of the local cortical cells. These currents should agree with the extracellular currents along the cortical column which are determined as the first-order differentials of the FPs.

The correspondence between the first-order differentials of ORs (Fig. 2.13B and dotted trace in Fig. 2.13D) and FPs (solid trace in Fig. 2.13D) supports this view. The differentials of the two responses closely resembled each other in laminar distribution. Both were greater in the superficial layer than in other layers. There was also good agreement in the time course, with the onset being longer in the superficial layer (2.7 and 2.7 ms for the differentials of ORs and FPs at 0.2 mm from the cortical surface), and shorter in the middle and deep layers (0.5 and 0.8 ms at 0.5 mm).

Altogether, agreement between the first-order differentials of the ORs and FPs which were obtained by quite different recording means was substantial. However, it is pointed out that the optical recording does not represent the IPSPs demonstrated by the intracellular recording. The cause of this discrepancy is not very clear. The IPSPs may be too small to be detected by the optical recording where cortical cells are not invaded by a microelectrode, and demonstrate deeper resting potentials than those for the intracellular studies, or alternatively the later optical responses corresponding to IPSPs may be concealed by the optical or light-scatter responses produced by non-neural elements such as glial cells. Still, optical recording with a photodiode array seems to be a reliable technique for multi-channel recording of the excitatory responses in slice preparations, and it may be useful in studying the spatio-temporal characteristics of visual cortical circuitry.

2.10 Concluding Remarks

Cross-correlation studies have provided a great deal of important information concerning the structure-function problem of visual cortical circuitry. However, the application of cross-correlation analysis has been restricted to the basic circuitry of the visual cortex which is probably constructed of focal, point-to-point, axonal projection. However, diffuse pathways projecting in a area-to-area fashion, suggested by morphological techniques (Gilbert, 1983), remained undetected probably because their effects are small and difficult to detect by the present technique of cross-correlation analysis.

Spike-triggered averaging was found to be a more sensitive method than cross-correlation analysis (Kirkwood, 1979) for the demonstration of synaptic connections (Komatsu et al. 1988). The study confirmed all intracortical connectivities suggested by cross-correlation studies (excitatory and inhibitory connections from juxtagranular to supragranular cells), and demonstrated two additional excitatory connections (from supragranular to layer V and from layer V to supragranular cells). Convergence and divergence were found in the known connections, but never in the newly-found pathways. The unitary EPSPs produced through the old connections had larger amplitudes (a few hundred microvolts vs. 100 μV or less), quicker rise times (1.7 vs. 2.4 ms) and shorter half widths (5 vs. 14 ms) than those for the new connections. Another interesting finding is that unitary inhibitory postsynaptic potentials (IPSPs) were smaller in amplitude (less than 100 μV), but exhibited much larger shunting conductance (probably more than 10 fold) than that for the unitary EPSPs of the old connections.

The signals conveyed by EPSPs would effectively be nullified by strong shunting of IPSPs, in cases where both would occur together. Therefore, the neuronal circuitry constructed of excitatory and inhibitory pathways provides an efficient gate which only opens for the "yes-and-no" combination (the presence of activities in the excitatory pathway and the absence of activities in the inhibitory pathway), which may serve as a efficient device for producing neuronal selectivity.

A difficulty common in these two techniques is the simultaneous recording from multiple neurons. Optical recording provides a promising solution to this problem, particularly in the study of the spatio-temporal aspects of cortical circuitry.

References

Aersten, M.H., Gerstein, G.L. (1985): Evaluation of neuronal connectivity: Sensitivity of cross-correlation. Brain Res. **340**, 341–354

Bach, M., Kruger, J. (1986): Correlated neuronal variability in monkey visual cortex revealed by a multi-microelectrode. Exp. Brain Res. **61**, 451–456

Blasdel, G.G., Salama, G. (1986): Voltage-sensitive dyes reveal a modular organization in monkey striate cortex. Nature **321**, 579–585

Cope, T.C., Fetz, E.E., Matsumura, M. (1987): Cross-correlation assessment of synaptic strength of single Ia fibre connections with triceps surae motoneurones in cats. J. Physiol. Lond, in press

Fetz, E.E., Gustafsson, B. (1983): Relation between shapes of post-synaptic potentials and changes in firing probability of cat motoneurones. J. Physiol. Lond. **341**, 387–410

Gilbert, C.D. (1983): Microcircuitry of the visual cortex. Ann. Rev. Neurosci. **6**, 217–247

Grinvald, A., Manker, A., Segal, M. (1982): Visualization of the spread of electrical activity in rat hippocampal slices by voltage-sensitive optical probes. J. Physiol. **333**, 269–291

Grinvald, A., Segel, M., Kuhnt, U., Manker, A., Anglister, L., Freeman, J.A., Hildesheim, R. (1986): Real-time optical mapping of neuronal activity in vertebrate CNS in vitro and in vivo. In *Optical methods in cell physiology*, ed. by P.D. Weer, B.M. Salzberg (Wiley, New York)

Hubel, D.H., Wiesel, T.N. (1962): Receptive fields, binocular interaction and functional architecture in the cat's visual cortex. J. Physiol. Lond. **160**, 106–154

Hubel, D.H., Wiesel, T.N. (1965): Receptive fields and functional architecture in two nonstriate visual areas (18 and 19) of the cat. J. Neurophysiol. **28**, 229–289

Humphery, A.L., Sur, M., Uhlrich, D.J., Sherman, S.M. (1985): Projection patterns of individual X- and Y- cell axons from the lateral geniculate nucleus to cortical area 17 in the cat. J. Comp. Neurol. **233**, 159–189

Jankowska, E., Roberts, W.J. (1972) Synaptic actions of single interneurones mediating reciprocal Ia inhibition of motoneurones. J. Physiol. (London) **222**, 623–642
Kirkwood, P.A. (1979): On the use and interpretation of cross-correlation measurements in the mammalian central nervous system. J. Neurosci. Methods, **1**, 107–132
Komatsu, Y., Nakajima, S., Toyama, K., Fetz, E. (1988): Intracortical connectivity revealed by spike-triggered averaging in slice preparations of cat visual cortex. Brain Res. **442**, 359–362
Kruger, J., Aiple, F. (1988): Multimicroelectrode investigation of monkey striate cortex: Spike train correlations in the infragranular layers, J. Neurophysiol. **60**, 798–828
Kruger, J., Bach, M. (1981): Simultaneous recording with 30 microelectrodes in monkey visual cortex. Exp. Brain Res. **41**, 191–194
Lee, B.B., Cleland, B.G., Creutzfeldt, O.D. (1977): The retinal input to cells in area 17 of the cat's cortex. Exp. Brain Res. **30**, 527–538
Levick, W.R., Cleland, B.G., Dubin M.W. (1972): Lateral geniculate neurons of cat: Retinal inputs and physiology. Invest. Ophthalmol. **11**, 302–311
Michalski, A.G., Gerstein, L. Czarkowska, J., Tarnecki, R. (1983): Interactions between cat striate cortex neurons. Exp. Brain Res. **51**, 97–107
Mitzdorf, U., Singer, W. (1978): Prominent excitatory pathways in the cat visual cortex (A 17 and 18): a current source density analysis of electrically evoked potentials. Exp. Brain Res. **33**, 371–394
Perkel, D.H., Gerstein, G.L., Moore, G.P. (1967a): Neuronal spike trains and stochastic point processes. I. Single spike train. Biophys. J. **7**, 391–418
Perkel, D.H., Gerstein, G.L., Moore, G.P. (1967b): Neuronal spike trains and stochastic point processes. II. Simultaneous spike trains. Biophys. J. **7**, 419–440
Poggio, G., Motter, B.C., Squatrioto, S., Trotter, Y. (1985): responses of neurons in visual cortex (V1 and V2) of the alert macaque to dynamic random-dot stereograms. Vision Res. **25**, 397–406
Poggio, G., Poggio, T. (1984): The analysis of stereopsis. Ann. Rev. Neurosci. **7**, 379–412
Ross, W.N., Krauthamer, N. (1984): Optical measurements of potential changes in axons and processes of neurons of a barnacle ganglion. J. Neurosci. **4**, 659–672
Saggau, P., Galvan, M., Bruggencate, G.T. (1986): Long-term potentiation in guinea pig hippocampal slices monitored by optical recording of neuronal activity. Neurosci. Lett. **69**, 53–58
Stone, J., Dreher, B. (1973): Projection of X- and Y-cells of the cat's lateral geniculate nucleus to areas 17 and 18 of visual cortex. J. Neurophysiol. **36**, 551–567
Tanaka, K. (1983): Cross-correlation analysis of geniculostriate neuronal relationships in cats. J. Neurophysiol. **49**, 1303–1318
Toyama, K., Tanaka, K. (1984): Visual cortical function studied by cross-correlation analysis. In *Dynamic Aspects of Neocortical Function*, ed. by G.M., Edelman, W.E. Gall, W.M., Cowan (Wiley, New York) pp. 67–86
Toyama, K., Kimura, M., Tanaka, K. (1981b): Organization of cat visual cortex as investigated by cross-correlation techniques. J. Neurophysiol. **46**, 202–214
Toyama, K. Matsunami, K., Onho, T., Tokashiki, S. (1974): An intracellular study of neuronal organization in the visual cortex. Exp. Brain Res. 21, 45–66
Toyama, K., Kimura, M., Tanaka, K. (1981a): Cross-correlation analysis of interneuronal connectivity in cat visual cortex. J. Neurophysiol. **46**, 191–201
Ts'o, D.Y., Gilbert, C.D., Wiesel, T.N. (1986): Relationships between horizontal interactions and functional architecture in cat striate cortex as revealed by cross-correlation analysis. J. Neurosci. **6**, 1160–1170
Tsumoto, T., Creutzfeldt, O.D., Legendy, C.R. (1978): Functional organization of the corticofugal system from visual cortex to lateral geniculate nucleus in the cat (with appendix on geniculo-cortical mono-synaptic connections). Exp. Brain Res. **32**, 345–364

3. Optical Imaging of Neuronal Activity in the Living Brain

R.D. Frostig[1], E.E. Lieke[2], A. Arieli[3], D.Y. Ts'o[4], R. Hildesheim[3], and A. Grinvald[4,5]

1 Department of Psychobiology, University of California at Irvine, Irvine, CA 92717, USA
2 Institut für Tierphysiologie, Neurobiologie, Konigin-Luise-Str. 28–30, D-1000 Berlin 33
3 Dept. of Neurobiology, The Weizmana Institute of Science, Rehovot 76100, Israel
4 Laboratory of Neurobiology, Box 138, The Rockefeller University, 1230 York Ave., New York, NY 10021, USA
5 IBM Research Division, Yorktown Heights, New York, NY 10598, USA

In this book questions regarding neuronal populations were primarily investigated using single- and multi-unit recordings with electrodes. The complexity of the functional organization which is revealed by these studies indicates a need for new methods which could obtain complementary information that is hard or impossible to obtain with microelectrodes. Imaging methods that provide high spatial or temporal resolution maps of the cortical organization of large cortical areas are of particular promise for studies of functional organization.

3.1 Why Optical Imaging?

There are already existing in vivo imaging methods, but most of them suffer from either limited spatial resolution, temporal resolution or both. These methods include radioactive imaging of changes in blood flow, electroencephalography, magnetoencephalography, nuclear magnetic resonance imaging (MRI), thermal imaging, metabolic mapping using 2-deoxyglucose and positron-emission tomography (PET).

In this chapter we will describe two optical imaging techniques: optical imaging using voltage-sensitive dyes and optical imaging using intrinsic signals. Each technique offers advantages over other techniques.

Optical recording using intrinsic signals offers:

a) high resolution mapping (50–100 μm) of surface areas of the brain;

b) ability to record from the same cortical region with many different experimental or stimulus conditions for extended periods of time;

c) study of the interaction between the neuronal system and the microvascular system supporting it.

The disadvantage of imaging with intrinsic signals is the relatively low time resolution (the rise-time of the intrinsic signal was more than 1 second).

Optical recording with voltage-sensitive dyes offers: a) the imaging of spatio-temporal patterns of the activity of neuronal populations with submillisecond time resolution both in vitro and in the living brain; b) the direct recording of the summed intracellular activity of neuronal populations, including fine dendritic

Springer Series in Synergetics
J. Krüger (Ed.)
Neuronal Cooperativity

and axonal arbors; c) the possibility of repeated measurements from the same cortical region with different experimental or stimulus conditions; d) the ability to combine simultaneously single unit recordings with optical imaging to selectively image a given neuronal assembly. The disadvantage of imaging with voltage-sensitive dyes is in the limited duration of the experiments and side effects that the dyes may have on the imaged tissue. Both optical imaging techniques are limited in their ability to image only surfaces of the brain. However, both techniques can be applied in the same experiment and thus they compliment each other substantially.

3.2 Optical Imaging of Neuronal Activity with Voltage-Sensitive Dyes

The development of suitable voltage-sensitive dyes has been the key to the successful application of optical recording, because different preparations often required dyes with different properties (Grinvald et al. 1988; Cohen and Lesher 1986; Ross and Reichardt 1979). The preparation under study is first stained by bath application of the dye. The dye molecules bind to the external surface of excitable membranes and act as molecular transducers that transform changes in membrane potential into optical signals (Grinvald et al. 1988; Cohen and Lesher 1986). The resulting changes in the absorption or the emitted fluorescence occur in microseconds and linearly correlate with the electrical activity of the stained neurons. These changes are then monitored with light measuring instrumentation. By using an array of photodetectors positioned in the microscope image plane, the activity of many individual targets can be detected simultaneously. Figure 3.1 depicts the optical monitoring of neuronal activity from 100 to 128 sites (Grinvald et al. 1988; Cohen and Lesher 1986; Grinvald et al. 1981). Optical imaging with voltage-sensitive dyes permits the visualization of cortical activity with a submillisecond time resolution and a spatial resolution of 50–100 μm. However, the instrumentation needed to record these fast optical signals with a high spatial resolution over a large area required fast detectors with many more pixels and is currently expensive. Other limitations of the use of dyes are briefly discussed in Sect. 3.6. It is important to note that optical signals recorded from the cortex are different from those recorded from single cells or their individual processes, and thus should be interpreted with care. In simpler preparations where the single cells are distinctly visible, the optical signal looks just like an intracellular electrical recording. In optical recordings from the cortex, however, single cell activity is not yet resolved and the optical signal represents the sum of intracellular membrane potential changes, in both pre- and post-synaptic neuronal elements, with a possible contribution from the depolarization of neighboring glial cells. Since the optical signals measure the integral of the membrane potential changes (integrating membrane area and time), slow subthreshold synaptic potentials in the extensive dendritic arborization are easily

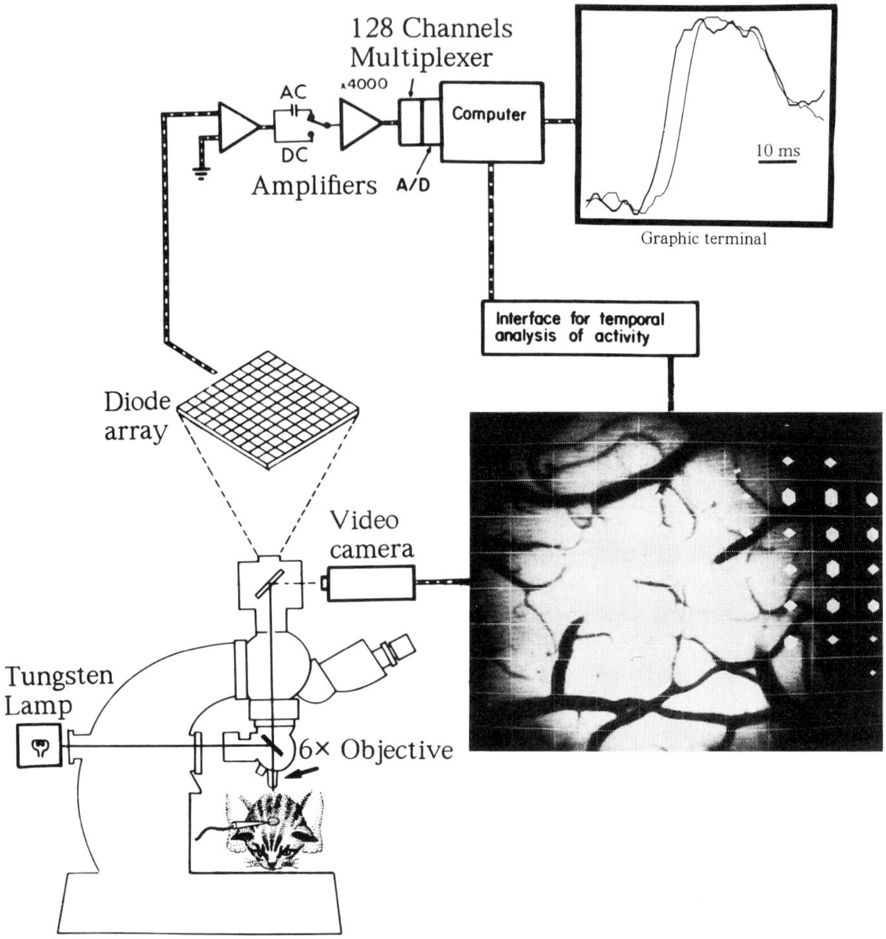

Fig. 3.1. Optical apparatus for real-time imaging of electrical activity with a photodiode array. The video monitor shows an image of the exposed cortical area. Time sections (frames) of the optically detected activity are superimposed on that video image and can be visualized in slow motion. The frame shown illustrates the development of seizure activity in cat visual cortex (after topical application of bicuculline). Inset: comparison between the time course of interictal event in the anterior (bold trace) and posterior (thin trace) portion of the imaged cortical area

detected by optical recording, as are action potentials in arbors of axons running horizontally in the cortex. Thus optical recording with voltage-sensitive dyes emphasizes the dendritic and axonal arbor domains of neurons, domains of high importance for neuronal communication and integration processes. Obtaining information from the axonal and dendritic domains of the living brain is a unique advantage of this imaging technique compared to other imaging techniques or single unit recordings.

3.3 Optical Imaging Based on Instrinsic Signals

Another imaging strategy based on slower intrinsic changes in the optical properties of active brain tissue permits visualization of active cortical regions at high spatial resolution, and without some of the problems associated with voltage-sensitive dyes. Possible sources for these activity-dependent intrinsic signals include either changes in physical properties of the tissue itself or changes in the absorption, fluorescence or other optical properties of intrinsic molecules having significant absorption or fluorescence. The existence of small intrinsic optical changes associated with metabolic activity in many tissues, has been known since the pioneering experiments of Kelin and Millikan on the absorption of cytochromes (Kelin 1925) and hemoglobin (Millikan 1937). The first optical recording of neuronal activity was made over forty years ago by Hill and Keynes (1949) who detected light scattering changes in active nerves. Changes in absorption or fluorescence of intrinsic chromophores were extensively investigated by Chance and his colleagues (1962), and Jobsis and his colleagues (1977). However, the intrinsic optical signals were usually small. Until recently, the optical measurement of intrinsic signals was not applied to the imaging of spatial patterns of neuronal activity (for reviews see Chance et al. 1962; Cohen 1973; Mayevsky and Chance 1982).

The remainder of this chapter will first provide examples of work that illustrate the types of experimental strategies and results that have been obtained from optical imaging with voltage-sensitive dyes. The next series of examples demonstrate the imaging of the functional organization of the cortex using the intrinsic signals. Finally, the relative merits of each imaging method will be discussed. The technical details of optical imaging techniques cannot be presented here, but are nevertheless crucial for optimizing these techniques and evaluating the feasibility of new types of experiments. For earlier reviews see (Grinvald et al. 1988; Cohen and Lesher 1986; De Weer and Salzberg 1986; Grinvald 1984; Loew 1987; Orbach 1987; Salzberg 1983; Salzberg et al. 1985; Tasaki and Warashina 1976; Waggoner 1979; Waggoner and Grinvald 1977).

3.4 Studies of the Organization and Interactions of Neuronal Populations in Vivo Utilizing Voltage-Sensitive Dyes

The detection of spatio-temporal patterns of activity in mammalian brain slices (Grinvald et al. 1982) suggested that optical imaging techniques could prove to be a useful tool for the study of the mammalian brain in vivo. However preliminary experiments in rat visual cortex in 1982 revealed several complications for in vivo optical imaging, which included large noise due to respiratory and heartbeat motion. In addition, the relative opacity and the packing density of the cortex limited the penetration of the excitation light and the ability of available dyes to stain deep layers of the cortex. Better dyes (e.g. RH 414) were developed at the Weizmann Institute and proved useful in extensive dye screening experiments on rat cortex. In addition, an effective remedy for large heartbeat noise was found by synchronizing the data acquisition with the EKG and subtracting a no-stimulus trial. These improvements facilitated the in vivo real-time imaging of the whisker barrels in rat somatosensory cortex (Orbach et al. 1985), the retinotopic responses in the frog optic tectum (Grinvald et al. 1984) and the experiments on the salamander olfactory bulb (Orbach and Cohen 1983; Kauer 1988). More hydrophilic dyes have also been developed that provide better performance in cat and monkey visual cortex (e.g. RH 704 and RH 795).

3.4.1 Imaging Evoked Responses

Among the first applications of in vivo real-time optical imaging using the voltage-sensitive dyes were experiments that visualized the topographic distribution of sensory responses. The frog retinotectal connections are one such system that is topographically organized. The optical signals obtained from the tectum in response to discrete visual stimuli corresponded well to the known retinotopic map of the tectum. When a flash of light was presented to the frog's eye, large fluorescence signals were detected from the exposed and topically stained contralateral optic tectum, even without signal averaging. Displays of light patterns to the frog's eye on an oscilloscope screen (moving or stationary spots, bars or annuli) evoked smaller and even more complex response patterns in the tectum. The optical signals were highly structured; the number of resolved components in a single optical trace was similar to the number of components in the corresponding electrically recorded evoked potential (14–21 components).

The retinotectal connections are topographically organized such that a point stimulus in the visual field evokes activity in a restricted area in the tectum. The spatial resolution and sensitivity of the optical technique was characterized by presenting discrete and weak natural stimuli. A spot of light was first positioned on a corner of a monitor screen, and then, after a delay, it was positioned in the opposite corner. Figure 3.2 illustrates that two excitation foci in the tectum (circled area) could be resolved. Both had a latency ~100 ms with respect to the onset of each light stimulus. The spatial distribution of the signals showed smaller, delayed activity (3–20 ms) covering a much larger area than expected from classical single unit mapping (Grinvald et al. 1984).

Similar mapping experiments were performed in the rat somatosensory cortex, where the simple organization of the whisker barrels offered a convenient preparation for testing new strategies. When the tip of a whisker was gently moved, optical signals were observed in the corresponding cortical barrel field. However a discrepancy was noted between the size of an individual barrel recorded optically (1300 microns) and the histologically defined barrel (300–600 microns in layer IV of the cortex). This difference is reasonable considering that most of the optical signal originates from the superficial cortical layers (I–III)

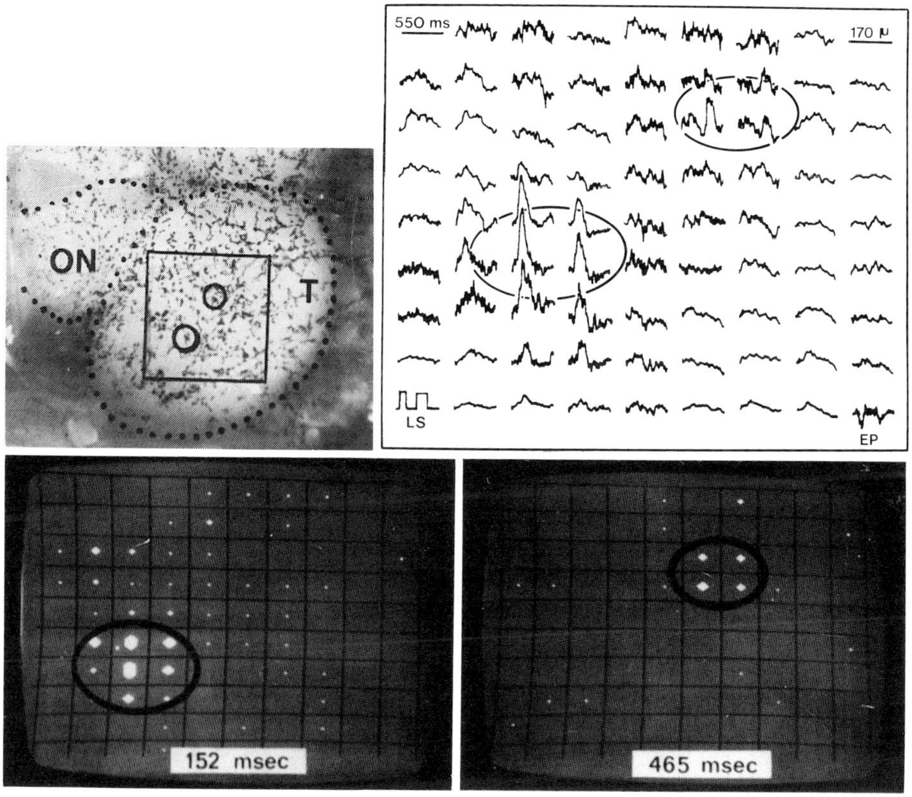

Fig. 3.2. Real-time imaging of visually evoked activity patterns in intact frog tectum. Top panel: evoked responses from a 9 × 9 photodetector positioned over the tectum. Two loci of activity could be resolved, corresponding to 2 different light spots presented one after the other to the contralateral eye (40 trials were averaged). Bottom panels: visualization of spatial patterns of activity at 2 different times depicting 2 loci of activity. Note that the response to the second stimulus was inhibited by the first response. Superimposed picture of tectum was omitted for clarity and is shown at left. EP: evoked potential; LS: timing of the light stimuli

and that barrel neurons extend long processes to neighboring barrels (Orbach et al. 1985). Retinotopic-imaging experiments in monkey striate cortex (Grinvald et al. 1985; Lieke et al. 1988) also showed activity over a cortical area much larger than predicted from standard receptive field analysis (Figs. 3.3,4), but consistent with the anatomical finding of long-range horizontal connections in the visual cortex (Gilbert and Wiesel 1983).

These optical imaging experiments on the topographic distribution of sensory processing all demonstrated a central focus of excitation corresponding to the applied stimulus, and a surrounding area of cortical activity larger than predicted by single unit recording (Fig. 3.3). Possible explanations for the expanded area of optically detected activity include: a) the optical pick-up of axonal activity in long-range horizontal connections that spread far beyond the directly acti-

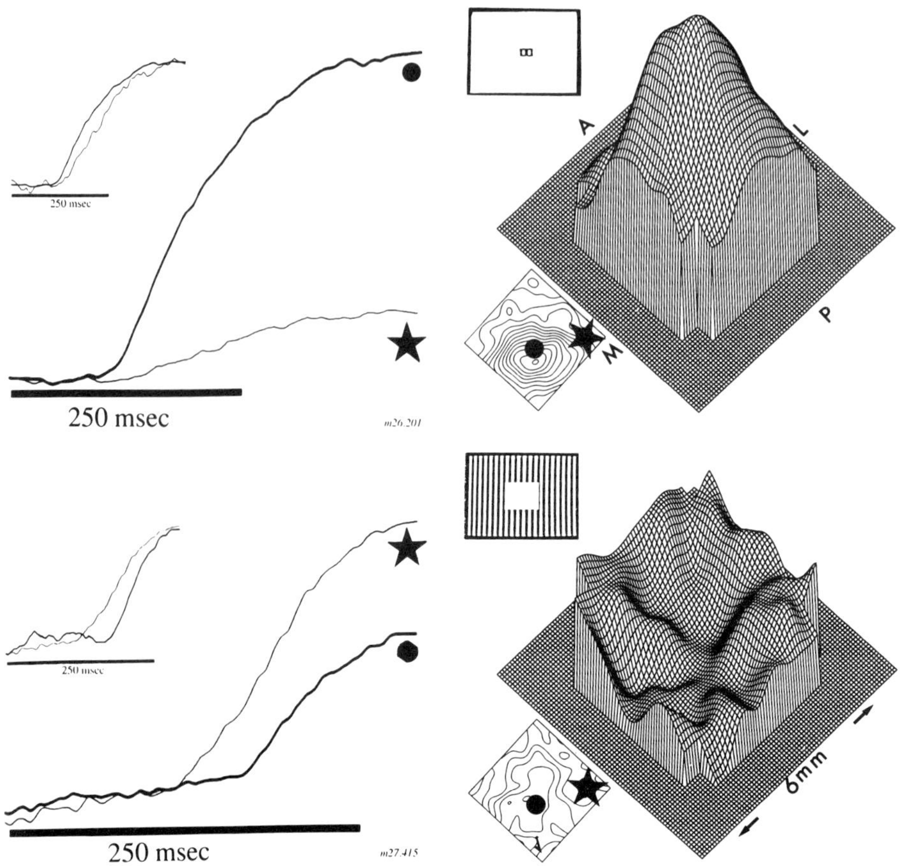

Fig. 3.3. The spatio-temporal spread of the optical signal beyond its retinotopic origin. Left top: Optical signals from the center (bold) of the photodiode array and the periphery elicited by a small (0.5 degree) moving grating ("center-only" stimulus). Inset: The same signals were normalized to equal height to show the increased latency of the optical signal as one moves away from the center of the site of peak activity. Bottom: Optical signals elicited by a large moving grating with a "hole" in the center ("surround" stimulus). The hole was 3 degrees in diameter, so that the cells activated by the stimulus were al least 6 mm away from the center of activity elicited by the "center-only" stimulus. The center of the monitored cortical area was activated 48 ms after the edge, consistent with a conduction velocity of 0.07 of 0.15 m/s (nearly identical in each direction). Right: 3-D picture of the spatial patterns of the two signals from one frame, corresponding to 350 ms after the stimulus onset. The insets display the slimuli (top) and contour maps of the signals (bottom) at that time. The star and the bold circle on the contour map show the optical recording sites for the traces shown in the left part of this figure

vated area; b) a large contribution from subthreshold excitation of post-synaptic dendrites at the regions where neurons are not actually firing in response to the given stimulus; c) a smearing of neuronal signals by slower glial signals (Lev-Ram and Grinvald 1986; Konnerth and Orkand 1986) that spread across the electrically coupled glial synsitium; d) a large contribution from layer 1 whose functional organization is not well characterized.

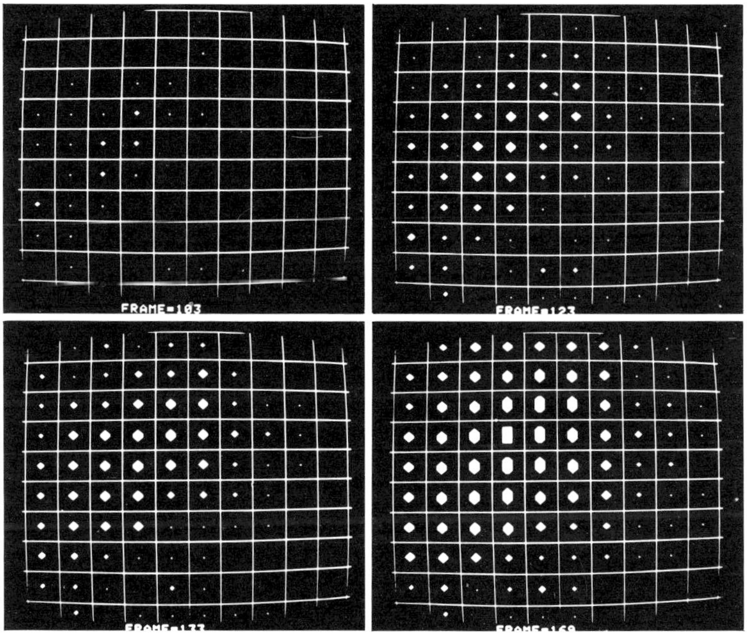

Fig. 3.4. Movie frames showing the time dependency of the signal spread. Four frames are shown. The symbol size reflects the amplitude of the signal. These frames show the activity resulting from a "center" stimulus, which spreads from the center of the imaged cortical area to its edges. Frame sequence alter stimulus presentation: 167 ms, 224 ms, 243 ms and 307 ms, each frame represents activity in one ms

3.4.2 Surround Inhibition Revealed by Optical Imaging

Although the amplitude of optical signals is linearly related to the membrane potential change, the detection of positive optical signals in a given cortical region does not necessarily imply that inhibition is not present. For example, a strong inhibition may not be accompanied by marked hyperpolarization and furthermore it may be masked by a larger depolarization at other neuronal sites. In addition, action potentials in inhibitory axons may also mask the small hyperpolarization in the post-synaptic cells. Thus, it was important to determine whether the surround activity in the topographic mapping experiments discussed above had any inhibitory component. The surround signals were tested with more revealing stimulus paradigms and with the use of pharmacological agents that block inhibitory synapses. Upon presenting a second spot of light that was delayed and displaced (26 degrees) from a preceding spot of light, optical imaging of the frog tectum showed that the activity evoked by a second flash was inhibited by the first flash (by 60%). When a single light spot was used, application of the GABA blocker bicuculline led to a 10-fold increase in the region of excitation, consistent with the notion that inhibitory interactions shape the excitation area. It was also noted that the conduction velocity of the bicuculline-

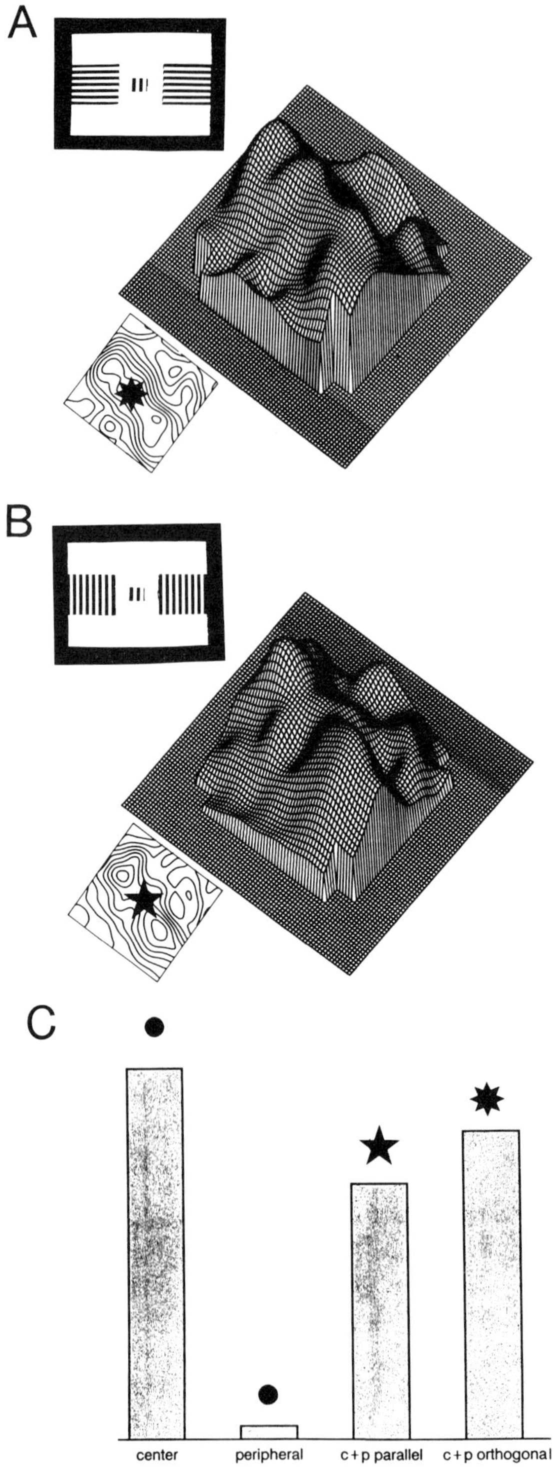
A
B
C
center
peripheral
c+p parallel
c+p orthogonal

induced excitation spread from the excitation focus was faster along the rostral-caudal axis than the medial-lateral axis, implying a possible asymmetry in the lateral surround connections (Grinvald et al. 1984).

In the studies of the rat somatosensory cortex, stimulation of two whiskers that were relatively far apart (e.g., A1 and D4) evoked responses in two circumscribed areas. A large overlap in the response area was obtained when neighboring whiskers were stimulated, but the responses could be resolved if they were activated with some delay (< 20 ms). Activity evoked by one whisker often inhibited the activity evoked by another whisker when the interstimulus interval was 20–120 ms, indicating a surround inhibition extending 2–4 mm in the rat somatosensory cortex (Orbach et al. 1985).

Inhibitory center-surround interactions were also revealed by optical imaging of a 6×6 mm region of monkey striate cortex. Four interlaced visual patterns of moving gratings were presented, stimulating either a small center field or the surround alone, or both together, with center and surround gratings having either matching or orthogonal orientation. The small (0.5–2 degrees) center gratings alone yielded activity in the surrounding cortical area that was delayed relative to the center and had a smaller amplitude. An annular surround grating, instead, produced activity in the center that was smaller and slower. Presenting the center and surround gratings together yielded center and surround cortical signals that were both smaller (~50%) and slower relative to signals from either grating alone, and were smallest when the center and surround gratings had the same orientaton. These results (Fig. 3.5) indicate that the interactions between center and surround have a strong inhibitory component (Lieke et al. 1988). However these experiments do not rule out the existence of lateral excitation between specific cortical sites as has been shown by other physiological and anatomical studies (e.g. Nelson and Frost 1985; Ts'o et al. 1986). This apparent discrepancy illustrates that different techniques may reveal different aspects of cortical interactions. The use of different stimulus conditions may help to clarify this issue. Local injections of drugs which block either excitation or inhibition may also improve such studies as well as nail down the contribution of intracortical connections to the observed interactions.

◁

Fig. 3.5. Center surround inhibitory interactions. 3-D pictures of the optical signals 350 ms after stimulus onset for each of the two different stimuli (both containing center and surround stimuli, but the surround orientation varies in the two experiments). When the surround orientation matches that of the center, there is greater inhibition than when the surround is orthogonal to the center. The histogram (C) at the bottom compares the amplitude of the responses at a position in the center of the exposed cortex. The top and the bottom pictures show the responses to combined stimuli. In A the surround stimulus is orthogonal to the center stimulus, while in B the surround stimulus is parallel to the center stimulus. The response to the center + surround stimulus is smaller than that to the "center only" condition (shown only in histogram). This histogram summarizes the results for the center detector of the array only. At peripheral cortical regions (corresponding to the mask, not shown), where none of the stimuli evoke a retinotopic response, the nature of the center/surround interaction depends on both the cortical site and orientation of the stimulus. The interactions also vary as a function of time. From each experiment which lasts 30 minutes we obtained 30,000 histograms. Such results are best shown in a movie

3.4.3 Selective Visualization of Neuronal Assemblies in Cat Visual Cortex

A neuronal assembly may be defined as a group of neurons that cooperate to perform a specific computation required for a specific task. The activity of cells in an assembly is time-locked (coherent). However, the cells that comprise an assembly may be spatially intermixed with cells in other neuronal assemblies that are performing different computational tasks. Therefore, techniques that can visualize only the average population activity in a given cortical region may not be adequate to study neuronal assemblies. What is needed, then, is a method to discriminate between the operations of several co-localized assemblies. A promising approach to solve this problem is to use the time-locked activity of the assembly neurons, to visualize selectively only that assembly. The firing of a cell that is part of the assembly can serve as a source of synchronization for this selective visualization.

We therefore combined single unit recordings and subsequent spike-triggered averaging of the optical recordings to study the spatio-temporal organization of neuronal assemblies. The visual cortex (area 18) of an anesthetized cat was stained with the dye RH 795, and on-going activity (without a stimulus) was recorded continuously for 70 seconds using simultaneous optical (from 124 sites), local field potentials (LFP) and single unit (from 1–3 isolated units) re-

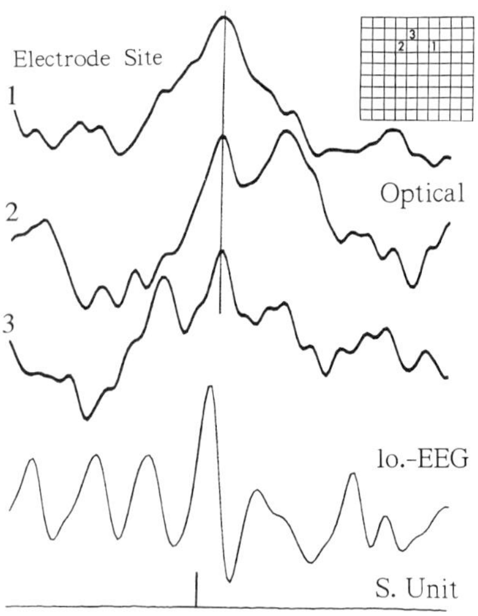

Fig. 3.6. Optical imaging of coherent activity in cat cortex. Single-unit activity and local field potentials were recorded via an electrode placed at site 1 (see inset). Spike-triggered averaging of spontaneous activity was performed on 78 spikes. The bottom trace shows the trigger event. The averaged local field potential (lo.-EEG), shows that this unit had a large probability of firing 7 ms prior to a large peak in the local field potential. Three optical traces from different cortical loci are shown at the top. The topmost trace shows the large coincident activity that was optically recorded from the same site as the single unit and lo.-EEG recordings. The two remaining optical traces show activity with both coincident firing, delayed firing and preceding firing from a neighboring cortical loci (200–400 microns apart). Time scale of the whole picture is 450 ms.

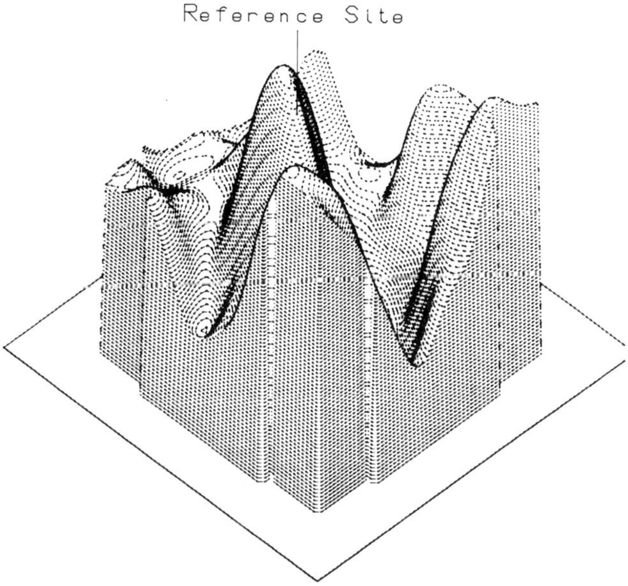

Fig. 3.7. A correlation map of a orientation columns. This 3 mm × 3 mm correlation map was made by dividing the optical map obtained with the horizontal stimulus, by the map obtained with the vertical stimulus. Then the correlation between the optical trace obtained with the photodiode imaging the electrode recording area and all the other optical traces, obtained with all the other photodiodes, was calculated. The neuron at the electrode site was selectively tuned for the vertical orientation. Thus, peaks in this correlation map represent areas in the cortex which tend to respond selectively more to vertical stimuli, while valleys represent areas which tend to respond selectively more to horizontal stimuli.

cordings. The spike-triggered averaging analysis confirmed that single units had a tendency to fire at the positive peak of the local EEG (e.g. Frost 1986). In addition, the averaged optical signal at the electrode site had a peak that also coincided with the occurrence of a peak in the LFP. This result indicated that many neurons next to the electrode site had coherent firing patterns (Arieli et al. 1987; Arieli and Grinvald 1988). However, the optically observed signals were heterogeneous in the field of view of 2 × 2 mm of cortex, (Fig. 3.6) indicating that optical recording provides a better spatial resolution relative to field potential recordings.

The orientation columns in cat visual cortex provide a convenient system to test if neuronal assemblies can be selectively visualized by the above strategy. We therefore employed this approach, performing the recordings for 70 seconds with multiple periods which included a visual stimulus having the optimal orientation for activating the single unit (the reference cell). Figure 3.7 shows a correlation map from such an analysis. A correlation map is obtained by correlating the optical signal at the electrode site with all other simultaneously obtained optical traces. The shape of this map suggests that we were able to selectively image the assembly processing for this stimulus orientation. An independent confirmation of this approach is still lacking, although orientation maps obtained with voltage-sensitive dyes were confirmed by optical imaging with intrinsic signals (see below).

3.5 Optical Imaging of Activity Using Intrinsic Signals

Although considerable progress has been made in the study of various neuronal systems with extrinsic signals originating from voltage-sensitive dyes, the use of intrinsic signals has several distinct advantages. Since no dye is required, the measurement of intrinsic signals is less invasive, can be made for an indefinite period of time and is free of several technical complications associated with the use of dyes (e.g. photodynamic damage, bleaching and pharmacological side effects, see Grinvald et al. 1988; Cohen and Lesher 1986). The primary disadvantage of the use of slow intrinsic signals is the lack of the millisecond time resolution possible with the use of voltage-sensitive dyes. However many applications of the optical imaging of activity are less demanding in time resolution and thus are candidates for the imaging of intrinsic signals.

3.5.1 Determining the Source and Nature of the Intrinsic Signals

Neuronal activity produces at least three types of optical changes in brain tissue which may affect the light intensity reflected from the active cortex (Kelin 1925; Millikan 1937; Hill and Keynes 1949; Chance et al. 1962; Jobsis et al. 1977). The first are light scattering signals, initially reported by Hill and Keynes (1949) who detected light scattering changes in active nerves. These signals have multiple origins (see Cohen 1973 for review). A second type of intrinsic signal originates from changes in the absorption and fluorescence of intrinsic chromophores like hemoglobin, cytochromes or NADH as investigated by several groups (Chance et al. 1982; Jobsis et al. 1977; Mayevsky and Chance 1982). Of particular interest is the extensively studied transition from oxy-hemoglobin to hemoglobin in response to an increased level of electrical activity (oxymetry). A third type of intrinsic signal originates from changes in the amount of light absorption by hemoglobin due to changes in blood volume.

To differentiate individual components of the intrinsic signal, we performed reflection measurements at different wavelengths from 450 to 930 nm. The maps of functional organization obtained were identical in all wavelengths even though different mechanisms dominate the intrinsic signal at different wavelengths (Frostig et al. 1988, 1990). At the blue-green part of the spectrum the dominant mechanism is the change in blood volume, probably due to capillary recruitment and venular dilation. At the yellow-red part of the spectrum the dominant mechanism is the oxymetry, while in the near-infra-red part the dominant mechanism in the cat was probably a mixture of the two mechanisms with a contribution from light scattering changes.

These experiments have also shown that proper control of depth of field is an important factor in optical imaging of cortical activity. Since near-infra-red light penetrates the cortex deeper than visible light it facilitates the imaging of deeper cortical structures. In addition, the confocal microscope (Egger and Petran 1967) may dramatically improve depth resolution. Thus infra-red imaging may prove more useful for three-dimensional confocal imaging than visible light.

3.5.2 Mapping of Functional Organization in Visual Cortex

A prominent feature of the visual cortex is the arrangement of cells having like-orientation preference into columns or bands (Hubel and Wiesel 1962). The optical imaging of the orientation columns with the aid of intrinsic signals required the separate collection of signals for each of two presented moving gratings of orthogonal orientation, and subsequent analytical comparison. A moving grating stimulus evoked intrinsic signals over a large region of cortex. However, at each cortical site the signals evoked by the optimal stimulus orientation were somewhat larger than the signals evoked by the orthogonal orientation (Fig. 3.8A). The spatial distribution of the regions preferring the horizontal stimulus is shown as a two-state coded map in Fig. 3.8B, and the raw data in the upper part of Fig. 3.8C. To obtain a more precise map of the orientation columns, the results from vertical and horizontal stimulus presentations were combined with results from presentations at orientations of 45 degrees and 135 degrees. The resulting map is shown at the lower left of Fig. 3.8C. The orientation map was confirmed, using a double-blind procedure, by making a series of 42 electrode penetrations (Grivald et al. 1986). Also in cat visual cortex, when a similar strategy to visualize the orientation columns was used with the fast voltage-sensitive-dye signals, the resulting orientation maps were identical to those obtained with the aid of the slow intrinsic signals. (Grinvald et al. 1985; Lieke et al. 1988). A comparison of the time courses of the voltage-sensitive dye signal and the intrinsic signal is shown in the lower right panel of Fig. 3.8C.

3.5.3 Increasing the Spatial Resolution of Imaging with Intrinsic Signals

Gross et al. (1984, 1985), Blasdel and Salama (1986) and Kauer (1988) have demonstrated, in imaging experiments with dyes, that standard video techniques may be used to economically permit higher spatial-resolution optical imaging at the expense of time resolution. However, in addition to the poorer time resolution relative to photodiode arrays, a standard video camera also has a poor signal-to-noise ratio, usually no better than 100:1. A slow-scan charge-coupled device (CCD) camera offers improved signal-to-noise ratios (e.g. 1400:1), while retaining the advantages of higher spatial resolution and moderate cost and com-

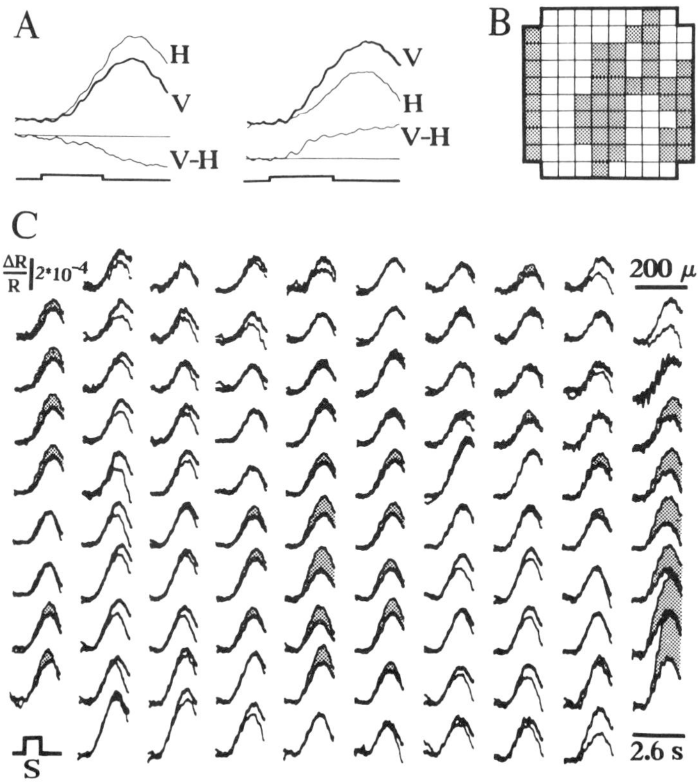
A
H
V
V
H
V-H
V-H
B
C
ΔR/R 2*10^-4
200 μ
S
2.6 s

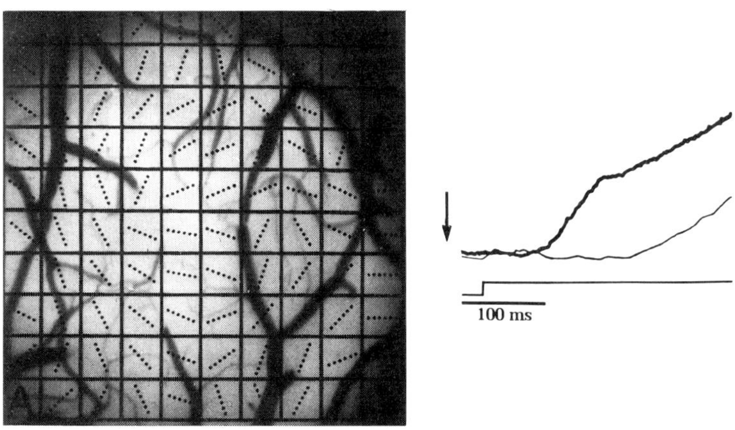
100 ms

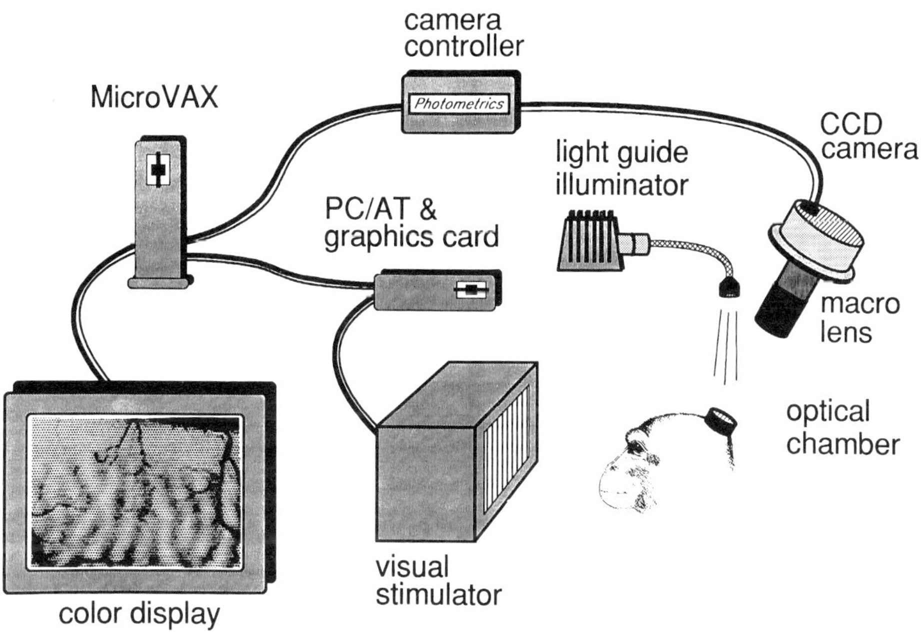

Fig. 3.9. A schematic of the optical imaging setup using the CCD camera. A MicroVAX II computer controlled the experiment and the acquisition of image frames from the camera. To produce a map of ocular dominance (example seen in the "color display" unit), the visual stimulus was gated to only one of the animal's eyes at a time using computer-controlled shutters (not shown)

plexity. We have employed such a CCD camera for the optical imaging at high spatial resolution of the intrinsic signals, where the millisecond time resolution of photodiode arrays is not required (Ts'o et al. 1988, 1990). A scheme of our apparatus is shown in Fig. 3.9.

CCD imaging of the intrinsic signals has enabled us to image directly *in vivo* several of the features of the functional organization of cat and monkey visual areas 17 and 18 (Ts'o et al. 1988, 1990). In each type of experiment, a separate set of frames from the CCD camera is collected for each of several opponent visual stimulus conditions (e.g. vertical vs. horizontal orientation, left eye vs.

Fig. 3.8. Imaging of the orientation columns in cat visual cortex with intrinsic optical signals. A: In one cortical site (left traces), single units preferentially responded to gratings with horizontal orientation. The corresponding intrinsic signal (thin trace) was larger than that obtained in response to the vertical stimuli (thick trace). In a second cortical site (right), single units had preferential responses to the gratings with vertical orientation and the opposite result was obtained. Stimulus duration: 1 s. B: The spatial distribution of cortical regions preferring the horizontal stimulus. C: The raw data showing the distribution of the signals in 96 cortical sites for horizontal (thin traces) and vertical stimuli (thick traces). Bottom left: Superimposed on the picture of the exposed cortex are dashed bars showing the optimal stimulus orientation for each cortical site, as determined by combining results using four different stimulus orientations. Botom right: A comparison between the time course of voltage-sensitive-dye signal (bold trace) and an intrinsic signal (thin trace). Note the expanded time course relative to A.

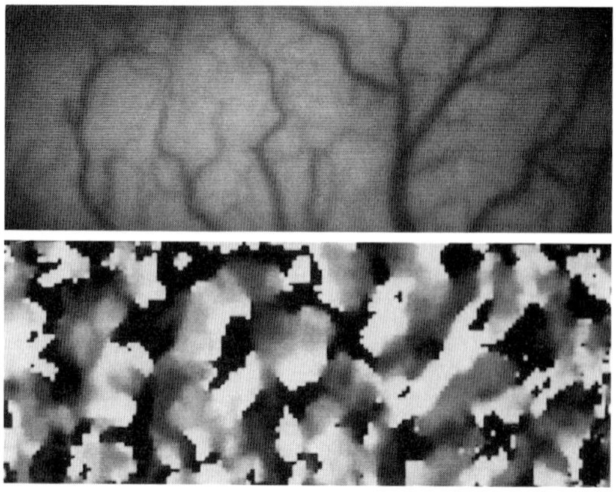

Fig. 3.10. High resolution CCD imaging of orientation columns in cat areas 17 and 18. Top: the CCD picture of the exposed cortex. Bottom: the orientation columns imaged at the same area. The orientation columns were obtained as explained in Fig. 3.8. Scale: 4 × 2.5 mm

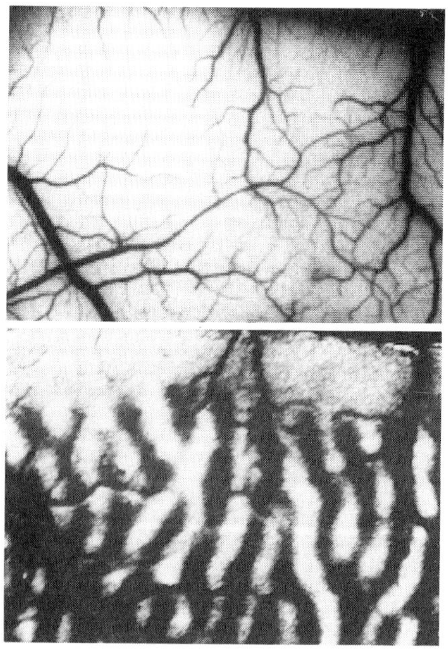

Fig. 3.11. Ocular dominance columns in monkey striate cortex. The frames show an area of 9 × 6 mm. Top: CCD picture of the exposed cortex showing the vasculature of the imaged brain area. The 17/18 border was about 2 mm from the top. Bottom: The ocular dominance columns, obtained by presenting moving gratings of various orientations to the right and left eye individually, averaging the optically recorded frames for each eye and subtracting the two sets of frames. Note the lack of columnar organization in area 18. This map was confirmed by comparison with cytochrome oxidase histology

right eye) and the resulting sets of frames are analytically compared. This procedure has revealed the orientation columns in the cat areas 17 and 18 (Fig. 3.10) to a higher spatial resolution than possible with photodiode arrays. Similar success has been obtained in the monkey visual cortex, where the CCD optical imaging has revealed the orientation and ocular dominance columns (Fig. 3.11), and the thick and thin stripes (Hubel and Livingstone 1987) of area 18 usually seen only with post-mortem cytochrome oxidase histology.

3.5.4 Imaging of Functional Organization Through the Intact Dura and Thinned Skull Preparation

The ability to image the functional organization using near infra-red light encouraged us to try to image cortical activity through the normally opaque dura covering the brain. Because of its penetrating power and reduced absorption by hemoglobin, near infra-red light has allowed us to image, using the CCD camera, the orientation columns of the cat and the ocular dominance columns of the monkey through the intact dura. The maps obtained proved to be identical to maps obtained after the dura was removed (Fig. 3.12). Similar results were obtained for imaging the orientation columns of a cat through a skull which was thinned to about 300 microns. This ability to image the functional organization of the cortex through the dura and thinned skull will be advantageous for research on chronic and awake behaving animal preparations (Frostig et al. 1988,

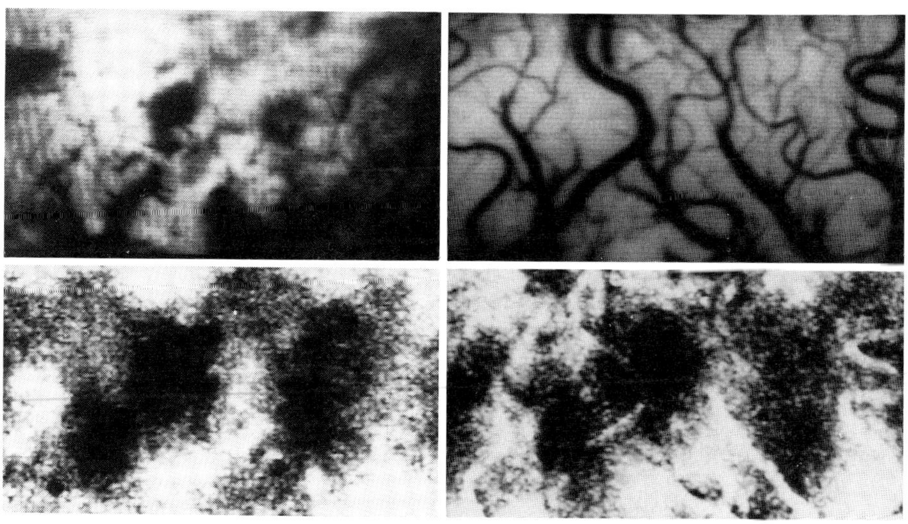

Fig. 3.12. Infra-red imaging of the orientation columns (a two-state map) through the intact dura. Left top: CCD picture of the intact dura illuminated with green light. Left bottom: imaging of the orientation columns through the dura using infra-red light (810 nm). Right top: underlying visual cortex under green light (after the dura was removed). Right bottom: the orientation map obtained with infra-red light after the dura was removed

1990). Not only would it enable mapping the functional organization of an unanesthetized cortex in developing and adult cortices, but it also makes the imaging with intrinsic signals far less invasive and overcomes potential problems associated with compromising the dura (infections, etc.).

3.5.5 Revealing Inhibitory Interactions Between Adjacent Cortical Regions

The visible world contains many richly textured patterns and thus the study of texture-related questions is critical to the understanding of vision processes. A related important question is: at which level does texture-related processing start? The CCD high spatial-resolution images enabled us to image even very small patches of activity in the cortex activated by small visual stimuli (~1/4 degree). The ability to image small areas of activation enabled us to study the interactions between simultaneously activated discreet, adjacent areas in the cortex (Forstig et al. 1989). Such studies compared the response elicited by the center area alone to the same response when the surround texture differs. These studies have demonstrated how any activity in the surround, or in parts of it, inhibits the activity in the center. Moreover, we found that there exists a dependency of these interactions on stimulus orientation. Not all orientations inhibited the center response equally. The dependency of surround inhibition on the orientation of the stimulus was first observed with the dyes, but at lower spatial resolution and unrelated to the functional organization (see above). Such texture-related processing obtained with both imaging techniques was found in the primary visual cortex (area 17) of a Macaque. Thus, texture-related processing appears as early as the primary visual cortex.

3.6 Comparison of Optical Imaging With and Without Voltage-Sensitive Dyes

The principal advantage of cortical mapping using voltage-sensitive dyes is its millisecond time resolution, in contrast to the time resolution of the intrinsic signals which is in the order of seconds. This time resolution is important for a detailed understanding of flow of information and its processing at different cortical sites. For example, in rat somatosensory cortex the area of the observed barrel at 15 ms after stimulation of a single whisker, is about half the size of that seen after 20 ms (Orbach et al. 1985). The trade-off for the good time resolution is a loss in spatial resolution. It is much easier to obtain a good signal-to-noise ratio if signals are slow (a measurement of small slow-signals with a rise-time of a second rather than a millisecond may yield a 33-fold improvement in the signal-to-noise ratio, due to the square root relationship between the number of samples and signal-to-noise ratio.) The spatial resolution of optical imaging of the cortex with the current voltage-sensitive dyes is not ultimately limited by the number of pixels of the photodetector or the available computer and instrumentation capabilities, but instead is limited by obtainable signal size and concomitant photodynamic damage. Without a significant improvement in the quality of the present dyes, we estimated that a 64×64 pixel array is close to the usable limit. In addition to the photodynamic damage associated with the dyes upon prolonged or intense illumination, the use of dyes has other difficulties such as bleaching, limited depth of penetration into the cortex and possible pharmacological side effects. Unfortunately, the extent of pharmacological side effects on cortical function have not been carefully evaluated.

Although the spatial resolution of optical imaging techniques clearly depends on the number of pixels and the resolution of the optics used, these elements may not always be the limiting factors in real experiments. There are sources of biological noise and artifacts that can be averaged out through multiple trials (e.g. heartbeat and respiratory movements). A more subtle source of artifacts may be the accumulation of systematic errors that are not effectively reduced by averaging (e.g. activity-related signals originating from the microvasculature that have spread beyond the activated cortical region, or an uneven distribution of the sources of the optical signals). Therefore, signal-to-noise ratio and reproducibility of the optical images of activity are necessary, but not sufficient, criteria. Controls for a lack of systematic distortions and a satisfactory signal-to-noise ratio may obviate such pitfalls.

Using small visual stimuli we have explored the spread of activity measured by imaging with intrinsic signals, and compared it with the spread of activity imaged with voltage-sensitive dyes. The activity maps obtained in the two cases are different. In the case of imaging with intrinsic signals, the images obtained were identical to areas where spiking neurons responding to the same stimuli were found using microelectrodes (Frostig et al. 1989). However, in the case of voltage-sensitive dyes the area of activation spreads beyond such regions. This difference is due to the fact that the dye also detects subthreshold pre- and postsynaptic activation and action potentials from axonal arbors running horizontally across the cortex. Thus, the two methods complement each other with the type of information that each method contributes.

Since the imaging of the intrinsic optical signals makes no use of dyes, the technique does not suffer from the possibility of photodynamic damage and can be used with much higher spatial resolution. The slow time course of the intrinsic signals further facilitates the use of more economical cameras (e.g., video, CCDs) to achieve higher spatial resolution. Optical imaging with intrinsic signals is less invasive and therefore can be repeated over a long period of time. As we have shown, the imaging of intrinsic signals may also be performed through the intact dura and thinned skull, a very promising prospect that would probably be quite difficult to implement with dyes. These advantages suggest that optical imaging with intrinsic signals may also have clinical applications.

Single- and multi-unit recording can significantly benefit from optical recording. Detailed high spatial-resolution maps of the functional organization of the area of interest, that can be related to surface markers such as blood vessels, enable a quick and precise penetration at carefully chosen loci, for microelectrodes. Obtaining images of functional organization is especially attractive for multiple electrode studies, which are aimed at simultaneous recordings from a certain system or sub-system (e.g., recording only from vertical orientation columns, red-blue blobs, whisker barrels, etc.) and it should reduce the number of electrodes required to get meaningful data.

3.7 Conclusions

Optical recording techniques offer both high temporal and spatial resolution images. Such information is complementary to information obtained by single- and multi-unit recordings. Thus it seems that the combination of optical and single-unit recordings is particularly promising in the study of the living brain, as demonstrated in the case of imaging neuronal assemblies. Moreover, the combination of both optical methods (i.e., voltage-sensitive dyes and intrinsic signals) is a promising combination, where using intrinsic signals would produce a high spatial resolution map (answering the question "where are the active modules"), and using voltage-sensitive dyes to get a high temporal resolution map of the same area (answering the question "how is information flowing between these active modules"). The technical development of such methods calls for a parallel development in the conceptual framework needed for design, analysis and interpretation of new experiments.

Acknowledgements. We thank Drs. L. Katz and T.N. Wiesel for their critical and useful comments, and Dr. Gilbert and Dr. Wiesel for collaboration in several experiments. Our work was supported by IBM and grants from the NIH (NS 14716), the U.S.-Israel Binational Science Foundation, The U.S. Air Force, Klingenstein Foundation and Dupont to AG.

References

Arieli, A., Frostig, R.D., Lieke, E., Hildesheim, R., Grinvald, A. (1987): Soc. Neurosci. Abstr. **13**, 52.
Arieli, A., Grinvald, A. (1988): Soc. Neurosci. Abstr. **14**, 1122
Blasdel, G.G., Salama, G. (1986): Nature **321**, 579
Chance, B., Cohen, P., Jobsis, F., Schonener, B. (1962): Science **137**, 499
Cohen, L.B. (1973): Physiol. Rev. **53**, 373
Cohen, L.B., Lesher, S. (1986): Soc. Gen. Physiol. Ser. **40**, 71
De Weer, P., Salzberg, B.M. (1986): Eds. Soc. Gen. Physiol. Ser. Vol. 40 (Wiley, New York)
Egger, M.D., Petran, M. (1967): Science **157**, 305
Frost, J.D. (1986): Electroenceph. clin. Neurophysiol. **24**, 434
Frostig, R.D., Lieke, E.E., Ts'o, D.Y., Grinvald, A. (1988): Soc. Neurosci. Abstr. **14**, 897
Frostig, R.D., Gilbert, C.D., Ts'o, D.Y., Grinvald, A., Wiesel, T.N. (1989): Soc. Neurosci. Abstr. **15**, 799
Frostig, R.D., Lieke, E.E., Ts'o, D.Y., Grinvald, A. (1990): Proc. Natl. Acad. Sci. USA **87**, 6082
Gilbert, C.D., Wiesel, T.N. (1983): Neurosci. **3**, 116

Grinvald, A. (1984): Trends in Neurosci. **7**, 143
Grinvald, A., Anglister, L., Freeman, J.A., Hildesheim, R., Manker, A. (1984): Nature **308**, 848
Grinvald, A., Cohen, L.B., Lesher, S., Boyle, M.B. (1981): J. Neurophysiol. **45**, 829
Grinvald, A., Frostig, R.D., Lieke, E., Hildesheim, R. (1988): Physiol. Rev. **68**, 1285
Grinvald, A., Gilbert, C.D., Hildesheim, R., Lieke, E., Wiesel, T.N. (1985): Soc. Neurosci. Abstr. **11**, 8
Grinvald, A., Lieke, E., Frostig, R.D., Gilbert, C.D., Wiesel, T.N. (1986): **324**, 361
Grinvald, A., Manker, A., Segal, M. (1982): J. Physiol. **333**, 269
Gross, D., Loew, L.M., Webb, W.W.(1985): Biophys. J. **47**, 270a
Gross, D., Webb, W.W. (1984): Biophys. J. **45**, 269a
Hill, D.K., Keynes, R.D. (1949): Physiol. **108**, 278
Hubel, D.H., Livingstone, M.S. (1987): J. Neurosci. **7**, 3378
Hubel, D.H., Wiesel, T.N. (1962): J. Physiol. **160**, 106
Jobsis, F.F., Keiser, J.H., LaManna, J.C., Rosenthal, M., (1977): J. Appl. Physiol: Respirat. Environ. Exercise Physiol.
Kauer, J.S. (1988): Nature, **331**, 166
Kelin, D. (1925): Proc. R. Soc. B **98**, 312
Konnerth, A., Orkand, R.K. (1986): Proc. Natl. Acad Sci. USA **83**, 6651
Lev-Ram, V., Grinvald, A. (1986): Proc. Natl. Acad Sci. USA **83**, 6651
Lieke, E.E., Frostig, R.D., Ratzlaff, E.H., Grinvald, A. (1988): Soc. Neurosci. Abs. **14**, 1122
Loew, L.M. (1987): Optical measurement of electrical activity (CRC Press, Boca Raton)
Mayevsky, A., Chance, B. (1982): Science **217**, 537
Millikan, G.A. (1937): Proc. R. Soc. B **123**, 218
Orbach, H.S. (1987): In Optical measurement of electrical activity. Ed. L.M. Loew (CRC Press, Boca Raton)
Nelson, J.I., Frost, B.J. (1985): Exp. Brain Res. **61**, 54 (1985).
Orbach, H.S., Cohen, L.B. (1983): J. Neurosci. **3**, 2251
Orbach, H.S., Cohen, L.B., Grinvald, A. (1985): J. Neurosci. 5, 1886
Ross, W.N., Reichardt, L.F. (1979): J. Membr. Biol. **48**, 343
Salzberg, B.M. (1983): In Current Methods in Cellular Neurobiology, Eds. J. Barber, J. McKelvy (Wiley, New York)
Salzberg, B.M., Obaid, A.L., Gainer, H. (1985): J. Gen. Physiol. **86**, 395
Tasaki, I., Warashina, A. (1976): Photochem. Photobiol. **24**, 191
Ts'o, D.Y., Frostig, R.D., Lieke, E.E., Grinvald, A. (1988): Soc. Neurosci. Abstr. **14**, 898
Ts'o, D.Y., Gilbert, C.D., Wiesel, T.N. (1986): J. Neurosci, **6**, 1160
Ts'o, D.Y., Frostig, R.D., Lieke, E.E., Grinvald, A. (1990): Science **249**, 333
Waggoner, A.S. (1979): Bioener. **8**, 47
Waggoner, A.S., Grinvald, A. (1977): Ann. N.Y. Acad. Sci. **303**, 217

4. Dynamic Aspects of Neuronal Cooperativity: Fast Stimulus-Locked Modulations of Effective Connectivity

Ad M.H.J. Aertsen[1] and George L. Gerstein[2]

[1] Max-Planck-Institut für Biologische Kybernetik, Spemannstrasse 38, D-7400 Tübingen
[2] Dept. of Physiology, University of Pennsylvania, Philadelphia PA 19104-6085, USA

Cooperativity among neurons is usually measured by cross-correlation of their simultaneously recorded firing activities under some appropriate stimulus condition. The ordinary 'raw' correlogram presents a time averaged count of (near) coincident spikes, and in general contains contributions from different origins: stimulus-induced modulations of single neuron firing rates and various types of (possibly nonstationary) interaction between the neurons. Appropriate normalization procedures are therefore essential in order to compare results from different experiments; only after such normalization may we interpret residual correlation as indicative of (possibly stimulus dependent) 'effective connectivity' among the neurons.

Recently we reexamined the possibilities for analyzing and interpreting the time course of correlation in two neuron spike trains; in particular we developed procedures to quantify and properly normalize the Joint Peri-Stimulus Time Histogram, a stimulus-locked temporal decomposition of the ordinary cross-correlogram. This normalization allows one to separate purely stimulus-induced correlation from intrinsic interneuronal correlation; moreover it enables one to study the dynamic characteristics of the interneuronal cooperativity.

Another extension of the usual cross-correlation analysis aims at providing a global description of the correlations among larger groups of neurons. The 'gravity representation' gives us a tool that examines the entire set of observed neurons as a single entity, and indicates with high sensitivity which neurons are showing signs of temporal correlations.

These new procedures are described and illustrated by results of their application to physiological multi-unit spike trains from the cat visual cortex (area 17). We present evidence for fast stimulus-locked modulations of 'effective connectivity', pointing at dynamic cooperativity as an emergent property of neuronal assembly organization in the brain.

4.1 Introduction

Since the time of Sherrington, neurobiologists have accepted the doctrine that neurons do not act in isolation, but rather that they form into assemblies for

Springer Series in Synergetics
J. Krüger (Ed.)
Neuronal Cooperativity

various computational tasks. A number of somewhat conflicting definitions of 'neuronal assembly' have been proposed, each implying different functions and properties. Thus, we distinguish (a) Sherringtonian '*neuron pools*', defined by a shared target, (b) *cortical columns*, defined by shared stimulus preference, (c) *Hebbian assemblies*, defined as a net held together by synapses strengthened via the Hebb rule, (d) *correlational assemblies*, defined through correlated time structure in the spike trains of their member neurons; a more exhaustive discussion with many references to the original literature is given by Gerstein et al. (1989).

Two special cases of correlational assemblies can be noted: (1) *pattern generating assemblies*, defined by output of repeating timing sequences in their spike trains (Abeles and Gerstein 1988), and (2) *oscillatory assemblies*, defined by shared coherent oscillatory firing among their member neurons, possibly in response to a particular stimulus (Eckhorn et al. 1988; Gray and Singer 1989; Gray et al. 1989; see also Gerstein 1970).

Correlated spike time structure is readily interpreted as 'effective connectivity' among the observed neurons. Thus, correlational assemblies can be a subset of Hebbian assemblies. On the other hand, neither neuron pools nor cortical columns imply anything about either time structure or effective connectivity. For the rest of this chapter we will be concerned with the physiologically observed properties of correlational assemblies.

In the recent past, it has become possible, in a relatively small number of laboratories, to study directly various phenomena within neuronal assemblies. This has required simultaneous but separable observation of spike trains from as many neurons as possible (currently up to about 30) in a working brain (for reviews see Krüger 1983; Gerstein et al. 1983). The new technology involved in such experiments includes novel types of electrodes and spike shape sorting devices (Schmidt 1984a, b). Above all, it has become necessary to invent new computational tools to analyze and interpret the enormous flow of information coming from such multi-neuron experiments.

The salient and somewhat surprising result of direct assembly observation is that the effective connectivity is *context dependent* and *dynamic* on several different time scales. This suggests that the usual concept of neurons with static interconnections of fixed or only slowly changing efficacy (during learning, for example) is no longer adequate. Instead, we should distinguish between *structural* (or anatomical) connectivity on the one hand, and *functional* (or effective) connectivity on the other. The former can be described as (quasi) stationary, whereas the latter may be highly dynamic, with time constants of modulation in the range of tens to hundreds of milliseconds. It appears that dynamic cooperativity is an emergent property of neuronal assembly organization in the brain, which could not be inferred from single neuron observations. In this chapter we will document these dynamic and context-dependent assembly properties using examples drawn from cat visual cortex recordings. Similar results have been observed from a number of other preparations (Gerstein 1987; Gerstein and Aertsen, in preparation).

4.2 Computational Methods

4.2.1 Cross-Correlation Analysis

The basic tool used to examine timing relations between two spike trains is the cross-correlogram (Perkel et al. 1967). Observed peaks (or troughs) in this measurement, when contrasted with the appropriate control or predictor computations, allow us to define an 'effective connectivity' between the neurons and to parse this into (a) shared modulation by the stimulus, (b) shared input from unobserved, other neurons, and (c) direct connections between the observed neurons (Moore et al. 1970). The 'effective connectivity' is thus equivalent to the simplest possible circuit diagram that would replicate the timing relations between the observed neurons; additional interneurons can obviously not be ruled out. We emphasize that the effective connectivity is a partial subset of whatever actual anatomical connectivity exists, since it represents only the interactions and connections that were sufficiently active to be detectable at the time of observation, and those in a simplified and stylized manner.

The ordinary cross-correlogram is an average measurement over the entire length of the available data. It is appropriate to inquire whether the underlying correlation process is indeed stationary in time (as is required for a time average) or instead has some as yet unknown but potentially significant time structure. There are a number of possible approaches to this problem; here we will use only the Joint Peri Stimulus Time Histogram (JPSTH) – a temporal decomposition of the ordinary cross-correlogram – which will allow us to study the dynamical aspects of the correlation structure that are time-locked to the stimulus events.

4.2.2 Joint Peri Stimulus Time Histogram

The Joint Peri Stimulus Time Scatter Diagram is constructed (Gerstein and Perkel 1969, 1972) as a two-dimensional display where the firing of one neuron relative to each repeating stimulus occurrence is measured along the x-axis while that of the other neuron is measured along the y-axis. Points are entered on this plane at positions corresponding to all the (delayed) coincidences of the firings of each of the two neurons for each successive stimulus repetition. The resulting scatter diagram may show increased point densities parallel to the axes, representing stimulus related firing modulations of one or both neurons. There may also be increased point densities along and near the diagonal of the plane, representing correlated firing of the two neurons. With appropriate binning of the scatter diagram one obtains a matrix histogram which measures the point density as a function of the location in the scatter diagram: the *Joint Peri Stimulus Time Histogram* (JPSTH; Aertsen et al. 1989). Various summations and marginal distributions recover the two ordinary single neuron *PST histograms* (along the x- and y-axes), the *PST coincidence histogram* (along the diagonal), and finally the *cross-correlogram* (perpendicular to the diagonal). The PST coincidence histogram is of particular interest here, and represents the stimulus time-locked average of near-coincident firing of the two neurons in the same sense that the ordinary PST histogram represents the stimulus time-locked average of firing by each individual neuron.

These are all 'raw' measurements, i.e. they do not take into account the effects of possible stimulus modulation of the two individual neuron firing rates. Changes in these rates through the repeating stimulus cycle would necessarily lead to corresponding changes in the near-coincidence rates. For the present purpose this is an uninteresting 'background' effect that we would like to eliminate from the correlation in order to study those components which indicate temporal changes in the underlying neuronal interaction. We have devised the appropriate normalizing calculations as well as a confidence test (see Aertsen et al. 1989; Palm et al. 1988 for details). The *normalization* involves subtracting from each bin in the JPSTH plane the product of the two corresponding bins in the individual PST histograms, and dividing the result by the product of the standard deviations of these same PST bins. Subsequent summations over appropriate diagonal and para-diagonal bins in the plane give us the corrected, 'neuronal interactions only', PST coincidence histogram and cross-correlogram. Thus we can examine fast, stimulus-related changes in the interactions between two observed neurons.

We stress that neither the normalized PST coincidence histogram nor the normalized cross-correlogram could have been obtained by any correction procedure applied to the ordinary cross-correlogram. The proper normalization fundamentally requires the temporal decomposition of coincident firings as done in the JPSTH, its subsequent dynamic correction (subtraction, followed by scaling) and reintegration across or along the diagonal. The nature of these operations means that their order cannot be interchanged. Consequently, the usual procedure for correcting the ordinary cross-correlogram (i.e. subtracting the so-called 'shift predictor') would, in general, lead to different results.

4.2.3 Gravitational Clustering

Unfortunately, once we simultaneously observe even some ten neurons, there is such a combinatorial increase of pairs for ordinary or joint cross-correlation analysis (in general, N neurons give rise to $N(N-1)/2$ different pairs) that the experimenter (not the computer) is overwhelmed. The *gravity representation* (Gerstein et al. 1985a, b; Gerstein and Aertsen 1985; Aertsen et al. 1986, 1987) images each of N observed neurons as a massless particle in an N-space; particles are initially placed at the corners of a hypercube so as to be equidistant. Each particle is given a charge (like electric or gravitational charge) which is a low-pass filtered version of the spike train from the neuron it represents. At each action potential the charge is incremented, and subsequently decays with an appropriate time constant. Suppose that there is a time-varying force along the line joining any two particles that is proportional to their instantaneous charge product. Now assume a viscous medium, so that the velocity of each particle is proportional to the vector sum of all forces upon it. All particles start equidistant from each other, but with these rules, those particles that experience a temporal correlation of high charge (because of correlated firing of the neurons the particles represent) will tend to aggregate in the N-space. Similarly, particles representing neurons which fire with less correlation than expected from their firing rates (i.e. from some form of inhibitory interaction) will tend to move apart. (For further details the reader is referred to the original literature.)

Using methods similar to those for the JPSTH (Aertsen et al. 1987), it is possible to make adjustments for direct stimulation-induced effects on the individual neuron firing rates, so that we may isolate the neural coordination effects (i.e. the effective connectivity) by observing the clustering process. Thus the gravity representation transforms temporal firing correlation among all the simultaneously observed neurons into a spatio-temporal clustering of the corresponding particles. These clusters are easily detected by standard methods, and identify any and all neuronal assemblies defined through correlated firing.

4.3 Results

4.3.1 General Results

With the above tools, we have examined multi-neuron recordings from a number of laboratories. The observations are thus drawn from a number of different preparations and sensory or association systems, and were recorded by somewhat differing techniques. (A more extensive survey of results will be published elsewhere; Gerstein and Aertsen, in preparation.) Notwithstanding the diverse origin of these data, they share a number of interesting properties regarding correlational assemblies.

1) Such assemblies are easily detected. Even with the relatively small numbers of neurons that it is currently practical to observe simultaneously, we frequently find several distinct assemblies.
2) The organization of an assembly may be context-dependent and highly dynamic. Different stimuli can create different correlational structure – interpreted as organization – among the observed neurons. The effective connectivity for some neurons (as measured by the gravity representation) can change both quantitatively and qualitatively. The membership of an assembly may change; some individual neurons may change their 'allegiance'. These measurements suggest that there is a particular organization in an assembly as averaged over many presentations of the same stimulus. But the organization may be different in a similar average over presentations of a different stimulus. Since stimuli are generally presented in an interleaved manner on a time scale of several seconds, these observations suggest that the effective connectivity (or corresponding organization) of an assembly is dynamic on this time scale. Such dynamic reorganization can also sometimes be demonstrated as a

function of recent stimulus history, so that the average organization during presentation of a particular stimulus may depend on the sequence of preceding stimuli. Note that the time scale for all these changes is far faster than that usually associated with development or learning.

3) Within the time of presentation of a single stimulus (now in the averaged sense of the JPSTH) there can be rapid modulations of effective connectivity. The time scale here can be in the tens of milliseconds.

4.3.2 Cat Area 17

We will illustrate these assembly properties with examples drawn from experiments with cat area 17 neurons performed in Krügers' laboratory (Krüger 1982). The electrodes were a 12-fold linear and organ-pipe-like array of glass-coated Pt-Ir wires, with a spacing of 160 microns. They were introduced into the cortex so that their tips fell on a line perpendicular to the cortical surface, i.e. within a single cortical column. The top electrode was near the pial surface, while the lower electrode was just into white matter.

The stimulus for these experiments consisted of a light bar (3° by 14′) moving at constant velocity in a direction perpendicular to its orientation. The distance travelled was 3° in 1.8 s, arranged to cover the entire receptive fields of the neurons under study. At the end of the movement the bar remained stationary for 0.4 s, after which it moved in the opposite direction. Finally the bar was rotated by 22.5° (duration 1 s) after which another cycle of movement back and forth, perpendicular to the new orientation was begun. This protocol was continued until, after 8 cycles of 5 s, the original orientation was reached again. The complete stimulus sequence, lasting 40 s was presented repeatedly (in the example shown, $n = 133$). We present results from an experiment in which the stimulus was presented to the ipsilateral eye only.

4.3.3 Gravitational Clustering

We first demonstrate the application of the gravity representation to these data. Figure 4.1 shows particle pair distances during the gravity condensation for the same set of 8 neurons during 25 repetitions each of two different stimuli (bar orientation 67.5°, direction of movement east-south-east vs. west-north-west, as indicated in the panels). All pairs start at the same distance at the beginning of each run, but the distance between particles of some pairs diminishes rapidly, while the distance between others remains (noisily) constant or even increases. The identities of the pairs are coded in color, according to the table shown at the top left of each panel, and are the same for both panels. The difference between the panels indicates that the correlation structure among the observed neurons was different in the two stimulus conditions, i.e. that the condensation proceded differently. The condensation was calculated with compensation for the direct effects of the stimuli on the firing of the individual neurons. Hence, we

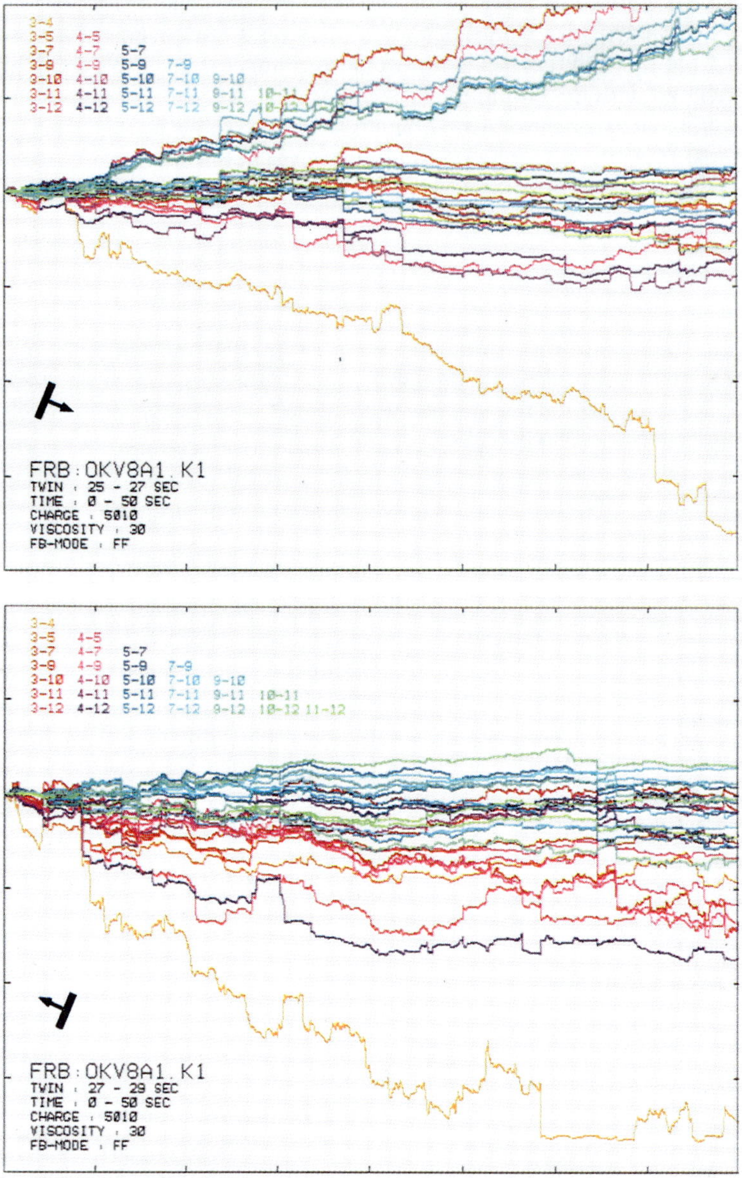

Fig. 4.1. Gravitational clustering: inter-pair distance (y-axis) as a function of time (x-axis) for the 28 pairs of neurons in an 8-neuron recording from the cat visual cortex (area 17) during repeated (n = 25) presentation of a moving bar stimulus. The two panels illustrate the dependence of condensation, and hence of assembly organization, on the stimulus condition: the bar and arrow in each of the two panels indicate the orientation of the light bar (which was the same in both cases) and the direction of motion (which was opposite). The colour coding for the pairs is displayed in the upper left corner of each panel. For further explanation see text

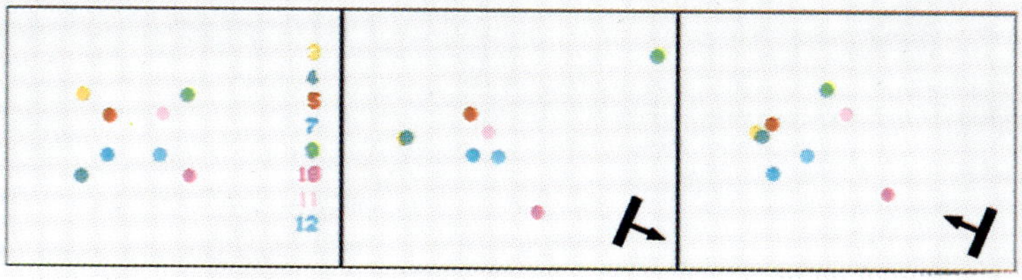

Fig. 4.2. Geometrical projection: 'snapshots' of the particle trajectories in 8-space, projected onto a 2-dimensional plane for visual inspection. The data are the same as shown for the two different stimulus conditions in Fig. 4.1. The leftmost panel shows the initial projected positions of all particles, which were the same in both cases. The colour coding for the particles, indicated in the leftmost panel, is the same in all three panels. The final projected clusterings for the two stimulus conditions are shown in the right-hand two panels. As in Fig. 4.1, the bar and arrow in each of the two panels indicate the orientation of the light bar (which was the same in both cases) and the direction of motion (which was opposite). Note that the final condensations are quite different in the two cases (and consistent with the distance curves in Fig. 4.1). For further explanation see text

may conclude that the effective connectivity among our observed neurons is different with the two stimuli.

Additional aspects of the condensation process are presented in Fig. 4.2 for the same data from two stimulus conditions. Here we see an appropriate projection from the 8-space (in which the particles actually move) to a plane which we can examine by visual means. The projection process necessarily loses information; in order to evaluate apparent clusterings in the projection, we must also examine the pair distances in Fig. 4.1. We show these projections simply as a convenient didactic tool; the clustering process should more properly be examined in the full 8-space in which the particles move, thus avoiding ambiguities and the need to consult the pair distance graphs. Initial projected positions of all particles are shown in the left hand panel of Fig. 4.2. The final projected clusterings shown for the two stimulus conditions in Fig. 4.2 (right-hand two panels) are clearly quite different (and are consistent with the distance information in Fig. 4.1). As noted before, the gravity condensation process has been carried out with compensation for the direct effects of the stimuli on the firing of the individual single neurons. Thus we again demonstrate that the effective connectivity is dependent on the stimulus conditions; the organization and membership of the neuronal assemblies that we have detected is significantly different for the two stimulus conditions shown here.

4.3.4 Joint Peri Stimulus Time Histogram

We now turn to modulations of effective connectivity on a faster time scale. Here we will examine the firing of two neurons from the same data set under one particular stimulus condition making use of the Joint Peri Stimulus Time Histogram. We will evaluate the stimulus time-locked average modulations of the correlations between the firing of these two neurons.

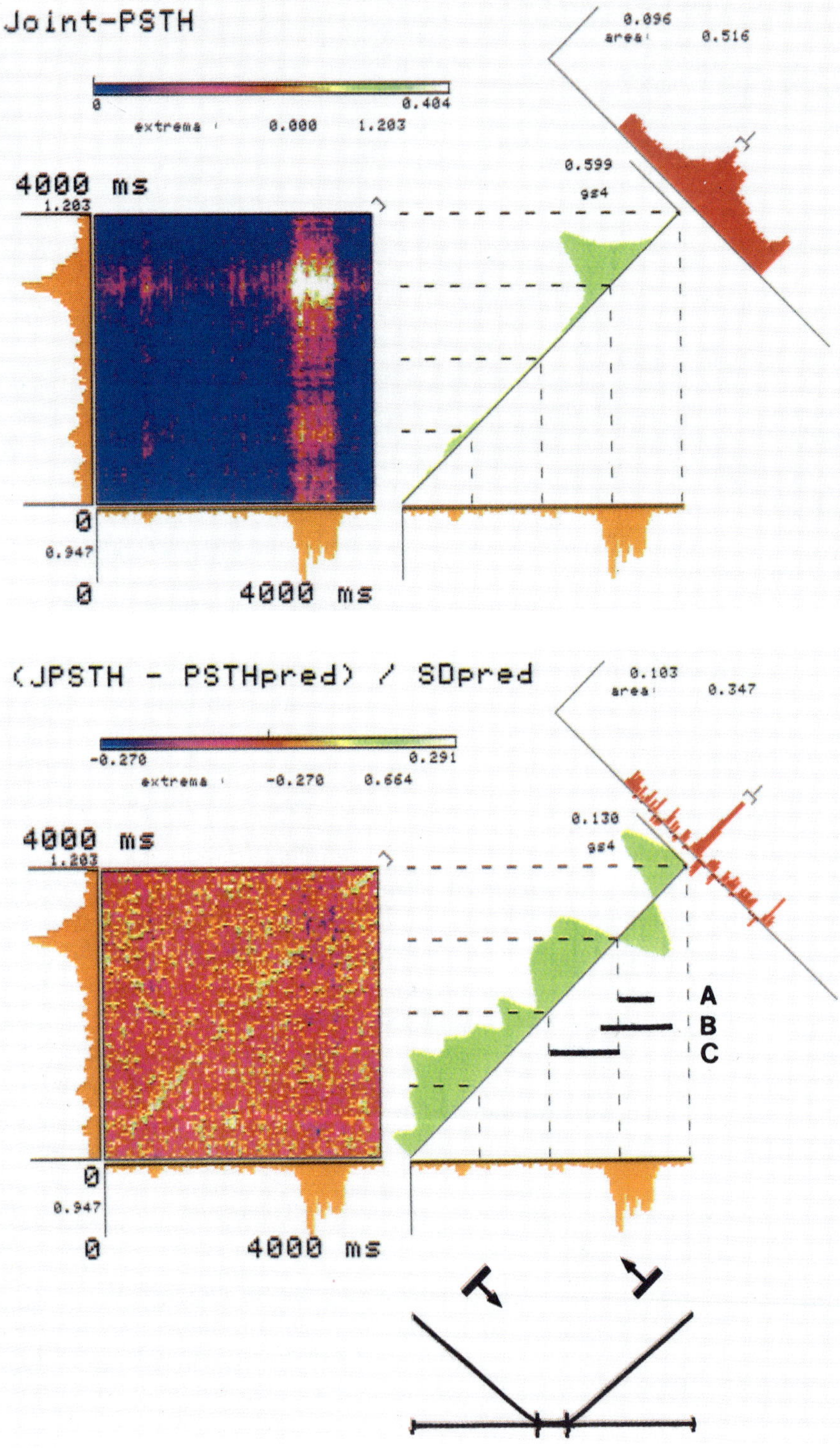
Joint-PSTH
0.096
area: 0.516
0
0.404
extrema: 0.000 1.203
0.599
gs4
4000 ms
1.203
0
0.947
0
4000 ms
(JPSTH - PSTHpred) / SDpred
0.103
area: 0.347
-0.270
0.291
extrema: -0.270 0.664
0.130
gs4
4000 ms
1.203
A
B
C
0
0.947
0
4000 ms

Figure 4.3 shows both the 'raw' (top panel) and the normalized JPSTH (bottom panel) and the associated marginal distributions for the two selected neurons. Within each panel, we show the individual PST histograms at the left of the matrix, and (upside down) twice below. The matrix is the JPSTH, with color used to indicate the density of coincident firing, according to the color bar above the matrix. At the right of each panel are respectively the PST coincidence histogram (along the diagonal rising to the right) and the cross-correlogram (perpendicular to the diagonal and descending to right). We pay particular attention here to the PST coincidence histogram which bins the counts near the diagonal of the joint matrix, and which represents the time-locked average of near-coincident firings of the two neurons in the same sense that the ordinary PST histogram represents the stimulus time-locked average of the individual neuron's firings. The time course of the light bar stimulus movement is indicated by the V-shaped line under the JPSTH matrix; the movement was perpendicular to the bar orientation, and the descending and ascending portions of the V correspond to the two directions of movement (bar orientation 45°, direction of movement south-east vs. north-west, as indicated along the V-shape).

In the 'raw' JPSTH panel at the top we note that the individual PST histograms show a strong time-locked increase in firing for both neurons as the stimulus enters their receptive fields, but only during the second direction of motion. These are direction- (as well as orientation-) selective neurons. The Joint PSTH matrix shows a considerable hill at a location matching the PST peaks, and a corresponding peak is visible in the PST coincidence histogram. In other words, individual, joint, and near-coincidence firings all are increased during a portion of the second direction of movement. In contrast, note that none of these quanti-

◁

Fig. 4.3. Stimulus-locked dynamic correlation of neuronal firing for two neurons recorded simultaneously from the cat visual cortex (area 17) during repeated (n = 133) presentation of a moving bar stimulus. The upper panel shows the 'raw' Joint-PST histogram; the lower panel shows the normalized Joint-PST histogram, calculated by subtracting from the 'raw' JPSTH the cross-product matrix of the individual PST histograms, and by dividing the resulting difference matrix (bin by bin) by the cross-product matrix of the standard deviations of the individual PST histograms. The display format of both panels is the same. The left-hand half of the panel shows the JPSTH matrix and the two ordinary PST histograms along its x- and y-axis (binwidth: 40 ms). Values in the JPSTH matrix are displayed by using color as indicated in the color bar above the matrix. The tic mark above the color bar corresponds to the value zero. All counts were divided by the number of stimulus presentations. The right-hand half of the panel shows the PST coincidence histogram (running along the diagonal from lower left to upper right) and the ordinary cross-correlation histogram (perpendicular to the diagonal, and running from upper left to lower right). The PST coincidence histogram was smoothed using a Gaussian with a sigma of four bins; this particular value (gs4), as well as the location and width of the selected diagonal band are indicated in each panel. The position of true coincidence (zero delay) in the cross-correlogram coincides with the intersection point of the PST coincidence histogram and the cross-correlogram; it is indicated by a tic mark near the diagonal band marker above the correlogram. Numbers of events: 2118 (neuron 1, x-axis), 3911 (neuron 2, y-axis), 133 (stimulus events). Spike trains were recorded during repeated presentation of a moving bar stimulus; the orientation of the bar and the direction and time course of movement are schematically indicated by the bars with arrows and the V-shaped line underneath the lower panel. In addition, the lower panel shows three different time windows (A, B and C) along the PST coincidence histogram for partial paradiagonal summation, leading to the three corresponding normalized cross-correlograms shown in Fig. 4.4. For further explanation see text

ties is increased as the stimulus passes through the receptive field in the first (and opposite) direction of movement.

The bottom panel of Fig. 4.3 contains the normalized JPSTH, i.e. corrected for the effects of stimulus time-locked modulation of the individual firing rates. We concentrate our attention on the normalized PST coincidence histogram along the diagonal. Note that during the first direction of movement, there is a positive level of residual near-coincidence firing, which is approximately constant, even when the stimulus passes through the receptive field region of the neurons. In contrast, during the second direction of movement, there is a lower than expected amount of near-coincidence firing (the correlation has gone negative) just as the individual firing rates go up in response to the stimulus. Thus, Fig. 4.3 demonstrates that near-coincidence firing can be strongly modulated by the stimulus presentation. These two neurons are repeatedly switching from a condition of ongoing co-firing to a stimulus-related period of strong anti-coincidence: when one fires, the other has a strong tendency to be more silent than expected from the individual firing rates.

In the example shown here, the switching from a condition of co-firing to one of anti-co-firing seems to be associated, at least roughly, with the transition of low to high firing rate of the two observed neurons (compare the time course of the PST histograms with that of the normalized PST coincidence histogram in Fig. 4.3). Note, however, also the brief, but strong reduction of co-firing (almost down to zero-level) slightly before 2000 ms, i.e. immediately after the stimulus bar stopped moving (as indicated by the short horizontal portion in the middle of the V-shaped line under the JPSTH matrix). Note also that in this case the modulation in co-firing was *not* associated with a corresponding change in either of the two neurons' individual firing rates. We have observed similar phenomena in other data. For example, the JPSTH in Chap. 11 by Vaadia et al. (Fig. 11.13), taken from two neurons in the prefrontal cortex of a behaving monkey, shows a strong, temporary increase from zero to a considerable degree of co-firing upon presentation of a task-related "go" signal, without any sign of firing rate modulation in the two neurons individually. We conclude that correlation of firing among neurons may change dramatically and fast, even without any appreciable sign in the probability of firing of any of the individual neurons. We have also often observed the converse phenomenon, i.e. strong modulation of single neuron firing rates without a significant modulation in the rate of coincidence firing.

The result shown in Fig. 4.3 has important implications for the use of ordinary cross-correlograms (even after subtraction of the so-called 'shift predictor'). The 'raw' cross-correlogram for the two observed neurons is shown at the extreme right of the top panel of Fig. 4.3, while its normalized version is in the corresponding position of the bottom panel. Both the 'raw' and the normalized cross-correlograms show positive peaks straddling the time origin. However, note that these correlograms are obtained by making sums in paradiagonal bins over the entire 'raw' and normalized JPSTH matrices. In the case of the strong stimulus modulation of near-coincident firing shown here, this is clearly a misleading procedure. If we choose three different time windows along the PST

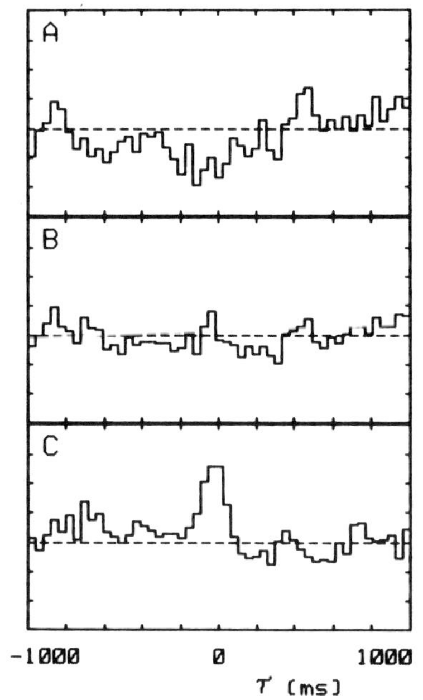

Fig. 4.4. Normalized cross-correlograms for different selections of time window over which to average the data. The three cross-correlograms (A, B and C) show the results for the corresponding time windows along the PST coincidence histogram in the lower panel of Fig. 4.3. Note the dramatic differences in correlogram shape in these three cases, due to the correlation process being dynamically modulated by the stimulus

coincidence histogram for partial paradiagonal summation (as indicated in the lower panel of Fig. 4.3), we obtain the three corresponding cross-correlograms shown in Fig. 4.4. Window A, taken during the time of strongest single neuron firing, coinciding with maximum stimulus-locked anti-correlation, produces a cross-correlogram with a strong valley, which would be interpreted as a strong *mutual inhibition* between the two observed neurons. By contrast, window B produces a flat cross-correlogram, which would be interpreted as *no interaction* between the two neurons. Finally, window C produces a strong peak in the cross-correlogram, which would be interpreted as a *shared excitation* from other, unobserved neurons. The point to be made here is that the ordinary cross-correlogram is an average measure over all the available data. If the correlation process is indeed being modulated by the stimulus (or any other source), such a global average is misleading. Rather, one should select appropriate time windows over which to average the cross-correlation measurement. These time windows, however, must be chosen intelligently, and only after examining the dynamic behavior of the correlation itself, as exhibited in the normalized PST coincidence histogram, rather than by mere inspection of the individual single unit firing rates, as measured by the PST histograms.

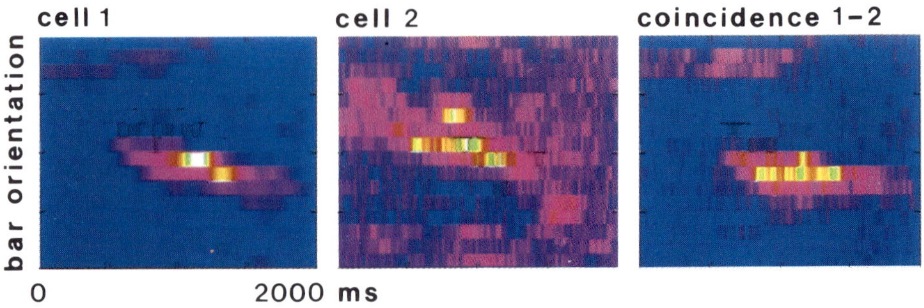

Fig. 4.5. Stimulus coding by correlated firing. The left-hand two panels show the response planes (composed of stacked PST histograms: each horizontal row corresponds to a single PST histogram, obtained for a particular stimulus condition) for the two cat area 17 neurons in Fig. 4.3. The coordinates of the plane are stimulus parameter (16 values for orientation and direction of bar movement) along the y-axis, and time relative to the stimulus along the x-axis. Color is used to indicate response magnitude (according to the same color key as used in the JPSTH in Fig. 4.3). The right-hand panel shows a comparable response plane for the *correlated* firing of these two neurons, made up of stacked PST coincidence histograms (such as in the upper panel of Fig. 4.3). Note that the spatio-temporal tuning domain of the near-coincident firing is more restricted (narrower) than the response domain of either of the two participating neurons alone

4.3.5 Stimulus Coding by Correlated Firing

Finally, we address the issue of stimulus coding by the use of correlated firing. The basic quantity to be examined is the tuning curve; we will compare the tuning of the individual neurons and the tuning of their near-coincidence firing as the stimulus parameter is varied (over bar orientation and direction of movement in this particular experiment).

Figure 4.5 shows three *response planes*; these contain averaged responses for each of the 16 different stimulus conditions (8 bar orientations, 2 movement directions). The averaged time-locked response to each stimulus condition is represented by a separate row in these planes, with color used to indicate response magnitude (according to the same color key as used in the JPSTH; cf. Fig. 4.3). The coordinates of the plane are therefore stimulus parameter along the y-axis, and time relative to the stimulus along the x-axis. Thus the plane exhibits spatio-temporal tuning domains (with spatio- referring to the stimulus parameter along the y-axis, in this case direction of bar movement) as an obvious generalization of a one-dimensional tuning curve in a compact manner.

The two left-hand response planes in Fig. 4.5 represent the stimulus time-locked activity of each of the two neurons individually; these are stacked ordinary PST histograms (cf. Gerstein et al. 1968). The right-hand panel shows a comparable plane made up of stacked PST coincidence histograms. It is clear from these pictures that the spatio-temporal tuning domain of the near-coincident firing is more restricted (narrower) than the response domain of either of the two participating neurons alone. Thus we demonstrate that near-coincident firing can have a more selective stimulus tuning than the individual neuron's firing. Since we represent 'raw' near-coincident firing in the right panel, this sharpening of tuning is potentially available as a neuronal code for transmitting information. Although we do not demonstrate it here, the normalized version of the coincidence panel in Fig. 4.5 can be used to show the changes of effective connectivity between the two neurons as a function of both stimulus parameter and time relative to stimulus presentation in a compact manner.

4.4 Discussion

In this paper we have examined some properties of real neuronal assemblies as observed by simultaneous but separable observation of spike trains from several neurons. Using appropriate computational tools, we have been able to examine both the 'raw' correlation structure of the spike trains and the effective connectivity. The latter measure gives a simplified model of the organization among the observed neurons, and defines the assembly structure.

Our principal observation is that the correlation structure, and hence the membership as well as the internal organization of the observed assemblies, is context-dependent and dynamic on several time scales ranging from 10 ms to 10 s.

The gravity representation allowed examination of the average organization of the observed assemblies over many presentations of the same stimulus, and allowed comparison with the average organization during repeated presentation of a different stimulus. Since stimuli are usually presented in an interleaved manner, these organizational changes are occurring on a time scale of seconds.

The Joint Peri Stimulus Time Histogram allowed us to examine the stimulus time-locked variations of correlation between the firings of two neurons. Here, by means of the appropriate normalization procedure, we could demonstrate stimulus time-locked changes on a time scale of fractions of a second. Similar results on other data not shown here bring the time scale for this kind of variation of correlation down to the 10 ms range (see e.g. Gochin et al. 1990). These fast modulations of firing correlation may occur with or without associated modulations in the individual neurons' firing rates. We have pointed out that ordinary cross-correlation is an average over all the data, and that, consequently, when the correlation is varying in a time-locked (or other) way, this is at best a misleading calculation. Both appropriate and inappropriate windows with respect to the stimulus structure can be chosen, so that inspection of the PST coincidence histogram is essential before deciding on time windows for the cross-correlation measurements.

Finally we have examined tuning curve structure as parameters of the stimulus are varied, showing both single neuron responses and near-coincidence firing. In some cases the correlated firing was more sharply tuned over the stimulus parameter variation, and hence, at least in principle, might be a useful code for the processing of information.

It is appropriate to seek the mechanisms that could underlie the observed context-dependent dynamics of assembly organization. One possibility is that there are rapidly changing synapses connecting the observed neurons, and that variations in the synaptic weights [of the kind proposed by von der Malsburg (1985)] account for the dynamics. Known mechanisms like pre-synaptic inhibition or NMDA synapses provide possible substrates for this type of explanation, but leave undefined the source of the neuronal activity which controls the synaptic modulation. Alternatively, we may suppose that the synaptic weights are constant, and that we are seeing different, partly overlapping Hebbian assemblies which are differentially ignited by the stimulus context, and which have a complex internal spatio-temporal structure (see e.g. Johannesma et al. 1986). A final class of explanation also assumes that synaptic weights are constant, but that we are seeing the effects of large numbers of unobserved neurons whose summed pool activity plays on our observed neurons, and hence modulates the apparent influence that the observed neurons exert on each other. Considerations of these alternatives are beyond the scope of this chapter, and are partially examined elsewhere (Gerstein et al. 1989; Boven and Aertsen 1989, 1990; Erb et al. 1986, 1989; Bedenbaugh et al. 1988, 1990).

Acknowledgements. We wish to thank Jurgen Kruger for providing us with the cat area 17 spike recordings, and Claudia Martin-Schubert for her help in making the final pictures. GLG was partially supported through ONR N-00014-87-K-0766 and AFOSR-88-0296, AA through grants from the BMFT and from GIF.

References

Abeles, M., Gerstein, G.L. (1988): Detecting spatiotemporal firing patterns among simultaneously recorded single neurons. J. Neurophysiol. **60**, 909–924

Aertsen, A., Gerstein, G., Johannesma, P. (1986): From neuron to assembly: Neuronal organisation and stimulus representation. In: *Brain Theory*, ed. by Palm, G., Aertsen, A. (Springer, Berlin, Heidelberg) pp. 7–24

Aertsen, A., Bonhoeffer, T, Krüger, J. (1987): Coherent activity in neuronal populations: analysis and interpretation. In: *Physics of Cognitive Processes*, ed. by Caianiello, E.R. (World Scientific, Singapore) pp. 1–34

Aertsen, A.M.H.J., Gerstein, G.L., Habib, M.K., Palm, G. (1989): Dynamics of neuronal firing correlation: modulation of "effective connectivity". J. Neurophysiol. **61**, 900–917

Bedenbaugh, P.H., Gerstein, G.L., Boven, K.-H., Aertsen, A.M.H.J. (1988). The meaning of stimulus dependent changes in cross correlation between neuronal spike trains. Soc. Neurosci. Abstr. **14**, 651

Bedenbaugh, P.H., Gerstein, G.L., Aertsen, A.M.H.J. (1990): Dynamic convergence in neural assemblies. Soc. Neurosci. Abstr. **16**, in press

Boven, K.-H., Aertsen, A. (1989): Functional connectivity in neuronal systems: fast modulation of synaptic efficacy reflects dynamics of network activity. In: Elsner, N., Singer, W. (eds.): Dynamics and plasticity in neuronal systems, 444 (Thieme, Stuttgart)

Boven, K.-H., Aertsen, A. (1990): Dynamics of activity in neuronal network give rise to fast modulations of functional connectivity. In: Eckmiller, R. et al. (eds.): Parallel processing in neural systems and computers (Elsevier, Amsterdam) pp. 53–56

Eckhorn, R., Bauer, R., Jordan, W., Brosch, M., Kruse, W., Munk, M., Reitboeck, H.J. (1988): Coherent oscillations: a mechanism of feature linking in the visual cortex? Multiple electrode and correlation analysis in the cat. Biol. Cybern. **60**, 121–130

Erb, M., Palm, G., Aertsen, A., Bonhoeffer, T. (1986): Functional versus structural connectivity in neuronal nets. In: Strukturbildung und Musteranalyse. Proc. 9th Cybernetics Congress (DGK). Göttingen, FRG, p. 23

Erb, M., Aertsen, A., Palm, G. (1989): Functional connectivity in neuronal systems: context-dependence of effective network organization does not require synaptic plasticity. In: Elsner, N., Singer, W. (eds.): Dynamics and plasticity in neuronal systems, 445 (Thieme, Stuttgart)

Gerstein, G.L. (1970): Functional association of neurons: detection and interpretation. In: The Neurosciences: Second Study Program, ed. by Schmitt, F.O. (Rockefeller University Press, New York) pp. 648–661

Gerstein, G. (1987): Information flow and state in cortical neural networks: interpreting multi-neuron experiments. In: Organization of Neural Networks: Structure and Models, ed. by Seelen, W. von, Leinhos, U., Shaw, G. (VCH, Weinheim) pp. 53–75

Gerstein, G.L., Perkel, D.H. (1969): Simultaneously recorded trains of action potentials: Analysis and functional interpretation. Science **164**, 828–830

Gerstein, G.L., Perkel, D.H. (1972): Mutual temporal relationships among neuronal spike trains. Biophys. J. **12**, 453–473

Gerstein, G., Aertsen, A. (1985): Representation of cooperative firing activity among simultaneously recorded neurons. J. Neurophysiol. **54**, 1513–1527

Gerstein, G.L., Butler, R.A., Erulkar, S.D. (1968): Excitation and inhibition in cochlear nucleus. I. Tone-burst stimulation. J. Neurophysiol. **31**, 526–536

Gerstein, G.L., Bloom, M.J., Espinosa, I.E., Evanczuk, S., Turner, M.R. (1983): Design of Laboratory for multi-neuron studies. IEEE-Trans. System, Man and Cybernetics **SMC-13**, 668–676

Gerstein, G., Perkel, D., Dayhoff, J. (1985a): Cooperative firing activity in simultaneously recorded populations of neurons: detection and measurement. J. Neurosci. **5**, 881–889

Gerstein, G., Aertsen, A., Bloom, M., Espinosa, I., Evanczuk, S., Turner, M. (1985b). Multi-neuron experiments: observation of state in neural nets. In: Complex Systems – Operational Approaches, ed. by Haken, H. (Springer, Berlin, Heidelberg) pp. 58–70

Gerstein, G.L., Bedenbaugh, P., Aertsen, A.M.H.J. (1989): Neuronal Assemblies. IEEE Trans. on Biomed. Engineering **36**, 4–14

Gochin, P., Gerstein, G., Kaltenbach, J. (1990): Dynamic temporal properties of effective connections in rat dorsal cochlear nucleus. Brain Res. **510**, 195–202

Gray, C.M., Singer, W. (1989): Stimulus-specific neuronal oscillations in orientation columns of cat visual cortex. Proc. Natl. Acad. Sci. USA **86**, 1698–1702

Gray, C.M., König, P., Engel, A.K., Singer, W. (1989): Oscillatory responses in cat visual cortex exhibit inter-columnar synchronization which reflects global stimulus properties. Nature **338**, 334–337

Johannesma, P., Aertsen, A., Van den Boogaard, H., Eggermont, J., Epping, W. (1986): From synchrony to harmony: Ideas on the function of neural assemblies and on the interpretation of neural synchrony. In: *Brain Theory*, ed. by Palm, G., Aertsen, A. (Springer, Berlin, Heidelberg) pp. 25–47

Krüger, J. (1982): A 12-fold microelectrode for recording from vertically aligned cortical neurones. J. Neurosc. Meth. **6**, 347–350

Krüger, J. (1983): Simultaneous individual recordings from many cerebral neurons: techniques and results. Rev. Physiol. Biochem. Pharmacol. **98**, 177–233

Malsburg, C. von der (1985): Nervous structures with dynamical links. Ber. Bunsenges. Phys. Chem. **89**, 703–710

Moore, G.P., Segundo, J.P., Perkel, D.H., Levitan, H. (1970): Statistical signs of synaptic interaction in neurons. Biophys, J. **10**, 876–900

Palm, G., Aertsen, A.M.H.J., Gerstein, G.L. (1988): On the significance of correlations among neuronal spike trains. Biol. Cybern. **59**, 1–11

Perkel, D.H., Gerstein, G.L., Moore, G.P. (1967): Neuronal spike trains and stochastic point processes II. Simultaneous spike trains. Biophys. J. **7**, 419–440

Schmidt, E.M. (1984a): Instruments for sorting neuroelectric data: a review, J. Neurosci. Meth. **12**, 1–24

Schmidt, E.M. (1984b): Computer separation of multi-unit neuroelectric data: a review, J. Neurosci. Meth. **12**, 95–111

5. Temporal Structure of Cortical Information Processing: Cortical Architecture, Oscillations, and Non-Separability of Spatio-Temporal Receptive Field Organization

Hubert R. Dinse[1], Katharina Krüger[2], Hanspeter A. Mallot[2], Johannes Best[2]

[1]Coleman Lab, UCSF, University of California, San Francisco, CA 94143, USA
[2]Lehrstuhl für Theoretische Biologie – Voustitut für Neuroinformatik, Ruhr-Universität Bochum, D-4630 Bochum

Since the work of Hartline in 1938 the concept of receptive fields (RFs) as one of the basic elements for information processing has been the guiding principle in the investigation of sensory systems. Although we perceive the visual world as a unitary whole, the neuronal machinery consists of elements that can analyse only small portions of the visual field. Knowing structure, organization and properties of RFs is indispensable for understanding their contribution for signal processing and, more generally, for brain function.

5.1 Introduction

5.1.1 Receptive Fields as Spatial Filters

Visual cortical neurons in area 17 have been described as responding in a rather specific way to the shape of a stimulus, especially to its orientation and length, known as orientation selectivity and hypercomplexity (end stopping). According to the RF concept proposed by Hubel and Wiesel (1962), the selectivity of cortical neurons for geometrical dimensions, i.e. to two-dimensional spatial distribution of brightness, is generated by a corresponding two-dimensional spatial distribution of excitatory and inhibitory inputs in the cortex. In this view, the spatial selectivity of a cell is essentially determined by the spatial arrangement of the afferent inputs onto a neuron. The distribution of excitation within RFs can be interpreted as a spatial filter or feature detector. On the whole, the RF concept is based on static response properties.

Fifty years after the Hartline concept of RFs, and about 25 years after the Hubel and Wiesel concept, there are still numerous questions even after repeated refinements of the original idea. For example, the dendritic field organization cannot account for the electrophysiological tuning characteristics of single cell responses (Gilbert and Wiesel 1979; Lin et al. 1979; Martin and Whitteridge 1984). There are controversial opinions concerning the existence and the extent

[1]present address: Lehrstuhl für Theoretische Biologie – Voustitut für Neuroinformatik, Ruhr-Universität Bochum, ND04, Box 102148, D-4630 Bochum

Springer Series in Synergetics
J. Krüger (Ed.)
Neuronal Cooperativity

of anisotropic dendrictic field organization (Collonier 1964; Mungai 1967; Wong 1967). A genuine anisotropic organization that matches orientation tuning is hard to discover. In contrast, Braitenberg (1981, 1986) favours a more probabilistic point of view, in which the dendritic field organization is described by statistical distributions of pre- and post-synaptic sites.

5.1.2 Functional Impact of Cortical Architecture

In spite of the enormous functional diversity of the neocortex in higher mammals, there is remarkable uniformity in its architecture. Relating function to structural features, we believe that some very fundamental aspects of this architecture determine its computational capacities and strategies.

Cortices are two-dimensionally extended sheets of neurons that are vertically segregated into laminae constituting a system, which can be regarded a type of multi-input/multi-output system. Although there is a multiplicity of shapes of neurons, the most frequent cell type are pyramidal cells that are exclusively excitatory. Inhibition is mediated by non-spiny stellate cells, which act in close vicinity to pyramidal cell somata. While the dendritic and axonal fields of stellate cells are of short range, pyramidal cells show an interesting overall pattern, in which their dendritic fields invariably reach from the site of their somata up to layer 1. This general anatomical pattern suggests that information from different layers can be combined in a very systematic way. The axonal fields of pyramidal cells also extend over several layers, suggesting distribution of output in multiple feedback loops. As the intrinsic elements outnumber the input fibers by a factor of 2 to 3, intracortical processing based on positive feedback is assumed to be crucial for functional performance and is supposed to dominate effects that are mediated by direct, afferent input (cf. Braitenberg 1978, 1986).

5.1.3 A Network Model of Cortical Information Processing Incorporating Some Basic Aspects of Cortical Architecture

All these anatomical constraints were incorporated into a computer model of cortical architecture based on a mean lamina-specific neuroanatomy of pyramidal and stellate cells (Krone et al. 1986). The main outcome of the model was:

1) processing is strongly layer specific, i.e. the properties of a given neuron are essentially determined by its position within the layered system;
2) low-frequency temporal oscillations occur, i.e. responses are structured in time that are layer and stimulus specific;
3) positive-feedback loops are responsible for the observed temporal response dynamics;
4) functional neuronal selectivity is derived through successive iterations within the feedback system, leading to variations of receptive field properties in time and
5) spatio-temporal behaviour is no longer separable into two independent terms.

According to these simulations and the briefly outlined neuroanatomical considerations, cortical neural networks are characterized by strong positive feedback. Cortical feedback architecture, however, has strong implications for the spatio-temporal organization of receptive fields (Krone et al. 1986; von Seelen et al. 1986, 1987).

5.4.1 Nonlinear Behavior vs. a Complex Temporal Response Pattern

Finally, there is a large body of evidence concerning nonlinear behaviour of cortical neurons. If a receptive field is static and linear, its behaviour can be predicted from a spatial distribution of weights. However, responses that are non-predictable in this sense may be produced in three different ways:

1) the cell is nonlinear in the strict meaning of systems theory;
2) the cell has an elaborated, spatio-temporally non-separable dynamic behavior;
3) both mechanisms might be present. Testing the spatial summation properties of cells alone cannot distinguish between these possibilities.

5.1.5 Hierarchical-Parallel Processing in the Cat Visual System

In the following, we report on experiments that were performed in the cortical visual areas 17, 18, 19 and PMLS (visual area in the postero-medial portion of the lateral suprasylvian sulcus according to the nomenclature of Palmer et al. 1978). These areas are part of a highly interconnected system of serio-parallel processing consisting of multiple areas (cf. Rosenquist 1985). Based on the pattern of thalamocortical connections, both areas 17 and 18 in cats are regarded as representing a parallel stage of primary areas directly receiving independent input from the LGN. The same is essentially true for area 19, but it is regarded as occupying a somewhat more advanced level than areas 17 and 18 because it additionally receives a substantial input from the LP-Pulvinar complex. This part of the visual thalamus located more medial from the LGN, in turn receives midbrain and back projections from areas 17, 18 and 19. Finally, the suprasylvian areas are clearly on a more complex and different level of processing, as they receive main input from the LP-Pulvinar complex and from areas 17, 18 and 19, while the LGN input remains marginal. While PMLS is discussed in respect to processing of motion and to its possible analogy to monkey mediotemporal cortex (MT), even less is known of areas 20 and 21, which some authors believe to be analogous to monkey inferotemporal cortex (IT). There is general agreement that the main input into the extrastriate areas is of the y- and w-type. It is therefore interesting to note that the x-type input is exclusively relayed via area 17. Based on these notions, it seems conceivable therefore that the areas studied in our experiments cover at least 2 to 3 different levels of information processing (for more details see Tusa et al. 1981; Rosenquist 1985; Spear 1989).

5.1.6 Synopsis

The experiments described in this chapter deal with the spatio-temporal response properties of cortical RFs and were influenced by our general assumption on cortical architecture and its implications of structure and function. We also gratefully appreciate some of the ideas provided by the performance of the above-described model, which were very fruitful and inspiring for the design of some of the experiments.

Firstly, we will characterize the low-frequency temporal oscillations of activity and describe their general and area-specific properties. We shall demonstrate that these oscillations are highly synchronized for closely spaced neurons, while synchronization of oscillation is absent for pairs of neurons separated by greater distances (2 mm). The existence of this type of oscillation initiated the investigation of the time course of RF organization during the oscillatory episodes, which will lead to the introduction of non-separability of spatial and temporal behaviour. We describe briefly some theoretical aspects of non-separability and the advantage of spatio-temporal filters for motion processing, and then discuss some implications arising from the introduction of non-separability. We will deal with time dependence of RF properties (orientation selectivity and hypercomplexity), and will discuss that selectivity is reached after a preceding non-selective state of excitation and therefore a high level of selectivity is no longer correlated with maximal spike count, but is much better described in terms of temporal modulation of response. We will briefly turn to time coding and feedback mechanisms and will then outline a general theory on dynamics of RF organization that incorporates what has so far been argued on oscillations, non-separability, feedback and time dependent selectivity. We discuss the consequences of non-separability for processing of moving stimuli that results in a new dynamic model of RFs, which is able to explain that RF properties become a function of temporal and spatial parameters of the stimulus such as velocity and overall shape. Finally, we discuss some evidence for time dependence of interareal cooperativity, by which assemblies or groups of neurons interact and cooperate even beyond areal borders.

5.2 Temporal-Spatial Aspects of Receptive Field Organization

5.2.1 Low-Frequency Oscillation in Neuronal Response Following Photic Stimulation

Brain activity comprises a broad spectrum of periodic phenomena ranging from high-frequency oscillations (30 – 90 Hz, the so-called gamma range) to very low frequencies in the range of seconds or minutes (besides ultradian and circadian rhythms). For several years we have investigated a highly significant and repro-

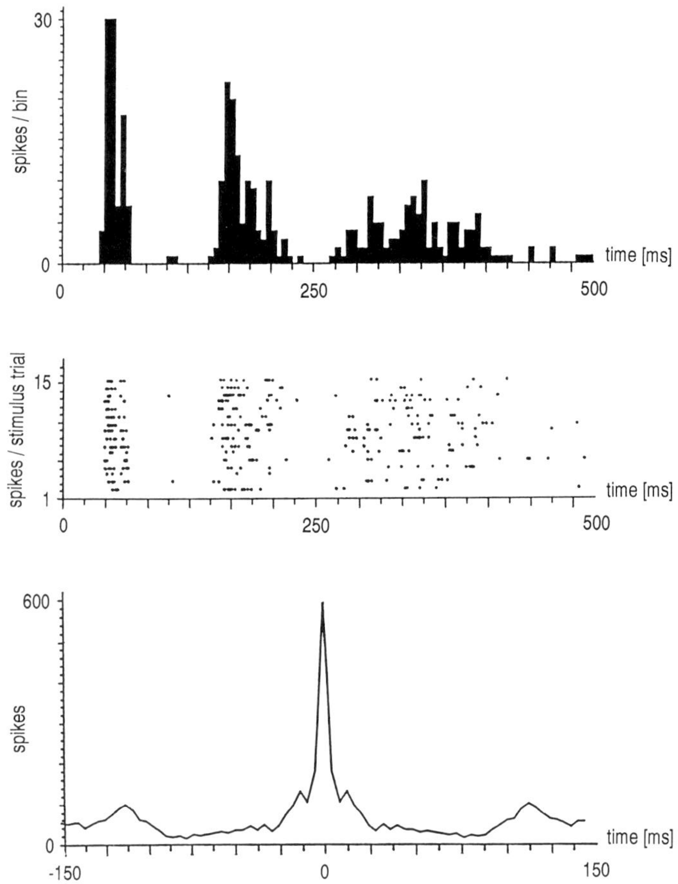

Fig. 5.1. a Characterization of low-frequency oscillations. Response to whole field stimulation, flickered with repetition rates of 2.5 s (1.25 s on, 1.25 s off). The off-response is shown. Top: PSTH, first 500 ms after stimulation (bin width 10 ms); middle: Dot-Display (same as PSTH); bottom: autocorrelation for +/− 150 ms. The characterstic oscillatory response pattern consists of a sequence of peaks of excitation at a mean frequency of 6–20 Hz. This low-frequency oscillation is well-documented in autocorrelation and appears to be superimposed by a smaller, high-frequency component in the range 50–60 Hz; b examples of PSTHs for neurons recorded in areas 17, 18, 19 and PMLS, respectively. Whole field on-off stimulation, which covered the entire receptive field. Repetition rate 2.5 s (1.25 s on, 1.25 s off). Abcsissa: 2.5 s; ordinate: spikes/bin. Bin width 10 ms. Time course of stimulation is indicated on the bottom

ducible temporal structure that can already identified in post-stimulus time histograms (PSTHs) using simple flashed stimuli. Figure 5.1A shows the characteristics of this temporal structure: a low-frequency oscillation in the range of about 10 Hz, which is superimposed by high-frequency oscillations and which might be supposed to be part of a much slower oscillatory pattern[1]. Any mutual relationship between high-frequency oscillations and those in the range 6–20 Hz as reported in this chapter has so far not been investigated. However, spectral

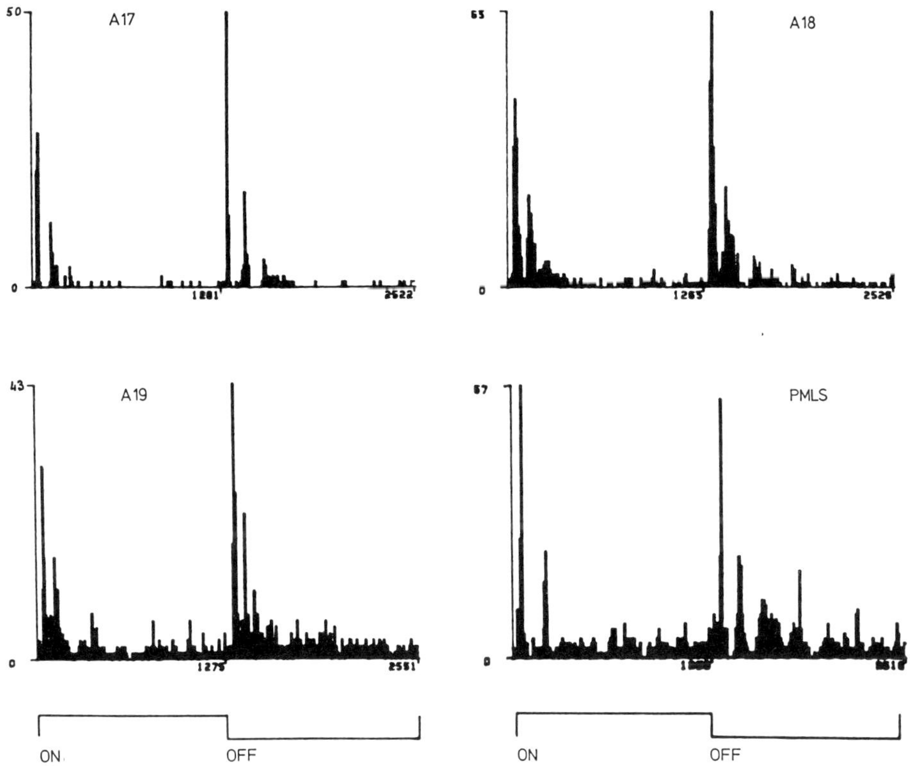

analysis of our data suggest that both types of oscillatory patterns are present in the neural discharge simultaneously, but with different amplitude characteristics.

This low-frequency type of oscillation has been described not only in cortical visual areas of cats, but also in monkey striate cortex (J. Krüger, private communication), rat somatosensory cortex (Dinse and Merzenich, unpublished) and cat auditory cortex (Schreiner and Joris 1986). Accordingly, oscillatory behaviour in the neuronal discharge in the range of about 10 Hz constitutes a fundamental attribute of cortical neurons and is found under a large variety of experimental conditions. Our investigations focused on areas 17, 18, 19 and PMLS in the cat, which cover 80% of the cat visual cortex and about 30% of the cat neocortex. This implies that our results on dynamic characteristics of discharge patterns and receptive field organization do not reflect sophisticated properties of neurons of highly specialized areas, but rather a common and general aspect of cortical information processing. Therefore we consider this

[1] High-frequency oscillations have been intensively investigated by Freeman and his coworkers in the rabbit olfactory system (cf. Freeman and Skarda 1985; Freeman and diPrisco 1986) and were more recently described for the cat visual system by Eckhorn et al. (1988) and Gray et al. (1989). (See also this volume)

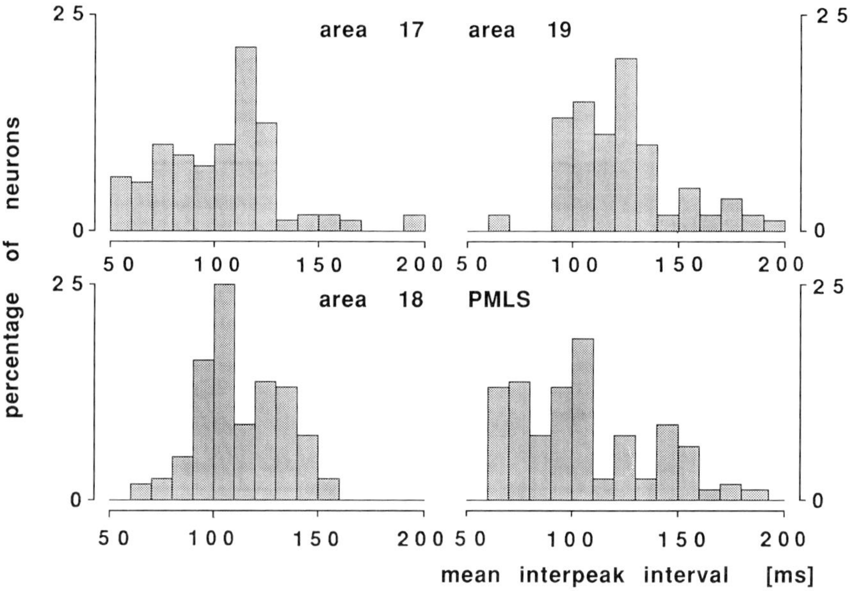

Fig. 5.2. Distribution of interpeak intervals for neurons of the areas 17, 18, 19 and PMLS. Mean interpeak intervals of all cells are between 50 and 170 ms

type of oscillation as the fundamental demonstration of a distinct temporal structure in the response pattern[2].

An example of oscillatory responses from neurons from all tested areas is given in Fig. 5.1B. While most of the cortical neurons show a clear oscillatory type of response, this is missing in our sample of geniculate cells[3]. At least two thirds of all cortical neurons of our sample, no matter to which area they belong, have an oscillatory type of response. There is a rather broad scatter of individual mean interpeak intervals, with the average for each area being between 50 and 170 ms, i.e. 6–20 Hz, respectively (cf. Fig. 5.2). However, the frequency of this oscillation is not constant, but decreases with the duration of oscillation, i.e. the interpeak interval is shortest for the first two peaks. Mostly, the amplitude of the first peak is also the largest. Accordingly, the response pattern seen is not a genuine periodic oscillation, but is best described as a *damped aperiodic oscillation*[4]. The maximal observed length of this oscillation is 500 to 600 ms during which 7 to 8 peaks may occur. The number of peaks observed is different

[2]These experiments were performed with so-called whole-field stimulation, in which the size of the stimuli were adjusted to cover the entire RF as detected by hand plotting.

[3]The results described in this paper are based on a total of more than 400 cortical neurons and 77 geniculate cells.

[4]A damped aperiodic oscillation might be the reason that in autocorrelograms we mostly obtain rather broad peaks although there are distinct oscillatory peaks of excitation in the PSTHs (cf. Fig 5.1).

in each area. Area 17 neurons resemble those of PMLS, as there is a considerable proportion of cells that show no oscillation, whereas all cells investigated in area 18 and the majority of cells in area 19 do show oscillations. The highest percentage of cells with long-lasting oscillation, i.e. five or more peaks, was found among the area 17 and PMLS neurons.

To evalutate mutual dependencies of parameters describing the oscillatory pattern, we performed linear regression analysis for the following variables: amplitude of the first peak, latency of the first peak, number of peaks, amplitude ratio of the first and second peak and time difference between the first and second peak. On the whole, we observed an overall lack of correlation between the different parameters, especially between response amplitude and latency, or between response amplitude and number of peaks. This indicates that the temporal aspects of the response patterns are not linked to measures of spike count or response strength. This is an important presupposition for the relation of temporal structure and information processing.

Including the oscillatory pattern as a significant portion of the response considerably prolongs the time over which neurons respond to even simple patterns. Assuming an average length of the temporal modulation in the range of 200 to 500 ms, this means that cortical neurons never reach a genuine steady state. As the dynamics of a visual scene[5] are clearly below these values, it follows that a continuous interaction between external stimuli in the visual field and the intrinsic temporal modulation must occur, which make neuronal responses highly dependent on the dynamics of stimulations. Indeed, we found abundant evidence for the kind of interaction that is described in the following sections.

Coincidence of Oscillations in Simultaneously Recorded Neurons. In order to clarify whether nearby neurons oscillate in phase, we recorded simultaneously single- and multiple-unit activity on one microelectrode in both area 17 and in PMLS. We analyzed the coincidence of oscillations in both recording channels at a more expanded time scale. Figure 5.3 illustrates a representative example of simultaneously recorded responses to on-stimulation. The single unit response shown at the bottom is not contained in the multiple unit recording. Comparing the time course of oscillations, we found that closely neighbouring neurons oscillate in phase: firstly, in all cases there was a high temporal coincidence of the peaks of oscillation between the single and multiple unit recording. Secondly, in the multiple unit recording we never observed any kind of blurring of the highly structured response pattern, which would have been an indication that the oscillation of the single neurons was out of phase. As a rule, we observed distinct peaks similar to those shown in the top of Fig. 5.3.

In addition, we initiated simultaneous recordings with two independent microelectrodes spaced at distances between 1 and 5 mm. In the paired recordings that are so far available of neurons of areas 17 and 19 and of areas 17 and PMLS, a quite different behavior was observed. Coincidence of the oscillatory pattern

[5] 100 ms are crucial for perceiving unblurred images (Burr 1981).

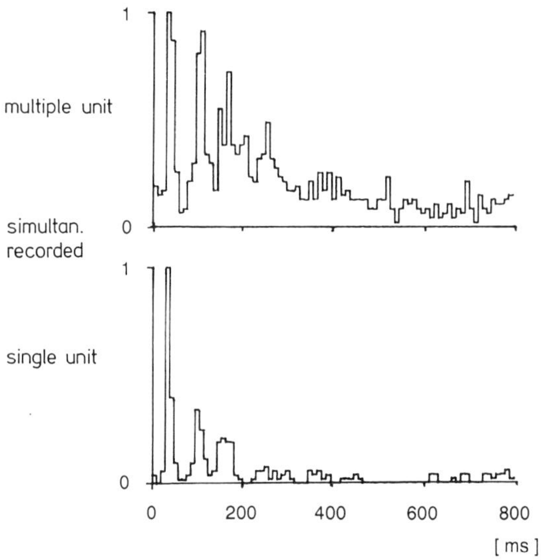

Fig. 5.3. Example of simultaneous recordings of multiple (top) and single unit activity (bottom) for a penetration site in area 17. Whole field stimulation, flickered with repetition rates of 2.5 s (1.25 s on, 1.25 s off). The on-response for first 800 ms is shown (bin width 10 ms). There is a high coincidence of oscillation between the single and multiple unit recording and a high in-phase oscillation within the multiple unit recording, which leads to distinct peaks of excitation, i.e. there is no indication of blurring due to out-of-phase oscillation

occurred only occasionally. As a rule, simultaneous recordings of this type across larger distances revealed a lack of in-phase oscillations.

We take the restriction of high coincidence of firing to local patches of the cortex as evidence that the oscillatory, temporally highly structured response reflects a local aspect of cortical information processing. So far we have no data on possible coincidence of oscillations for neurons which were recorded over larger distances, but would match with regard to other functional properties, e.g. preferred orientations or ocular dominance. It remains to be seen that if under these conditions a coincidence in oscillatory frequency can be demonstrated[6].

There are several theories of cortical information processing that rely on oscillations of the type reported here. Fuchs (1988), for example, proposed a neural network model for pattern recognition and associative memory based on nonlinear dynamics. The decision-making process involved in pattern recognition is modelled as the approach of one of a number of fixed points in the systems phase space. Temporal oscillations occur as the system approaches the fixed point. A model for the storage of temporal patterns in a non-local ("holographic") way was proposed by Longuet-Higgins (1968), which he called "holophon". This model predicts temporal bandpass behaviour of the type recorded here. Information about temporal patterns could be stored in the details of the oscillatory behaviour. Finally, the "trion model" of cortical organization

[6]For somatosensory cortex it was reported recently that a substantial reorganization of cortical skin representation was paralleled by a dramatic modification of neural assembly interaction. This was indicated by an increase in correlated activity, which was recorded even for pairs of neurons separated by up to 1 mm (Dinse and Merzenich 1989; Dinse et al. 1990). Interestingly, this overall alteration of correlated activity patterns included low-frequency oscillatory episodes in the neuronal discharge.

put forward by Shaw et al. (1986) is based on temporal oscillations in the range 50–100 ms, which the authors believe represents a significant clock-like temporal scale of cortical information processing.

5.2.2. Dynamics of Receptive Field Organization

a) Temporal Distribution of Excitation Within Receptive Fields

So far we have demonstrated a distinct temporal response structure of damped aperiodic oscillations in cortical neurons recorded in four visual areas. In these experiments, we used whole field stimulation which provides, however, little information on the internal receptive field organization because of spatial averaging and integration across the entire receptive field. In this and in the following section, we will relate RF properties to the temporally structured neuron responses.

In order to evaluate the distribution of excitation within the receptive field as a function of both space *and* time, we subdivided receptive fields into 5 by 5 to 8 by 8 subfields and stimulated them in random order with spots of light, which were of the same size as the diameter of the small subfields. We obtained 25 to 64 single PSTHs, which were used to compile temporal sequences of response planes (cf. Stevens and Gerstein, 1976). As these phenomena are basically dynamic, a more adequate illustration can be achieved by viewing this sequence on a video in slow motion.

The response planes obtained in this way were different for the on and off response of the same cell in approximately 80% of all neurons. A distinct temporal response structure may be restricted to either on or off responses.

Our paradigm for compiling sequences of response planes differs from conventional response planes in that it avoids integration of the response over time or the use of a peak response measure. Due to this, we can demonstrate that the spatial distribution of excitation and thus RF organization is not static, but changes over time. These changes are related to the oscillation of response, described in detail for whole field stimulation. However, the oscillation of excitation is not spatially homogeneous throughout the entire RF. Each subfield within the receptive field exhibits its own temporal characteristics, i.e the oscillatory pattern elicited at different locations within the receptive field may be different. In the sequences of response planes, the excitation appears as a "wave" spreading over the entire receptive field. The observation that each spatial location has a temporal characteristic of its own, is a clear indication that space and time are not separable[7].

[7]This notion is rather equivalent to the findings of concentric arrangements of visual latencies within RFs, which has been described recently for area 17 complex cells (Li and Nothdurft 1987) and earlier for extrastriate cells of area 18 and 19 (Dinse 1981). The temporal "singularity" of each location within RFs corresponds to what has been called "active points" by Reinis et al. (1988). In our terminology these are the "mechanisms" or "processing submodalities" that lead to non-separability of space and time as soon as a certain number of "mechanisms" is exceeded.

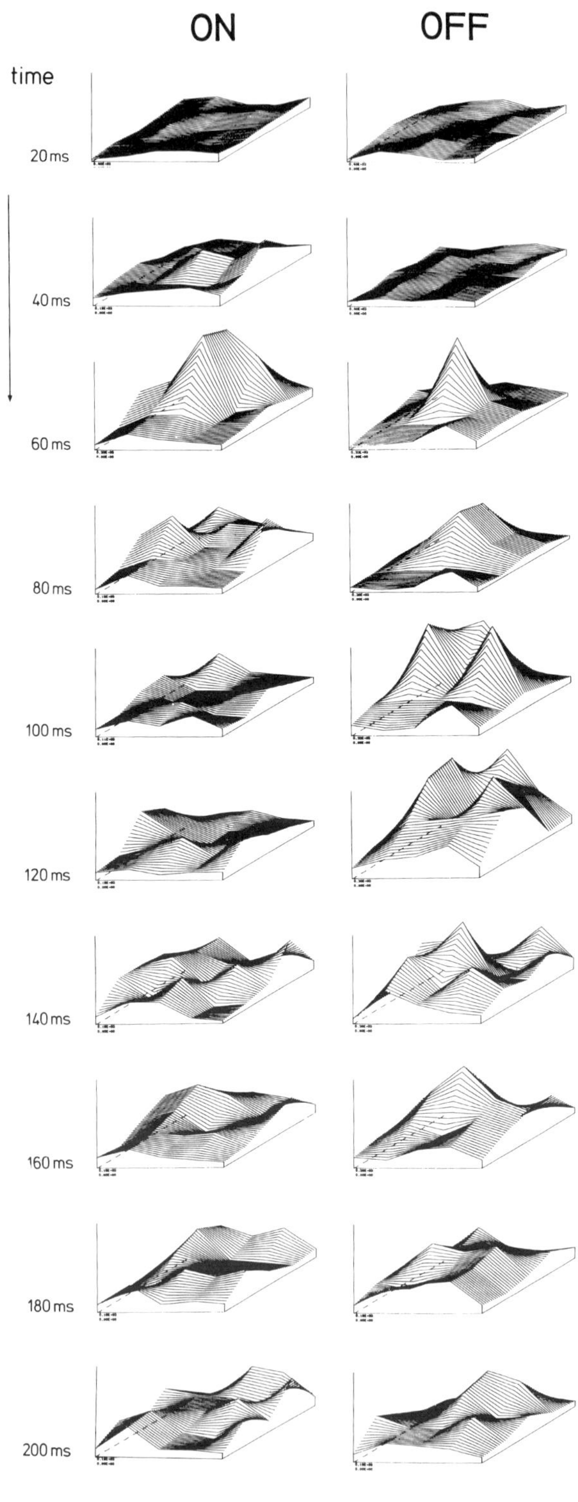

Fig. 5.4. Sequences of response planes for a neuron recorded in area 17. Time steps of 20 ms. On-off stimulation within a grid of 5 × 5 subfields was used. Size of the stimulation matrix was 2.5 deg (each subfield 0.5 deg). Ordinate is spikes/time step (in this example spikes/20 ms). This sequence illustrates the two-dimensional RF organization for each moment (i.e. each 20 ms time step) over the first 200 ms after photic stimulation. The RF organization changes dramatically over time. As each spatial location exhibits a temporal characteristic of its own, space and time are not separable in cortical cells that show such or similar behaviour

The RF dynamics described so far were obtained with stimuli that allowed the system to reach a steady state. In addition, no other clues were available, such as optical flow information, e.g., large textured background stimuli. Therefore we assume that under these conditions the intrinsic dynamics of cortical neurons become apparent. It seems conceivable, however, that these dynamic processes are subject to adaptive or plastic changes as soon as additional contextual information is available.

b) Non-Separability of Spatial and Temporal Behaviour

In a simple description, RFs may be regarded as spatial filters weighted with a temporal low-pass characteristic, i.e. both spatial and temporal filtering can be treated as independent factors. The experimental data, however, suggest that this separability might not be maintained in cortical neurons. Unless the complete temporal behaviour is taken into account, any statements concerning RF organization are either time-averages hiding important aspects or snapshots obtained at given moments revealing only partial information. For a complete description of RFs, it is necessary to obtain information on the response characteristics for every spatial location within RFs at every moment in time. This conclusion has considerable consequences on the understanding of RFs, as it makes RF organization time-dependent, a property that is already obvious from the sequences of response planes as they display different RF properties for different moments after stimulation.

To clarify the discussion of spatio-temporal interactions in receptive fields, we will make use of the following definitions. Firstly, a function $f(x,y,t)$ (e.g., a spatio-temporal response plane) is called separable if it has a spatial part $h(x,y)$ and a separate temporal part $g(t)$, such that

$$f(x,y,t) = h(x,y) * g(t). \quad (1)$$

If no such separation exists, the function is called non-separable. Secondly, we introduce the term piecewise separability for a function satisfying the equation

$$f(x,y,t) = \sum_{n=1}^{N} h_n(x,y) * g_n(t), \quad (2)$$

where N is a small integer (usually 2 or 4). Note that the term piecewise refers to a segmentation into mechanisms rather than spatial subfields. Both separability and piecewise separability are preserved by the Fourier transform[8].

The different types of separability may be illustrated by the following example. Suppose a set of spatial stimuli is presented with different temporal modulations, e.g., briefly flashed, flickered or turned on or off. In a linear-separable

[8]Most supposedly non-separable receptive field models introduced so far (e.g. von Seelen, 1970; Korn and von Seelen 1972; Marr and Ullman 1981; Enroth-Cugell et al. 1983; Dawis et al. 1984; Fleet et al. 1985) propose center-surround type functions, which are, according to our definitions, piecewise separable.

system, temporal behaviour can be factorized out of the summation process performed by the linear system. Therefore, the spatial characteristic of the system would be the same for all temporal modulations of the stimulus. In a piecewise separable system, the temporal characteristics g_n together with the temporal modulation of the stimulus impose a weight on each spatial mechanism h_n. The set of weights will be different for each temporal modulation and the overall spatial filter will thus be modified by the temporal modulation of the stimulus. The system may then "prefer" one special pattern among the flickered stimuli and a different one among the flashed ones. The more "mechanisms" there are, i.e. the larger the number N in (2), the more pronounced will be the influence of the temporal modulation of the stimuls on the spatial filtering. The most interesting case is reached when the system is entirely non-separable, in which case one should no longer talk of distinct "mechanisms". Theoretical studies have indicated that temporal characteristics that are not piecewise separable are likely to exist in the visual cortex. They arise from distributed feedback coupling, which is frequently found in the visual cortex (Krone et al. 1986; von Seelen et al. 1987). More generally, given the conditions of a highly interconnected network, the concept of conventional RFs might become obsolete (cf. Johannesma et al. 1986).

From our experimental data on non-separability we have to conclude that there is a smooth transition from non-separability involving a low number of "mechanisms" (a condition close to piecewise non-separability) to complete non-separability comprising a large number of "mechanisms". Accordingly, we can not characterize the separability state of a cell by a fixed number of "mechanisms". Instead, we observe a rather continuous distribution. This is consistent with the assumption that the position of a cell even within a highly interconnected network can be more or less close to the site of the afferent input. In addition, the degree of feedback that a cell receives might not be uniform throughout the network (Johannesma et al. 1986). Both conditions are supposed to act on the degree of non-separability. The occurrence of a graded manifestation of non-separability predicts a considerable variability of time dependence of RF properties (cf. Figs. 5.5, 5.7, 5.14 and Table 5.1). This prediction was fully substantiated by our experimental results and is discussed in detail in the following sections.

c) Earlier Findings

Although the interest in spatio-temporal receptive fields has been somewhat inhibited by the search for spatial feature detectors, the spatio-temporal frequency characteristics of cortical neurons and psychophysically assessable "mechanisms" have been studied intensively. Wilson (1980) psychophysically tested a transient visual mechanism with sinusoidal gratings flickered sinusoidally in time. He found that the threshold sensitivity curves are in fact separable into a spatial and a temporal part. Burr et al. (1986), however, using drifting instead of flickering gratings, found that the spatial frequency characteristics depend on the stimulus velocity, which is proportional to the temporal frequency of the stimu-

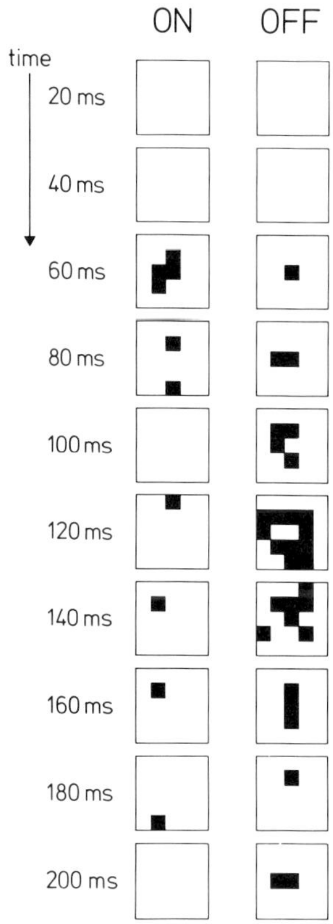

Fig. 5.5. Sequences of contour planes for the neurons shown in Fig. 5.4 recorded in area 17, illustrating the *dynamic receptive field size – DRFS*. Time steps of 20 ms. On-off stimulation within a grid of 5 × 5 subfields was used. Horizontal sections through the planes were applied at 50% maximal activity. Dark regions indicate spatial extent of excitatory regions within the receptive fields. Size of the stimulation matrix was 2.5 deg (each subfield 0.5 deg). These sequences illustrate RF diameter for each moment. Again there are dramatic changes over time that include systematic shift of RF location within the visual field, also expansion and contraction. A functional interpretation is suggested by the terms *motion type* and *looming type* (cf. Fig. 5.6)

lus. In a neurophysiological study with drifting sinusoidal gratings, Tolhurst and Movshon (1975) found the spatial frequency characteristic to be independent of the stimulus velocity; recently Galli et al. (1988) reported speed dependence of spatial frequency sensitivity for area 18 neurons. This result indicates spatio-temporal separability, as does the more recent study of Baker and Cynader (1988) who tested flashed and moving light bar stimuli in the cat striate cortex.

Recent studies on X-type retinal ganglion cells (Enroth-Cugell et al. 1983; Dawis et al. 1984) have resulted in receptive field models with different temporal behaviour for center and surround (cf. von Seelen 1970; Korn and von Seelen 1972). In our terminology, these fields are piecewise separable. True spatio-temporal receptive fields in the visual cortex have been discussed by Eckhorn and Pöpel (1980) and the "receptive field for motion", introduced by Krause and Eckhorn (1983), also suggests non-separability of space and time. As can be seen from the work of Stevens and Gerstein (1976), cells at the lateral geniculate level exhibit a piecewise separability of 3 to 4 mechanisms. According to these experiments, geniculate cells may be closer to non-separability than has been assumed earlier. They show a behaviour intermediate between those described for retinal ganglion cells (Enroth-Cugell et al. 1983; Dawis et al. 1984) and cortical cells. A comparable complex spatio-temporal behaviour was described for geniculate cells by Gielen et al. (1981) using white noise analysis. This raises the question of thalamo-cortical interactions and cortical involvement in establishing cortex-like non-separable spatio-temporal behaviour, which is currently under investigation (cf. Heinke et al. 1990)[9].

As to different sensory modalities, it is interesting to note that in the acoustic system of the grass frog, a non-separable spectro-temporal behaviour has been demonstrated (Eggermont et al. 1981) that was present at a higher level of processing, but not at first- or second-order neurons.

5.2.3. Impacts of Non-Separability for RF Organization

a) Time Dependence of RF Size

So far we have demonstrated non-separability of spatial and temporal behavior as a characteristic feature of cortical cells, suggesting that RF properties are to a considerable degree controlled by dynamic factors. This section deals with a series of experiments designed to show the time dependence of RF organization. Again we used stationary flashed stimuli in order to relate the resulting temporal response structure to a highly reliable and controllable time course of stimulation, which is not possible when using moving stimuli. We will demonstrate a time dependence of RF size and of RF properties such as orientation selectivity and hypercomplexity.

One important aspect of receptive field organization is its size in the visual field. It is generally believed that the obtained dimensions of RFs are essentially determined by the type of afferent input to the cell and the pattern of convergence. To test this assumption and to verify the possibility that non-separability may cause time dependence of RF size, we compiled temporal sequences of contour planes[10]. Figure 5.5 illustrates such sequences of contour planes obtained from horizontal sections at 50% maximal neuronal activity, which display the actual spatial extent of the receptive field for each moment, i.e. in time steps of 20 ms. We introduce the term *dynamic receptive field size* (DRFS). The dimensions of the receptive field determined in this way clearly differ over time, changing between 0.5 and 2.0 deg in diameter. When a time-averaged RF size is calculated, one obtains a diameter of 2.5 deg. This latter value corresponds to the size obtained with handplotting. What normally is considered the RF size is, according to these findings, caused by temporal integration, which masks the changes that occur over time.

---

[9]For several years, a Soviet group used a comparable approach to study dynamics of RF organization of cortical and geniculate cells and reported wave-like processes within RF organization; again these authors found mostly quantitative rather than qualitative differences between cortical and geniculate neurons (Podvigin et al. 1974; Shevelev et al. 1982; Volgushev et al. 1983; Shevelev, private communication).

[10]Conventional contour planes are regarded as a temporally integrated measures of receptive field size (cf. Stevens and Gerstein 1976).

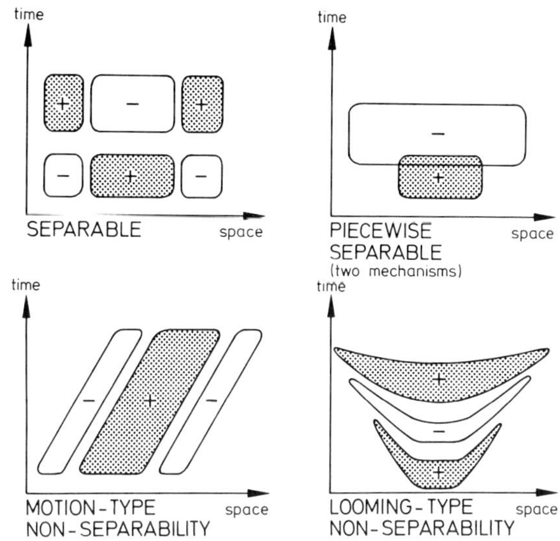

Fig. 5.6. Different types of spatio-temporal separability and non-separability. Piecewise separability has been found for retinal ganglion cells (e.g., Enroth-Cugell et al. 1983), which reflect the basic center-surround organization. Spatio-temporal *orientation* or *motion-type* non-separability has been reported for cortical simple cells by McLean et al. (1987). The cells from our recordings with mixed on-off discharge regions (complex-like cells) are of both the *motion type* and the *looming type* (cf. Figs. 5.4, 5.5)

We distinguished two different types of RF dynamics, based on the observed changes of RF size and organization. Type I exhibits little change of RF size over time, but shows a shift or motion of RF discharge centre (Fig. 5.5, on response). Type II shows no motion of RF discharge centre over time, but changes its dimensions. Type II RFs usually show the smallest size during the beginning of the response, followed by an expansion and subsequent contraction of the RF (Fig 5.5, off response). A more schematic illustration is given in Fig. 5.6. While separable and piecewise separable organizations are rather straightforward definitions, *motion-* and *looming-type* non-separabilities are two examples for the general case[11]. The terms *motion-type* and *looming-type* non-separability were chosen for spatio-temporally RFs to indicate their possible applicability in motion detection.

b) Time Dependence of RF Properties

Orientation selectivity and hypercomplexity (end-stopping property) can be regarded as the most prominent RF properties of cortical visual neurons. They describe specific neuronal responses to orientations or lengths of an elongated stimulus. Accordingly, we used stationary flickered bars of lights (0.2–0.4 Hz) of different orientations and lengths positioned in the RF centre. As with the time course of RF size, we investigated the time course of orientation selectivity and hypercomplexity. This was done by compiling sequences of tuning curves for time steps of 20 ms. We introduce the terms *dynamic orientation selectivity*

[11] While the so-called *looming type* of non-separability is different from the spatio-temporal orientation type reported by Burr et al. (1986) and McLean et al. (1987), the second so-called *motion type* seems comparable.

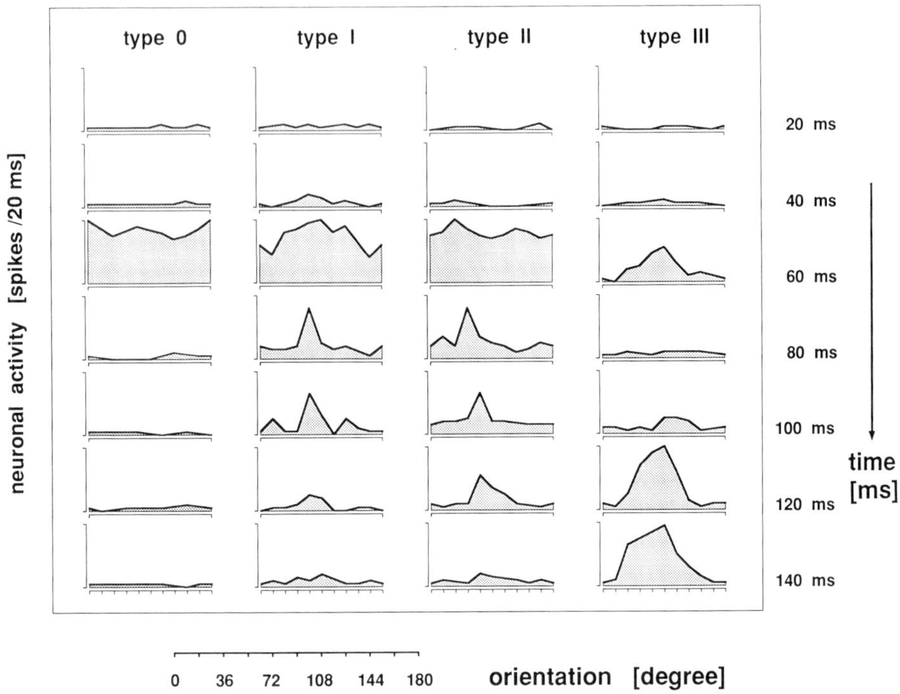

Fig. 5.7. Sequences of orientation tuning curves for four different response types found in area 17, illustrating *dynamic orientation selectivity – DOS*. Time step 20 ms. Stimulus was a bar of light, flickered at 0.4 Hz and set at different orientations. Abscissa: stimulus orientation in deg; ordinate: number of spikes elicited during each 20 ms time step. Type 0 is not stimulus selective (17%); type I is selective after a previous state of non-selective excitation (24%); type II is similar to type I, but changes its selectivity over time (34%); type III is selective without time dependence (25%)

(DOS), and *dynamic end-stopping sensitivity* (DES), respectively. In this case, each tuning curve describes the actual selectivity of a cell at that moment. In this way we avoid the temporal average found when compiling conventional tuning curves, which uses a measure of maximal spike rate. We distinguished four different types of response (Fig. 5.7). Type 0 is not at all orientation selective. Types I and II are the most intriguing cases, comprising more than 50% of our sample of area 17 neurons. Their selectivity to different orientations is clearly time dependent. Following on-off stimulation with flashed bars of light presented at different orientations, these neurons exhibit a first initial state of brisk excitation that does not exhibit a selectivity for any orientations. After this inital stage, orientation selectivity evolves, indicated by a high level of selectivity for a given orientation. Types I and II differ in that type II changes the selected orientation over time. Finally, type III neurons resemble most closely the conventional classification, in which selectivity is expressed without time dependence. However, type III neurons were found in only one quarter of our sample. The time dependence of RF properties, as seen in types I and II, has not only been observed for orientation tuning, but also for hypercomplexity (Fig. 5.8). It is noteworthy

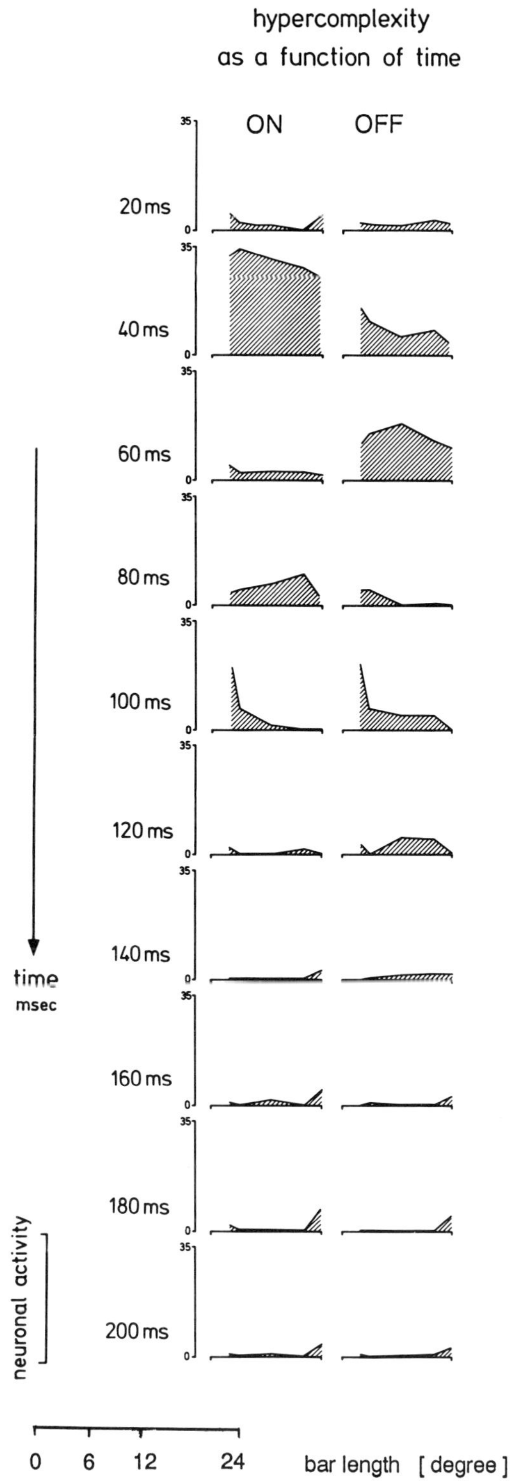

Fig. 5.8. Sequences of length-response tuning curves for on and off response of an area 18 cell illustrating *dynamic end-stopping sensitivity – DES*. Time step 20 ms. Stimulus was a bar of light, flickered at 0.4 Hz (1.25 s on, 1.25 s off) and set at different lengths. Abscissa: stimulus length in deg; ordinate: number of spikes elicited during each 20 ms time step. Time dependence of selectivity according to type I shown in Fig. 5.7

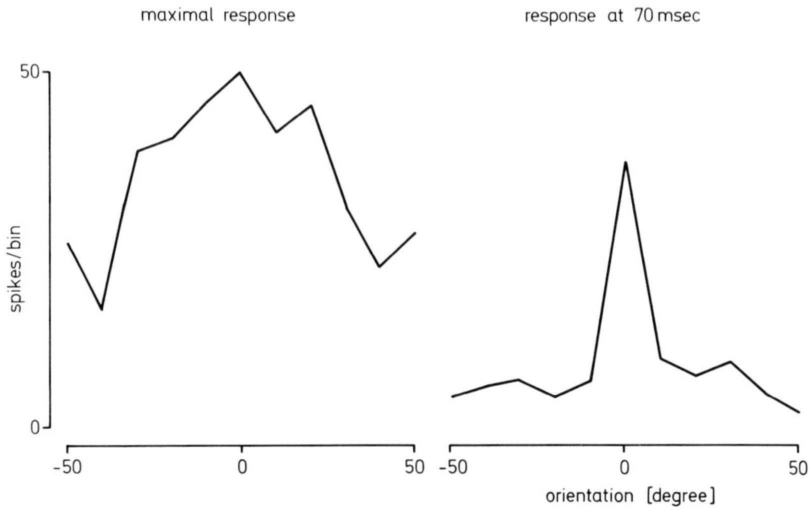

Fig. 5.9. Comparison of orientation tuning curves for a cell in area 17 based on two different measures of neuronal activity: maximal response amplitude (left) and number of spikes during a time window between 70 and 90 ms (right). High level of selectivity is not associated with maximal spike rate

that in all cases *selectivity evolves from an early state of non-selectivity*. Selectivity during the initial period of excitation and development of activity into an unspecific state was not observed.

These findings raise two important questions that will be treated briefly before we discuss the implications of non-separability for the processing of moving stimuli. One aspect deals with the obvious notion that states of maximal activity do not correlate with expression of high levels of selectivity, which questions a simple model representing stimulus information in terms of maximal or average spike counts. In addition, we will attempt to outline a theory on RF organization that makes use of the observation that a high level of selectivity evolves delayed after an initial state of stimulus-invariant excitation.

5.2.4. Evidence for the Use of a Time Code

Conventional tuning curves are compiled by plotting a given measure of neuronal activity against stimulus parameters. In most cases, the maximal spike rate obtained from PSTHs is used. However, as can be seen from the sequences of tuning curves (Figs. 5.7 and 5.8), maximal excitation can be completely stimulus invariant, whereas a high level of selectivity occurs delayed at an activity level much below that of the initial state. Figure 5.9 contrasts two different tunings of the same cell for different measures of neuronal activity, by comparing the maximal response amplitude against a measure of activity that is based on activity during some arbitrary time window, in this case the number of spikes elicited

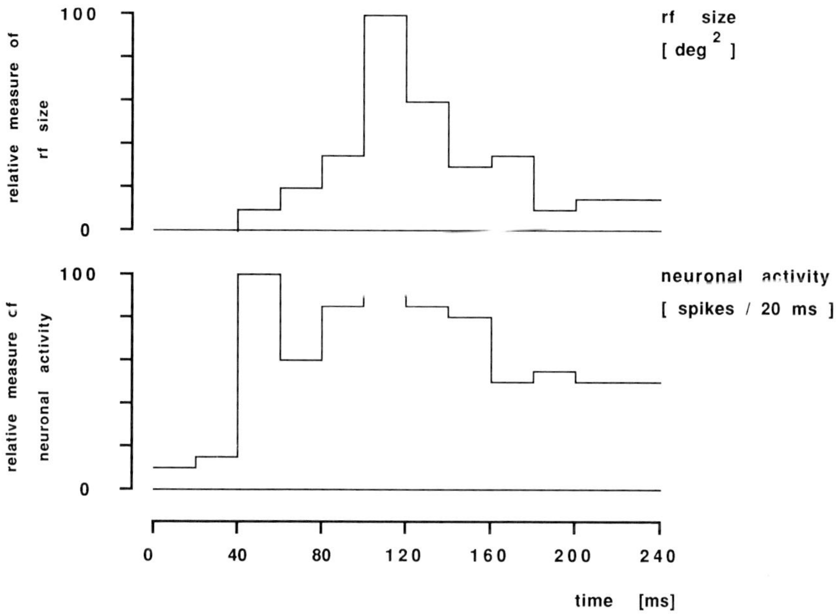

Fig. 5.10. Comparison of the time course of spike rate (bottom) and receptive field dimension (top). Off response of the area 17 cell shown in Figs. 5.4 and 5.5. Ordinate is an arbitrary measure of spike rate in number of spikes/20 ms (bottom) and of RF dimension in deg^2 (top). Note time-dependent association of the two parameters. While in the first 100 ms there is no correlation between spike rate and RF dimension, it might be assumed during the second half of response

during 20 ms between 70 and 90 ms after the onset of stimulation. Orientation tuning based on the first measure lacks a sharp selectivity, due to the fact that a broad spectrum of orientations elicits a viable response, which results in a broad tuning width. In contrast, orientation tuning based on the second measure reveals an extremely sharp selectivity, as indicated by the narrow tuning width. This supports the view that high level of selectivity is not always linked to measures of maximal spike count. Assuming such a relationship in the case of time dependence of RF properties may result in misrepresenting the selectivity of cortical neurons.

A similar observation could be made in the case of evaluation of RF size, in which maximal activation was also not correlated with RF size. A comparison of the time course for both spike rate and receptive field size is illustrated in Fig. 5.10. There is no simple, linear relationship between maximal spike rate and receptive field size. Rather, the correlation between the two variables is time dependent. During the first period of excitation, 50 to 100 ms after onset of stimulation, both variables are independent of each other but there is a clear correlation during the second period of excitation 100 to 200 ms after onset of stimulation.

This is in agreement with the initially described observation that measures of response amplitude and response latency are not correlated (cf. Sect. 5.2.1). We

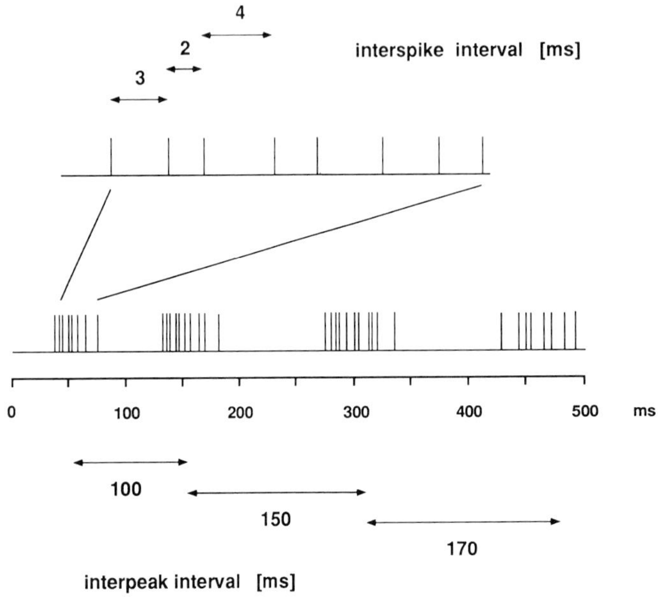

Fig. 5.11. Schematic illustration of a short and long time scale, based on short interspike intervals (range 1–20 ms) and interpeak intervals (range 50–500 ms), which might correspond to the so-called *psychological* time scale

interpret the lacking or weak association between temporal parameters and measures of maximal spike count as an indication of the biological significance of a time code, in which a temporal modulation of neuronal responses carries crucial information[12]. The fact that temporal modulation of response takes place over a range of about 500 ms, argues in favor of the existence of two important time scales: a short, spike-to-spike time scale in the ms range (interspike intervals) and a second, much longer, peak-to-peak time scale in the range of tenth of seconds (interpeak intervals) that might correspond with what have been called *psychological time scales* (Fig. 5.11).

The functional significance of the time dependence of RF properties is possibly reflected in human reaction time measurements. It is well established that simple detection of stimuli can be performed within the shortest range of reaction times (about 200 ms), whereas recognition and discrimination of even simple geometric patterns leads to considerable lengthening in reaction times to the range of 400 to 500 ms. The initial first state of excitation seen in the sequences of tuning curves might thus reflect the fact of just recognizing that a stimulus

[12] A similar conclusion was reached by Optican et al. (1987), Richmond et al. (1987) and Richmond and Optican (1987), who were able to demonstrate by means of principal component analysis of the wave forms of the temporal modulation elicited by an orthogonal set of complex geometrical patterns (Walsh stimuli), that the properties of modulation can carry a considerable amount of information independent of maximal spike count.

has appeared. The stimulus evaluation, however, takes a significantly longer time, which is reflected in the fact that a high level of selectivity in neuronal response occurs at a much later state. Similarly, it is well known from auditory psychophysics that frequency discrimination improves with longer observation periods.

5.2.5 Evidence for Positive Feedback

As already discussed, cells that showed time dependence of RF properties reach a maximal high level of selectivity after a previous non-selective state of excitation. Moreover, the activity level of the initial response state was in most cases above that containing a high level of selectivity. We argue that this behaviour can most appropriately be explained in terms of positive feedback mechanisms. According to this assumption, the first state of excitation obtained after onset of stimulation is the initial trigger, elicited by visual stimuli that activate the underlying cortical network. During this initial state, the activity is distributed across the entire system. Reverberating activity provides information that contributes to an increasing state of high level selectivity of the local circuitry. As these processes need time, we observe evolution of selectivity over a considerable time. In this view, selectivity evolves over time by feedback processes that consist of multiple iterative steps, which sculptures selectivity (cf. Sect. 5.1.2). This assumption is in agreement with the basic properties of an *average cortical neuroanatomy*, including its vertical lamination, average axonal and dentritic domains and intrinsic feedback loops via pyramidal cell axon collaterals, all providing the requisite 'wetware' background. As has been described in the introduction, all these constraints were incorporated in the computer model of cortical architecture and supported the notion of time dependence of RF organization (Krone et al. 1986)[13].

Further experimental evidence for the existence of iterative algorithms implemented in cortical processing was provided by Best et al.(1986): if excitation is indeed distributed throughout a network consisting of multiple feedback loops, one should be able to trace the course of excitation at different spatial and temporal locations across the proposed iterative process. The scatter diagram in Fig. 5.12 depicts visual latencies following on-off stimulation for cells in area 17 separated according to their lamina origin. We argue that both the considerable time delay between the cortical layers and the wide scatter of visual latencies can be interpreted as a snapshot of the dynamics of the excitation distribution due to iterative steps. It should be noted that the observed lamina-specific distribution does not match with the reported termination of x-, y- and w-input and can thus not been explained in terms of the afferent input pattern (Best et al. 1986).

---

[13] For details of an analytic approach to feedback systems see (von Seelen et al. 1987; Mallot and von Seelen 1989).

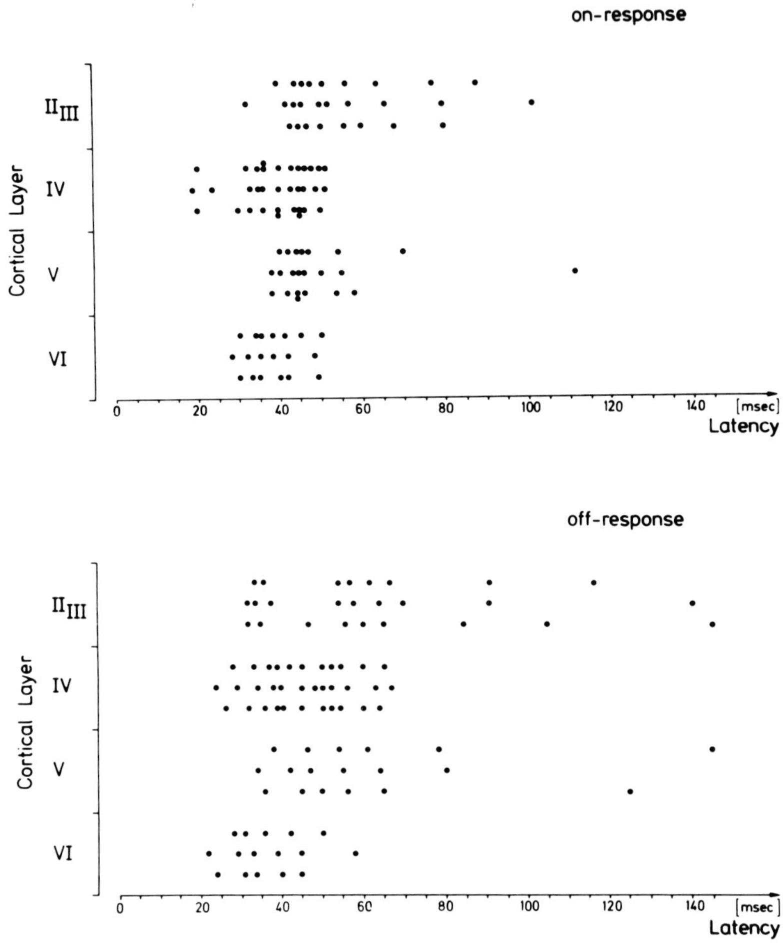

Fig. 5.12. Scatter diagram of lamina distribution of visual latencies following photic stimulation in area 17 (according to Best et al. 1986). It has been argued that the main flow of information passing through the incoming layers IV and VI is distributed throughout the other layers by way of multiple positive feedback loops. Since these loops need several iterations, they are suspected of generating both the pronounced differences between latencies of all layers and the wide scatter of latencies found within each layer

5.2.6. General Outline of a Theory of RF Organization

Based on what has been argued so far on the implications of temporal behaviour, time dependence and feedback for RFs, we will outline a general thesis on dynamic RF organization.

Starting points are neuroanatomical observations on average dendritic field dimension, as found in cat and monkey striate cortex. According to Lund (1973), Lund et al. (1979) and Jones (1974), the diameter of dendritic fields does not exceed 200 to 400 microns. There is also agreement that the average RF size in area 17 subserving foveal vision is about 1 to 2 deg visual angle. Utilizing the available information on retinotopic mapping studies (cf. Tusa et al. 1981), one can recalculate the cortical surface in terms of the visual field parameter. Taking for granted that 1 square mm of cortex represents 1 to 2 square deg, the observed dendritic fields cover at least 0.2 to 0.5 square deg (for more details see Dinse 1989a). However, as stated above, the observable average RF size is 1 to 2 deg, for which we introduce the term *apparent RF size*.

We now recall that the apparent RF size is due to considerable temporal averaging. The *dynamic receptive field size* is illustrated in the sequences of contour planes, which provide detailed information about RF size at every moment after stimulation. As can be seen from Figs. 5.5 and 5.10, the smallest extent of RFs was most frequently observed during the initial state of excitation. This is equivalent to the initial state of excitation that lacks selectivity when stimulus parameters such as orientation or length were varied. Based on what has been argued in respect to positive feedback, the smallest RF dimension seen during this initial state is the expression of the first direct, stimulus-triggered activity. Indeed, RF size at this time is about 0.5 square deg or less. We therefore suggest that this state, and only this, is mediated directly by the underlying dendritic fields. In this case, during this first state of excitation, we obtain an optimal match between measured dendritic field dimensions and apparent RF size. In addition, this state lacks a high level of selectivity. Both observations are compatible with the assumption that the underlying dendritic field is responsible for this part of the response. Next, the contributions from positive feedback become apparent, i.e. the non-local network is operating and acts on the cell under investigation. It is in this state that we observe a high level of selectivity and expansion of RF diameters or motion of RF location (both lead to what is normally measured with hand plotting or with quantitative averaging techniques). Accordingly, both apparent RF size and a high level of selectivity are considerably dynamic in nature and may be assumed to be based on both the underlying dendritic field organization *and* the organization of a distributed network. This explains time dependence of RF organization and the mismatch between apparent RF size and dendritic field diameter. It also explains that a high level of selectivity is always reached after a previous unselective state.[14]

In this view inhibition plays a relatively unspecific role. A high level of selectivity is not due to inhibitory processes alone, but arises from multiple iterations within a system of positive feedback. However, systems consisting of abundant positive feedback—such as cortical networks—require a very meticulous control in order to maintain stable operation. We therefore suggest that inhibitory action is primarily used to stabilize the positive feedback system. Under these assumptions, blocking of inhibition would result in exceeding of excitation and loss of a

[14] In addition, it overcomes the still controversial question as to what extent anisotropic organized dendritic fields are necessary to generate the sharp tuning seen in elctrophysiological measurements. A weakly biased input would be sufficient to produce a high level of selectivity on the basis of intracortical feedback mechanisms during successive iterative steps.

high level of selectivity due to an abnormal state of the entire positive feedback system. Interestingly, based on results from intracellular recordings, the concept of a rather unspecific role of cortical inhibition has been advocated by Martin (1988).

It should be noted that PMLS neurons, although basically displaying a very similar behaviour, are somewhat different in that the initially observed RF diameters are already larger by a factor of 3 to 5 than those described for area 17 neurons. However, this is compatible with the assumption that, due to massive convergence of inputs onto PMLS neurons, even initial and not temporally averaged RF dimensions are much larger than those observed in area 17[15].

5.2.7. Implications of RFs with Non-Separable Spatio-Temporal Behaviour for Processing of Moving Stimuli

In all experimental findings so far discussed, RFs were analysed with stationary stimuli that were flashed on and off. This method was used because the time course of stimulation can be controlled and monitored with high precision. These experiments revealed not only the highly structured response pattern, but also that RFs are composed of a set of multiple "mechanisms" or "processing submodalities" that are temporally independent, i.e. the time course of excitation is different. In terms of temporal modulation of the response pattern this means that each location or each "mechanism", respectively, may oscillate with a somewhat different frequency.

However, considering biotope and environmental conditions, there is little doubt that moving stimuli are more adequate for the visual system than stationary flashed stimuli. It may be tempting to try to predict neuronal response to moving stimuli with the knowledge of RF organization based on analysis with stationary signals. So far, successful predictions can be made only in neurons with very simple (i.e. non-separable) RF organization. However, considering a RF organization as described above in which each location displays a temporal behaviour of its own, it is clear that when stimuli are moving across such RFs, a

[15] While we are dealing with RF dynamics in the range of hundredths to tenths of seconds, there is an interesting link to dynamics in the minute range that stems from observations of reorganizational capacities of sensory maps. Representation of skin in somatosensory areas has been shown to be alterable by use (Merzenich et al. 1988). Dinse and Merzenich (1989) and Dinse et al. (1990) were able to demonstrate significant representational changes after several minutes of intracortical microstimulation (ICMS), a method that has been shown to produce changes in topographic respresentations that are comparable in extent to those found after behavioural training (Recanzone and Merzenich 1988, Dinse et al. 1990). As to the underlying mechanisms of this reorganization of cortical representations, it has been suggested that input selection occurs by cooperation between groups of neurons. These groups have to be regarded as dynamic, i.e. group constitution is the subject of continuous remodeling. Similarly to what we have described so far for RF organization in visual cortices, the functional maps are generated by actions of the entire network instead of direct, dendritic field-mediated effects. The anatomical projections are believed to provide an extended, rather unspecific input repertoire that is selected from in the generation and remodelling of sensory and functional maps.

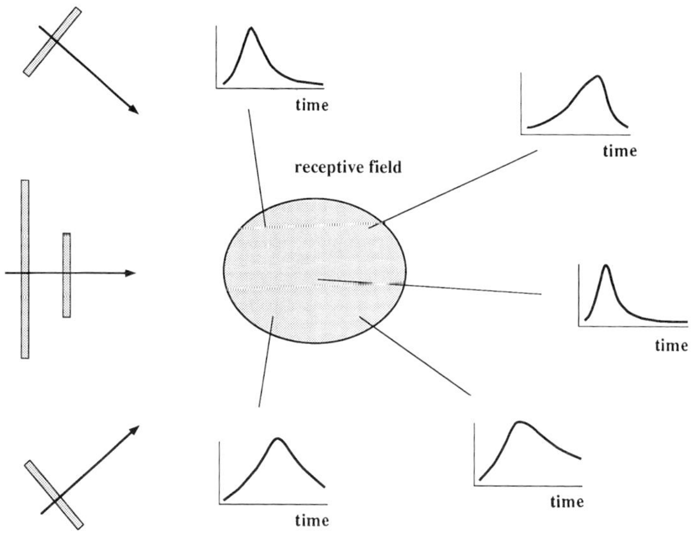

Fig. 5.13. Scheme of RF organization consisting of a number of single "mechanisms" that all differ in their time course. If stimuli are moving across such a RF type, the overall response depends on the interaction of adjacent mechanisms and is thus supposed to reveal dependency on speed, direction and stimulus shape, such as orientation and length

very complex response pattern may emerge. One should expect a strong interaction between neighbouring "mechanisms" that will depend on stimulus speed, i.e. the temporal sequence in which adjacent "mechanisms" are activated. Due to the temporal properties of each "mechanism" there will be suppression or facilitation of a response. The input history will interact with the internal temporal organization. Moreover, this interaction should not exclusively depend on stimulus speed, but also on the set of "mechanisms" actually utilized. Which "mechanisms" are involved first depends on the direction of motion and stimulus shape (cf. Fig. 5.13).

This hypothesis was tested by computer simulations (Dinse 1989b), in which the effect of systematic variation of several important parameters that are characteristic for individual "mechanisms" was studied. The parameters tested were the number of involved "mechanisms", the time course of excitation and inhibition of each "mechanism" and their spatial arrangement within RFs. In addition, using electrophysiological methods, the tuning properties of single cells in areas 17, 18 and 19 were investigated. Variation of stimulus parameters such as orientation or length was interpreted as changing stimulus shape. Under this assumption, tuning curves were calculated for different speeds and directions of moving stimuli presented with different orientations and lengths (Fig. 5.14).

The simulations indicated that relatively small numbers of "mechanisms" can generate the above-described mutual dependency of RF characteristics on stimulus parameters such as speed, direction, orientation and length, provided that there are sufficient differences in the time course of the involved "mechan-

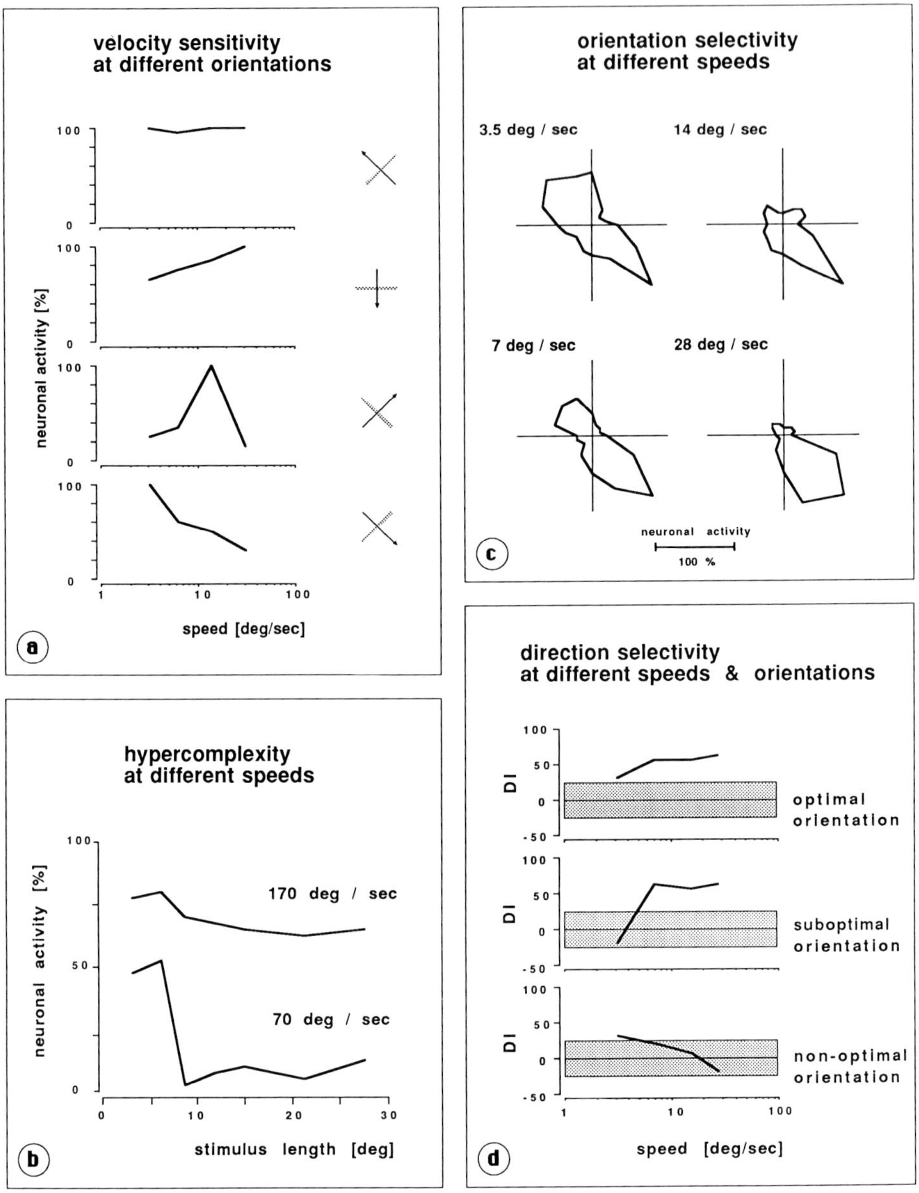

Fig. 5.14. a Velocity sensitivity for an area 17 cell measured at four different orientations. Optimal velocity or preferred range of speed can be considerably different for the same cell, when velocity tuning is tested at different orientations of the stimulus; b hypercomplexity for an area 18 cell measured at two different speeds. Distinct inhibition of response seen during elongation of the stimulus at low speed nearly disappears when the cell is tested with higher speed, indicating loss of hypercomplex properties; c orientation tuning foran area 17 cell measured at four different speeds; polar plots. Increasing speed is associated with considerable broadening of tuning half-width; d direction selectivity for an area 17 cell measured at four different speeds and for optimal, suboptimal and nonoptimal orientations. DI is an arbitrary index of direction selectivity for two directions, separated 180 deg (preferred response – non-preferred response/preferred response * 100); positive values indicate preference of 0 deg direction, negative values indicate preference of 180 deg direction. Zero means that the cell is not direction selective. There is a clear speed dependence of DI, which is, however, a function of orientation

Table 5.1. Percentage of neurons in areas 17, 18 and 19 in which RF properties such as orientation selectivity, hypercomplexity and direction selectivity are a function of stimulus speed

	area 17	area 18	area 19
orientation selectivity	42	56	47
hypercomplexity	55	60	59
direction selectivity	43	75	78

isms". These effects were even more pronounced when an asymmetric spatial arrangement of the "mechanisms" within RFs was introduced. Increasing the number of "mechanisms" affects the sensitivity to different stimulus shapes rather than speed dependence. Direction selectivity revealed the highest sensitivity to changes in internal RF organization.

The experimental data fully confirmed these observations, as summarized in Fig. 14. Expression of stimulus selectivity can be extremely speed dependent. However, this dependency is always a function of direction of stimulus motion. On the whole, orientation tuning was somewhat more robust than hypercomplexity (end-stopping). However, while in the example of Fig. 14c the overall preferred orientation is not affected, the half-width of the tuning doubles at high stimulus speeds. In terms of orientation discrimination, this might indicate loss of orientation selectivity. Table 5.1 summarizes the percentage of neurons in areas 17, 18 and 19 that showed speed dependence for different kinds of manipulations.

So far the experimental data revealed little support for any type of stimulus-invariant discrimination, because the tuning characteristics of most neurons are alterable when further stimulus parameters are changed simultaneously (cf. Dinse and von Seelen 1981). In contrast, they are best described in terms of *stimulus representation.* Under the described conditions, i.e. testing the mutual dependencies for a large number of stimulus parameters, RFs are *time* and *stimulus* dependent. While this behaviour is difficult to explain in terms of Hubel and Wiesel RF types, the simulations have shown that such behaviour can easily be generated and even quantitatively predicted by a dynamic model of RF organization, which is based on differences in the time course of excitation and inhibition for different locations within RFs. It should be noted that when more complex stimuli such as moving textured backgrounds were introduced, even more complex response patterns are observed that are not predictable from responses to moving bars of light presented without background. Again, direction selectivity is most sensitive to the introduction of complex stimuli (Dinse and Krüger 1987; Orban et al. 1987; Krüger and Dinse 1988; Allman et al. 1985, 1989). It is well known that considerable areas *outside the classical RF boundaries* are responsible for mediating such effects (Allman et al. 1985; Dinse and Krüger 1987). It seems conceivable that including the dynamic properties of remote locations outside the classical RFs will lead to a more complex response repertoire that is in full agreement with the assumption of context-bound representation of stimulus in the discharge of cortical cells.

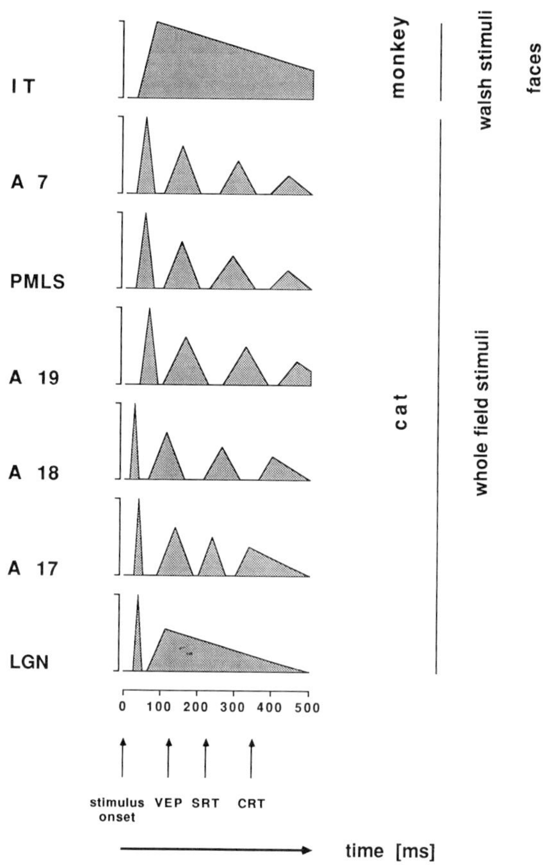

Fig. 5.15. Cumulative PSTHs constructed on the basis of mean values of the parameters describing the oscillatory response pattern. Data for monkey IT according to Richmond et al. 1987. All cat data obtained with whole field stimulation; IT data obtained with so-called Walsh stimuli. The time axis is scaled using averaged values for visual evoked potentials (VEP), simple reaction times (SRT) and choice reaction times (CRT). There are little differences in the onset of response, even for areas of an advanced level according to their number of input stages. The temporal modulation of response therefore considerably extends the possible range in which neuronal cooperativity can take place, to up to 500 ms, and provides a substrate for parallel processing beyond areal borders

5.2.8. Evidence for Parallel Processing Across Different Hierarchical Levels

So far we have restricted our discussion to issues relating to the dynamic properties of RF organization and to the time course of information processing performed by single neurons or small groups of neurons within a given cortical area. In this last section we discuss some implications that arise for *interareal* interactions. A distinct temporal modulation of responses longer than 500 ms counteracts the assumption of neural responses as a short event. Based on data on the characterization of oscillation of response in different areas presented in Sect. 5.2, we reconstructed a *mean average response* to whole field stimulation in the form of *cumulative PSTHs* (Fig. 5.15) for each substrate so far investigated (LGN, area 17, 18, 19, PMLS and area 7). In addition, we utilized the available date for neurons in monkey inferotemporal cortex (IT) as described by Richmond et al. (1987)[16]. Interestingly, the latencies of IT neurons are within the same range.

It is striking feature of the *cumulative PSTHs* the differences in latency to the first peak of excitation are small (Fig. 5.15). This is especially true for neurons of PMLS, area 7 and IT. Obviously, there is no time for any type of sequential processing within the different stages of the visual system, provided that "sequential" means that a subsequent stage is waiting for the end of the entire response of a preceding stage. Whereas areas 17, 18 and 19 of the cat predominantly process information in parallel by receiving independent inputs, areas PMLS, 7 and IT markedly differ in their level of processing. Several more input stages are involved regarding their thalamocortical connections, which should lead to longer time delays of response in comparison to those of areas 17, 18 or 19. However, the distribution of latencies of the former areas suggests that neurons in each of these areas independently start their processing. One has to assume a continuous interaction between all stages that already takes place during the very first state of excitation and continues during the complete temporal modulation of response. This is also true for cells of the A/A_1 lamina of the LGN and area 17, which show a considerable temporal overlap in response. Accordingly, there is no true preprocessing, but interaction between different levels is better described in terms of *temporally distributed processes*[17]. It seems therefore conceivable that there is also a strong time dependence of neuronal cooperativity of thalamocortical and corticocortical interactions. Obviously, this interaction never reaches steady-state conditions. As with *intracortical* feedback loops, this *interareal* time dependent interaction could be a functional substrate for positive feedback acting *beyond* areal borders throughout the entire visual system. It could be the substrate that enables the formation of assemblies or functional groups across long distances. Moreover, the temporal modulation of response considerably extends, up to 500 ms, the possible range in which neuronal cooperativity can take place.

5.3 Summary

5.3.1 Low-Frequency Oscillation of Neuron Response

The vast majority of neurons in the visual cortex of the cat (areas 17, 18, 19, 7 and PMLS) exhibit complex temporally modulated responses that can be characterized as damped aperiodic oscillations in the range of 6 to 20 Hz. This type of oscillation is highly synchronous in closely spaced neurons (<500 microns), but coincidence of oscillations is usually lacking for pairs of neurons separated by a greater distance.

[16] These authors reported that neurons responded with short latencies and strong temporal modulation to a set of geometrical black and white patterns, which were flashed on and off.

[17] One could argue that the fastest neurons in a preceding stage cooperate with the slowest of a subsequent stage. However, such an overlap can be constructed for only a small set of neurons, neglecting the major part of the cell sample.

Investigation of the internal RF organization during the oscillatory episodes revealed, however, that each location within RFs is temporally independent, i.e. its onset latency and oscillatory pattern are different. Accordingly, space and time are non-separable.

5.3.2 Non-Separability of Spatial and Temporal Behaviour

Among spatio-temporal interactions in the cat visual cortex, we frequently observed so-called looming-type and motion-type non-separability. The concept of the generation of RFs by a small number of distinct visual "mechanisms" (e.g. centre-surround) appears to fail for cortical information processing. There are several implications of non-separability: when using stationary flashed stimuli, RF properties will become time dependent, i.e. RF organization and RF size are not static entities, but change over time. High levels of orientation selectivity or hypercomplexity were observed to occur with a delay after a previous, non-selective state.

This spatio-temporal response structure enables neurons to analyze environmental stimuli in terms of elementary events rather than spatial features. These elementary events are the "primitives" of early natural vision such as movement, accretion and deletion of points, and looming. However, the described dynamics were obtained under conditions which allowed the neurons to reach steady state, and no contextual clues were available. Clues such as optical flow information can be assumed to have a strong impact on the observed dynamics. When moving stimuli were used, speed and direction of stimulus motion became crucial variables that determined the overall response characteristics, thus rendering RFs time- and stimulus-variant. These effects were essentially due to the existence of multiple, temporally independent "mechanisms" that contribute to the composition of RFs.

5.3.3 Further Implications of the Oscillatory Response Pattern

Including temporal modulation as a significant portion of the response considerably prolonges the time over which neurons resond to even simple patterns. Assuming an average length of temporal oscillation in the range of 200 to 500 ms, it follows that cortical neurons never reach a genuine steady state. Accordingly, a continuous interaction between external stimuli in the visual field and intrinsic temporal modulation must occur, which makes neuronal responses highly dependent on the dynamics of stimulation.

Both the considerable length of the temporal modulated response and the observation that a high level of stimulus selectivity is reached delayed and independent of maximal spike counts argue in favour of an essential role for a time code.

Finally, comparison of the time course of the temporal modulated response in different cortical areas provides little evidence for serial processing. In contrast, information seems to be processed in parallel but temporally distributed. Interareal time-dependent interaction could be a functional substrate for positive feedback acting beyond areal borders throughout the entire visual system. It could be the substrate that enables the formation of assemblies or functional groups across long distances.

5.3.4 The Origin of the Oscillatory Pattern

The occurrence of fully non-separable spatio-temporal RFs and of time dependence of RF properties in the visual cortex is evidence for the relevance of intracortical feedback. As shown theoretically by Krone et al. (1986) and von Seelen et al. (1987), spatially distributed feedback systems generally result in this type of spatio-temporal interactions. This is in agreement with notions of the impact of cortical architecture on cortical processing strategies. Accordingly, cortices provide the framework that leads to abundant positive feedback, which in turn favours generation of the oscillatory response pattern, as has been found in four cortical areas investigated so far. A manifestation of multiple positive feedback systems are non-separability of space and time, which makes RF organization time- and stimulus-dependent.

When positive feedback mechanisms and time dependence of RF properties are incorporated into a general model of RF organization, the importance of cortical network dynamics on the evolution of stimulus selectivity becomes apparent. Consequently, a high level of selectivity is to a considerable degree independent from the underlying dendritic field organization.

5.4 Perspectives – Anything is Temporal

We have described a temporal structure of neuron response following on off stimulation which consists of damped oscillations in the range of 10 Hz. What are the impacts of this temporal structure, which is causally related to the existence of positive feedback in the cortex, on classical concepts such as RFs? How is the temporal structure linked to fundamental problems of processing and coding of information? Figuratively, anything is temporal. In this scenario a temporal structure emerging during the first 500 ms after onset of stimulation plays the key role. Clearly that means that temporal averaging is no longer reasonable. Temporal averaging will erase the temporal structure to the observer. Therefore, we introduced descriptions in terms of *dynamic RF properties* which were illustrated by temporal sequences of RF descriptors.

Receptive fields are considered to be the basic elements for information processing. However, the original idea of feature detectors has lost its attraction. There is increasing evidence from several laboratories that the concept of RFs has to be extended or reformulated (at present there seems to be no coherent, generally accepted conceptual standard).

Our data that cortical RFs are characterized by non-separability of their spatial and temporal behaviour provided additional support for this view. Non-separability, of course, implies a negative definition: RFs *cannot* be described by a simple cascade of spatial and temporal filters. The notion of non-separability predicts time and stimulus dependence of RFs which could be experimentally verified. Based on these observations, we have introduced the contrasting concepts of *stimulus discrimination* vs. *stimulus representation*. What is still missing, however, is a new conceptual framework of RFs. This becomes even more evident as more complicated types of stimulation are used, such as moving background. Background effects described so far are clearly mediated by regions "outside the classical RF", a term already requiring revision of the old concept.

Our demonstration that cortical neuron response is not a short event, but lasting over some 300 to 500 ms, suggests that under normal conditions the visual system will never operate in a true steady state. However, the observed temporal structure may be ascribed to the intrinsic dynamic properties of these neurons, as it was obtained under conditions that allowed the system to reach steady state. This implies ongoing interactions between the intrinsic temporal structure of the neurons and the dynamics of the stimulus. For example, stimulus repetition faster than 1 Hz must elicit interference between the intrinsic temporal structure and the stimulus dynamics. This long-lasting intrinsic temporal structure also mediates contextual interaction based upon spatial cues (i.e. stimulus dependence). This is possible because the temporal interaction depends not only on the speed of the stimulus and its direction (i.e. the temporal sequence, in which adjacent subfields are activated) but also on which set of subfields is actually utilized. The selection of sets is largely determined by parameters describing the overall shape of the stimulus, such as its orientation or length.

The observation of *dynamic RF properties* and their time and stimulus dependence was a strong argument that stimulus information is indeed processed in this temporal structure, indicating that stimulus information is mapped into the entire temporal structure of the spike train. In the case of temporal encoding of information, temporal averaging of neural responses will not preserve such encoded information. Simplistic measures such as maximal firing rate are then not sufficient to understand the messages transmitted. This is consistent with our finding of a lack of correlation between the points in time where firing rate is maximal and the occurrence of a high level of selectivity. Interestingly, the length of the temporally modulated (oscillatory) response, which we suppose to be the substrate for this type of temporal encoding, links the short, "neuronal", interspike interval timescale to the much longer psychological time scale, in the range of tens of seconds.

Finally, in a more general sense, we believe that a temporal structure of the type described in this paper is optimally adapted for dealing with an environment that is itself essentially dynamic. As entire stabilized images fade after several seconds, genuinely stationary stimuli must be viewed as synthetic. Consequently, the environment must be analyzed in terms of elementary events rather than spatial features, an assumption that should be incorporated into new dynamic RF concepts.

There are obviously a number of open questions. It remains unclear how the striking difference in temporal structure between small-field and whole-field stimulation occurs. One might postulate a synchronizing effect acting across the entire RF and ask for its mechanisms. Moreover, are there general rules relating an idiosyncratic temporal structure to the type of stimulus used? Also, the significance of layer-specific and area-specific temporal patterns is not understood at present. We are left with the certainty, however, that any conclusive conceptual framework on cortical information processing must take the striking temporal structure of neuron response into account.

Acknowledgements. This study was supported by DFG grants (Se 251/* and Di 334/1-1). The experiments were performed at the Neurophysiological Laboratory of Prof. Dr. W. von Seelen's group at the Department of Zoology III at the University of Mainz. The manuscript was prepared while one of us (H.D.) was in receipt of a grant from the Max Kade Foundation, New York. We are very grateful to Prof. Dr. W. von Seelen and Dipl. Biol. W. Kiefer for extensive and fruitful discussions. We appreciate very much Dr. Th. Bullock's interest in this work and his helpful comments on the manuscript. We are also grateful to Dr. P. Johannesma and Dr. Ch. Schreiner for suggestions and discussion of the manuscript. We thank Mr. W. Hoch for his care with the animals, Ms. K. Rehbinder for preparing the illustrations and Ms. M. Grosz, Mr. T. Muller and Mr. G. Hradek for preparing the photographs.

References

Adelson, E.H., Bergen, J.R. (1985): Spatiotemporal energy models for the perception of motion. J. Opt. Soc. Am. A2, 284–299

Allman, J., Miezin, F., McGuiness, E.L. (1985): Direction- and velocity-specific responses from beyond the classical receptive field in the middle temporal visual area (MT). Perception **14**, 105–126

Allman, J., Miezin, F., McGuiness, E.L. (1990): "The effect of background motion on the responses of neurons in the first and second cortical visual areas", in *Signal and Sense*, NRP (Alan Liss, New York)

Baker, C.L., Cynader, M.S. (1988): Space-time separability of direction selectivity in cat striate cortex neurons. Vision Res. **28**, 239–246

Best, J., Reuss, S., Dinse, H.R.O. (1986): Lamina-specific differences of visual latencies following photic stimulation in the cat striate cortex. Brain Res. **385**, 356–360

Braitenberg, V. (1978): "Cortical architectonics: General and Areal", in *Architectonics of the Cerebral Cortex*, ed. M.A.B. Brazier, (Raven Press, New York) pp. 443–465

Braitenberg, V. (1981) "Anatomical basis for divergence, convergence and integration in the cerebral cortex. Adv. Physiol. Sci. **16**, 411–419

Braitenberg, V. (1986): "Two views of the cerebral cortex", in *Brain Theory*, eds. G. Palm, A.M.H.J. Aertsen, (Springer, New York) pp. 81–96

Burr, D.C. (1981): Temporal summation of moving images by the human visual system. Proc. R. Soc. London B **211**, 321–339

Burr, D.C., Ross, J., Morrone, M.C. (1986): Seeing objects in motion. Proc. R. Soc. London B **227**, 249–265

Colonnier, M. (1964): The tangential organization of the visual cortex. J. Anat. **98**, 327–344

Dawis, S., Shapley, R., Kaplan, E., Tranchina, D. (1984): The receptive field organization of x-cells in the cat: spatiotemporal coupling and asymmetry. Vision Res. **24**, 549–564

Dinse, H.R.O. (1981): The spatio-temporal organization of cells in the cat's extrastriate cortex. Neurosci. Lett. **S7**, 41

Dinse, H.R.O. (1989a): *Informationsverarbeitung im visuellen System der Katze. Neuronale Grundlagen*. (Thieme, Stuttgart)

Dinse, H.R.O. (1989b): "A new dynamic model of receptive field organization: stimulus representation vs. stimulus discrimination", in *Proceedings of the 16. German Neurobiology Meeting: Dynamic and Plasticity in neuronal systems*, eds. N. Elsner, W. Singer (Thieme Verlag) pp. 358

Dinse, H.R.O., Krüger, K. (1987): On the representation of background in the cat visual cortex areas 17, 18, 19 and PMLS-Global vs. local processing revisited. Soc. Neurosci. Abstracts **13**, 1624

Dinse, H.R.O., Merzenich, M.M. (1989): Alterations in correlated activity parallel ICMS-induced representational plasticity. Soc. Neurosci. Abstracts **15**, 1223

Dinse, H.R.O., Recanzone, G.H., Merzenich, M.M. (1990): In *Parallel Processing in Neural Systems and Computers*, ed. by R. Eckmiller, G. Hartmann, G. Hauske (Elsevier, Amsterdam) pp. 65–70

Dinse, H.R.O., von Seelen, W. (1981): On the function of cell systems in Area 18 of the cat, Part I and II. Biol. Cybern. **41**, 47–69

Eckhorn, R., Pöpel B. (1980): The spatio temporal receptive fields of visual neurons. Neurosci. Lett **S5**, 494

Eckhorn, R., Bauer, R., Jordan, W., Brosch, M., Kruse, W., Munk, M., Reitboeck, H.J. (1988): Coherent oscillations. A mechanism of feature linking in the visual cortex. Biol. Cybern. **60**, 121–130

Eggermont, J.J., Aertsen, A.M.H.J., Hermes, D.J., Johannesma P.I.M. (1981): Spectro-temporal characterization of auditory neurons: redundant or necessary? Hearing Res. **5**, 109–121

Enroth-Cugell, C., Robson, J.G., Schweitzer-Tong, D.E., Watson, A.B. (1983): Spatio-temporal interactions in the cat retinal ganglion cells showing linear spatial summation. J. Physiol. **341**, 279–307

Fleet, D.J., Hallett, P.E., Jepson, A.D. (1985): Spatiotemporal inseparability in early visual processing. Biol. Cybern. **52**, 153–164

Freeman, W.J., Skarda, C.A. (1985): Spatial EEG patterns, non-linear dynamics and perception: the neo-Sheringtonian view. Brain Res. Rev. **10**, 147–175

Freeman, W.J., di Prisco, G.V. (1986): "EEG spatial pattern differences with discriminated odors manifest chaotic and limit cycle attractors in olfactory bulb of rabbits", in *Brain Theory*, eds. G. Palm, A.M.H.J. Aertsen, (Springer, New York) pp. 97–119

Fuchs, A., Haken, H. (1988): "Computer simulation of pattern recognition as a dynamical process" in *Neural and Synergetic Computers* ed. H. Haken Springer Ser. Synergetics Vol. 42 (Springer, Berlin, Heidelberg) pp. 16

Galli, L., Chalupa, L., Maffei, L., Bisti, S. (1988): The organization of receptive fields in area 18 neurons of the cat varies with the spatio-temporal characteristics of the visual stimulus. Exp. Brain Res. **71**, 1–7

Gielen, C.C.A.M., van Gisbergen, J.A.M., Vendrik, A.J.H. (1981): Characterization of spatial and temporal properties of monkey LGN Y-cells. Biol. Cybernetics **40**, 157–170

Gilbert, C.D., Wiesel, T.N. (1979): Morphology and intracortical projections of functionally characterized neurons in the cat visual cortex. Nature **280**, 120–125

Gray, C.M., Konig, P., Engel, A.K., Singer, W. (1989): Oscillatory responses in cat visual cortex exhibit inter-columnar synchronization which reflects global stimulus properties. Nature **338**, 334–337

Hartline, H.K. (1938): The response of single optic nerve fibres of the vertebrate eye to illumination of the retina. Am. J. Physiol. **121**, 400–415

Heinke, D., Giannakopoulos, F., Best, J. (1990): "The thalamo-cortical feedback-system: a nonlinear network model for binocular interaction" in *Parallel Processing in Neural Systems and Computers*, ed. by R. Eckmiller, G. Hartmann, G. Hauske (Elsevier, Amsterdam) pp. 113–116

Hubel, D.H., Wiesel, T.N. (1962): Receptive fields, binocular interaction and functional architecture in the cat's visual cortex. J. Physiol. Lond. **160**, 106–154

Johannesma, P.I.M., Aertsen, A.M.H.J., van den Boogaard, H.F.P., Eggermont, J.J., Epping, W.J.M. (1986): "From synchrony to harmony: ideas on the function of neural assemblies and on the interpretation of neural synchrony", in *Brain Theory*. eds. G. Palm, A. Aertsen (Springer, New York) pp. 25–47

Jones, E.G. (1974): "The anatomy of extrageniculostriate visual mechanisms", in *The Neurosciences, Third Study Program*, eds F.O. Schmitt, F.G. Worden (MIT Press, Cambridge) pp. 215–227

Korn, A., von Seelen, W. (1972): Dynamische Eigenschaften von Nervennetzen im visuellen System. Kybernetik (Biol. Cybern.) **10**, 64–77

Krause, F., Eckhorn, R. (1983): Receptive fields for motion stimuli (RFM) determined for different types of cat visual neurons. Neurosci. Lett. **14**, Suppl. S 209

Krone, G., Mallot, H.A., Palm, G., Schüz, A. (1986): Spatio-temporal receptive fields: a dynamical model derived from cortical architectonics. Proc. R. Soc. London B **226**, 421–444

Krüger, K., Dinse, H.R.O. (1988): Background dependence of direction selectivity in cortical neurons of areas 17, 18, 19, 7 and PMLS of the cat. Soc. Neurosci Abstracts **14**, 602

Li, C., Nothdurft, H.C. (1987): Some observations on dynamic properties of receptive field organization of complex cells in cat visual cortex. Sci. Sin. Ser. B **30**, 44–54

Lin, C.S., Friedlander, M.J., Sherman, S.M. (1979): Morphology of physiologically identified neurons in the visual cortex of the cat. Brain Res. **172**, 344–345

Longuet-Higgins, H.C. (1968): The non-local storage of temporal information. Proc. R. Soc. London B **171**, 327–334

Lund, J.S. (1973): Organization of neurons in the visual cortex, area 17, of the monkey (Macaca mulatta). J. Comp. Neurol. **147**, 455–495

Lund, J.S., Henry, G.H., MacQueen, C.L., Harvey, A.R. (1979): Anatomical organization of the primary visual cortex (area 17) of the cat. A comparison with area 17 of the Macaque monkey. J. Comp. Neurol. **184**, 599–618

Mallot, H.A., von Seelen, W. (1989): "Why cortices? Neural networks for visual information processing", in *Visuomotor Coordination: Amphibians, Comparisons, Models, and Robots*, eds J.P. Ewert, M. Arbib, (Plenum, New York) pp. 357–382

Marr, D., Ullman, S. (1981): Directional selectivity and its use in early visual processing. Proc. R. Soc. London B **211**, 151–180

Martin, K.A.C. (1988): From single cells to simple circuits in the cerebral cortex. Quart. J. Exp. Physiol. **73**, 637–702

Martin, K.A.C., Whitteridge, D. (1984): Form, function, and intracortical projection of spiny neurons in the striate visual cortex of the cat. J. Physiol. Lond. **353**, 463–504

McLean, J., Raab, S., Palmer, L. (1987): Spatiotemporally oriented simple receptive fields: Local linear motion detectors. Soc. Neurosci. Abstracts **13**, 1623

Merzenich, M.M., Allard, T., Jenkins, W.M., Recanzone, G. (1988): "Self-organizing processes in adult neo-cortex", in *Neural Networks. Structure and Models*, eds W. von Seelen, G. Shaw, U.M. Leinhos (VCH, Weinheim) pp. 285–297

Mungai, J.M. (1967): Dendritic patterns in the somatic sensory cortex of the cat. J. Anat. **101**, 403–418

Optican, L.M., Richmond, B.J. (1987): Temporal encoding of two-dimensional patterns by single units in primate inferior temporal cortex. III. Information theoretic analysis. J. Neurophysiol. **57**, 162–178

Orban, G.A., Gulyas, B., Vogels, R. (1987): Influence of moving texture background on direction selectivity of cat striate cortex. J. Neurophysiol. **57**, 1792–1812

Palmer, L.A., Rosenquist, A.C., Tusa, R.J. (1978): The retinotopic organization of the lateral suprasylvian areas in the cat. J. Comp. Neurol. **177**, 237–256

Podvigin, N.F., Kuperman, A.M., Chuyeva, I.V. (1974): Space-time properties of excitation and inhibition and wave processes in the receptive fields of the external geniculate body of the cat. Biofizika **19**, 341–346

Recanzone, G., Merzenich, M.M. (1988): Intracortical microstimulation in somatosensory cortex in adult rats and owl monkeys results in large expansion of the cortical zone of representation of a specific cortical receptive field. Soc. Neurosci. Abstracts **14**, 223

Reinis, S., Weiss, D.S., Landolt, J.P. (1988): Lack of homogeneity of receptive fields of visual neurons in the cortical area 18 of the cat. Biol. Cybern. **59**, 41–48

Richmond, B.J., Optican, L.M. (1987): Temporal encoding of two-dimensional patterns by single units in primate inferior temporal cortex. II. Quantification of response waveform. J. Neurophysiol. **57**, 147–161

Richmond, B.J., Optican, L.M., Podell, M., Spitzer, H. (1987): Temporal encoding of two-dimensional patterns by single units in primate inferior temporal cortex. I. Response characteristics. J. Neurophysiol. **57**, 132–146

Rosenquist, A.C. (1985): "Connections of visual cortical areas in the cat", in *Cerebral Cortex Vol 3. Visual Cortex* eds. A. Peters, E.G. Jones, (Plenum, New York) pp. 81–117

Schreiner, C.E., Joris, P.X. (1986): Intrinsic oscillations in the primary auditory cortex of cats. IUPS Satellite Symposium on Hearing. UC, San Francisco, pp. 81

Shaw, G.L., Silverman, D.J., Pearson, J.C. (1986): "Trion model of cortical organization: Toward a theory of information processing and memory", in *Brain Theory* eds. G. Palm, A.M.H.J. Aertsen (Springer, New York) pp. 177–191

Shevelev, I.A., Sharaev, G.A., Volgushev, M.A., Pyshny, M.F., Verderevskaya, N.N. (1982): Dynamics of receptive fields of the visual cortex and lateral geniculate neurons in cat. Neurofisiologia **14**, 622–630

Spear, P. (1990): "Functions of extrastriate visual cortex in nonprimate species", in *The neural basis of visual function* ed. A. Leventhal, (Macmillan, London)

Stevens, J.K., Gerstein, G.L. (1976): Spatiotemporal organization of cat lateral geniculate receptive fields. J. Neurophysiol. **39**, 213–238

Tolhurst, D.J., Movshon, J.A. (1975): Spatial and temporal contrast sensitivity of striate cortical neurons. Nature **257**, 674–675

Tusa, R.J., Palmer, L.A., Rosenquist, A.C. (1981): "Multiple cortical visual areas: visual field topography in the cat", in *Cortical Sensory Organization*, Vol II, ed. by C.N. Woolsey, (Humana Press, New York) pp. 1–31

von Seelen, W. (1970): Zur Informationsverarbeitung im visuellen System der Wirbeltiere. I. Kybernetik (Biol. Cybern.) **7**, 43–60

von Seelen, W., Mallot, H.A., Krone, G., Dinse, H.R.O. (1986): "On information processing in the cat's visual cortex", in *Brain Theory* eds. G. Palm, A.M.H.J. Aertsen, (Springer, New York) pp. 49–79

von Seelen, W., Mallot, H.A., Giannakopoulos F. (1987): Characteristics of neuronal systems in the visual cortex. Biol. Cybern. **56**, 37–49 (1987)

Volgushev, M.A., Shevelev, I.A., Dec, K., Sharaev, G.A., Verderevskaya, N.N. (1983): Differences in dynamics of visual receptive field and its summation zone in cat. Neurofisiologia **15**, 466–473

Wilson, H.R. (1980): Spatiotemporal characterization of a transient mechanism in the human visual system. Vision Res. **20**, 443–452

Wong, W.C. (1967): The tangential organization of dendrites and axons in three auditory areas of the cat's cerebral cortex. J. Anat. **101**, 419–433

6. Spike Train Correlations on Slow Time Scales in Monkey Visual Cortex

J. Krüger

Neurologische Universitätsklinik, Hansastr. 9 D-7800 Freiburg

The brain is usually viewed as an organ in which neurons respond to signals from the environment, and other neurons control muscles acting upon the environment. The link between the two is formed by an enormous neuronal network within which a complex signal processing goes on, involving memory storage and retrieval. This view implicitly presupposes the knowledge of the notion of "environment", together with some of its fundamental properties, such as containing objects. The role of neuronal activity is then to generate a representation of individual situations occurring in that environment.

If, however, one takes a viewpoint from *within* the brain which is enclosed in a black box, the environment including its fundamental properties appear as unknowns. The network only obeys general principles of evolution and selection, and it has to find ways, by means of self-organization, to contribute to the survival of the species of its host. It does so by grouping correlating information together so that data processing and storage becomes more economical. The general strategy is to handle compact concepts instead of large quantities of individual nerve signals. The environment is not taken as pre-existing but rather as a concept being elaborated.

Thus, the conventional view "from without" stresses the role of nervous signals as such, whereas the view "from within" emphasizes *correlations* between such signals. Indeed, the elements of the neuronal network are well-suited for handling correlations: neurons are correlation *detectors* because dendritic depolarizations occurring at different input synapses of the same neuron are much more efficient in eliciting an output spike if they are synchronized (Abeles 1982). Clearly, neurons are also *generators* of correlation because their axones may bifurcate so that the same signal is sent to several neurons.

However, the latter property is a very common one in the biological domain, since any cell secreting substances will usually influence several other cells simultaneously. If a cell is round or compact, it can either influence a small number of direct neighbors, and if nothing is lost the interaction with at least one cell must necessarily be rather strong. It can also act indiscriminately upon a larger number of cells by extracellular diffusion. The speciality of nerve cells lies rather in how the correlations they generate are distributed: by virtue of their long cell processes, neurons can *avoid* limiting their strongest interactions to the nearest

Springer Series in Synergetics
J. Krüger (Ed.)
Neuronal Cooperativity

neighbors. Instead, each cell can select (in the cortex) among about 100,000 cells located essentially within a hypercolumn, and establish individual contacts with 1000 or 10 000 of them (Braitenberg 1978). The activities of the latter cells will then correlate, but since the input from one given cell is only a small fraction of the total inputs impinging on each target cell, not only the excitation but also the correlation generated by one neuron is weak. In this respect, neuronal networks are different from other tissues. In addition, two correlating neurons can be separated by other neurons not correlating with them.

Some well-known instances where the role of correlations is obvious are

- the concept of the solid object, formed by correlating activity from remote edges,
- the generation of complex movements from correlating signals of motor neurons,
- the activity of playing, where a motor signal is produced repeatedly and whose effect in the outer world, detected by the senses, is correlated with the former signal.

Finally, Hebbian synapses (Hebb 1949), which are thought to form the basis of memory, react to correlating pre- and post-synaptic activity with an increase of synaptic efficiency.

In our experimental work over the last years we were primarily concerned with neuronal signals belonging together. We wanted to know the spatial and temporal relationships, and other conditions, of neuronal correlation within a small volume of the intact neocortex, in order to discover principles of cortical signal processing, and we also wondered about the functional roles of correlation. Furthermore, it has been our long-standing aim to apply the "view from within the brain" experimentally. This implies that we have to take neuronal activity as the primary event (in a logical, not a temporal sense), and first try to determine its internal structures. The structures are then related to events, considered as secondary, occurring in the outer world. We believe that this is the more natural way of viewing the brain, in contrast to that of primarily considering events in the outer world, and to determine "reactions" in the brain taken as secondary events. In the latter case, no urgent need is apparent for recognizing structures in the neuronal signal flow.

If one part of the brain needs information from another part, it receives a bundle of fibres which is installed essentially permanently. One of our methodological aims is an approximation of this "natural recording technique", namely the long-term implantation of many microelectrodes. At present we are close to this goal but we could not yet apply it to scientific questions. The insights we have gained so far are based on recordings obtained with an array of 30 microelectrodes from the striate cortex of anaesthetized, paralyzed monkeys. Most of the results presented here are based on the recordings from four animals. In our 5×6 square lattice arrangement, the electrodes had mutual separations of 160 μm. Simultaneous recordings were taken during one to two hours from a cortical layer while various stimuli were presented. Four or five layers were recorded sequentially, with the recording loci situated one below another.

We could not individually adjust the electrode positions, so well-isolated single spikes were recorded only at 20 to 50% of the electrodes. The remaining electrodes usually showed one dominating spike with a mostly minor admixture of a further spike. Unresolvable mass activity was rare. We want to emphasize that the impressive stability of the recordings was much more important for the conclusions drawn here than a perfect single-spike isolation. We may also note that all simultaneously recorded neurons were exposed to the same complete set of stimuli. There was no inadvertent interruption of the recordings due to mechanical instability, as this usually occurs after variable time spans when only one electrode is used.

6.1 Spike Train Correlations

Long before the term "correlation" was applied in a strict sense to spike trains (Perkel et al. 1967a,b), researchers having recorded simultaneously from two spiking neurons, have looked for correlations. They simply counted how often exactly synchronous spikes (or spike pairs with a particular mutual delay) occurred in comparison with spike pairs with other mutual delays. A correlogram, as shown in several articles of the present volume, is a systematic way of presenting these counts in the form of a histogram, with the mutual delay being the abscissa. The histograms may differ in bin width, i.e. in the range within which the delays are accepted to be equal.

Since the systematic presentation of the correlation technique (Perkel et al. 1967a, b), it is known that an examination of structures in high-resolution correlograms can reveal details of functional connectivity including time dependency (see the contribution of Aertsen et al. in this volume). The relevant range of delays is between about 0.5 and 10 ms. We have published an analysis along these lines (Krüger and Aiple 1989), but here we will consider two types of temporally more inaccurate correlations. Firstly, when a spike is emitted in a cortical neuron we found that in another neuron the likelihood of spike emission is increased within about $\pm$ 25 ms around the former spike. Secondly, some years ago we found that neurons in small groups tend to covary, i.e. to raise and lower their firing rates together within seconds (Bach and Krüger 1986). This is another variety of spike train correlation on a very slow time scale.

Both these types of correlation cannot immediately be interpreted in terms of underlying connectivity but, since the phenomena are very prominent, an inquiry into their origin and functional role seemed necessary. A synthesis of relevant findings and discussion points will be presented here.

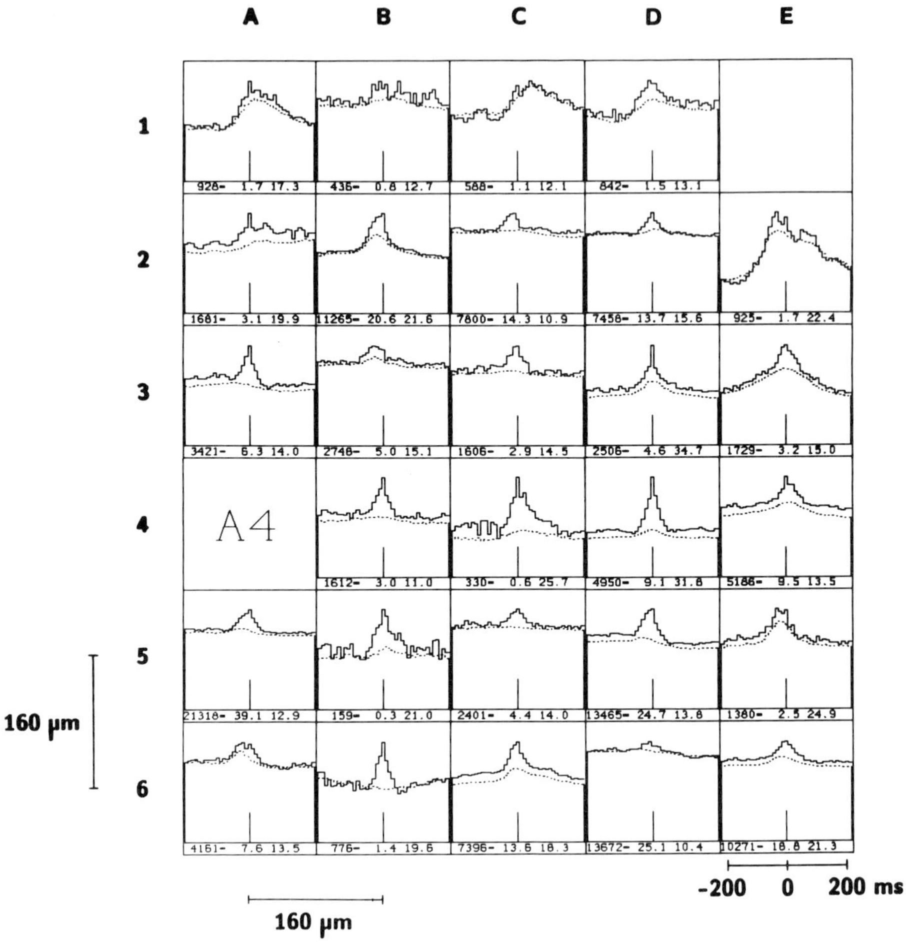

Fig. 6.1. Correlations of the spike train recorded at the electrode A4 with the trains of the remaining 29 electrodes. No suitable recording was obtained at E1. The arrangement of the correlograms corresponds to the array of electrodes. The dashed lines are the shift predictors. The correlogram bin width is 11 ms. Numbers below each histogram indicate the maximum bin count as absolute number (left), as number relative to the total A4 spikes (= 550; middle) and as number relative to the total spikes at each running electrode (right). The total accumulation time was 5200 ms. Data are from layer VI. It can be seen that most correlograms have maxima with half-widths of about 50 ms

6.2 Correlations in the 50 ms Range

The basic observation is shown in Fig. 6.1: the correlograms are depicted for the spike train from electrode A4 paired with that of each of the remaining electrodes. No activity was recorded at E1, but nearly all remaining correlograms have central peaks with half-maximum widths of about 50 ms. This is true even if the so-called "shift-predictor" (dashed curve; Perkel et al. 1967b) is subtracted, which describes the stimulation-induced correlation. The recordings are from layer VI but similar correlograms are obtained in all layers, and also between layers (in preparation). Thus, this type of correlation is clear cut and ubiquitous in area 17.

A priori one might think that many circuits, involving many synaptic stages and even remote brain areas, could be involved in these correlations so that clear insights into the underlying mechanisms could not easily be expected. However, one of the first observations made in layer IVc gave us a hint. Figure 6.2 exemplifies this finding: in Fig. 6.2a a distribution of ocular dominances is shown, and in Fig. 6.2b and c it can be seen for the left eye-dominated neuron A4 and for the right eye-dominated neuron A5, that (besides a decay with distance) the correlogram peaks are largest for pairs of neurons having similar eye dominances. The larger maximum of the pair A5-D2 shows that this is true even when two neurons are separated by a stripe of neurons linked to the opposite eye. This finding was confirmed in all layer IVc recordings, and it led to the idea that correlations of this type originate in the retinae. It would not be surprising because in the cat Mastronarde (1983) has observed that neighboring retinal ganglion cells have correlating spike trains. Four further findings corroborated our assumption. Firstly, receptive field (i.e. retinal) distances explain the variance of peak sizes more than do intracortical distances. Despite cortical retinotopy, the two kinds of distance are not equivalent for nearby loci because two cells with neigboring receptive fields may be separated by a stripe of cells linked to the opposite eye, and in addition in most cortical layers there is a considerable scatter of receptive field position (Hubel and Wiesel 1974).

Secondly, the shift-predictors (dashed curves in Figs. 6.1 and 6.2) have peaks that are often rather flat but have similar widths as those of the original correlograms. An explanation can be that the loci where the correlations are generated are very close to each other: in the case of the raw correlograms, it is one (or both) of the intraretinal plexiform layers interconnecting parallel afferent pathways. In the case of the shift-predictor, it is the level of the photoreceptors where synchrony is generated by common stimulation. For both types of correlation the relaxation of accuracy, introduced by several synaptic stages, has a comparable extent because it occurs on the common pathway to the cortical recording sites.

Thirdly, we found that the correlogram peak areas are proportional to the product of the spike rates. This is not trivial: such a relationship is obtained if common spike sources with similar, rather elevated spike rates are assumed, and the lower but individually different spike rates of cortical cells are produced by individual degrees of random spike *losses* (Aiple and Krüger 1988). This can be seen as follows: assume two neurons A and B each receiving some independent input, but also some shared input that causes the outputs to be correlated. If the rate of A-spikes is now reduced by a random removal of spikes, then whatever the degree of reduction is, the rate of *correlated* spikes is lowered by the same factor as the rate of A-spikes. Analogously, reducing the rate of B-spikes too, the rate of correlated spikes is further lowered by the factor of B-spike reduction.

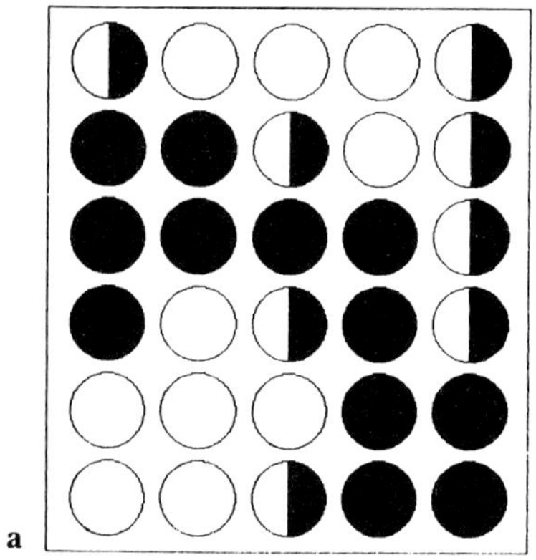

Fig. 6.2. Correlations in layer IVc. The format of the figure is as in Fig. 6.1. **a** Ocular dominance distribution (black = right eye, white = left eye); **b** correlations of the right-eye-dominated neuron A4 with the remaining ones; **c** correlations of the left-eye-dominated neuron A5 with the remaining ones. It can be seen that the correlogram peaks are largest for correlations between neurons dominated by the same eye. This is true even when an opposite-eye stripe has to be bridged, e.g. in (**c**) the correlation A5-D2

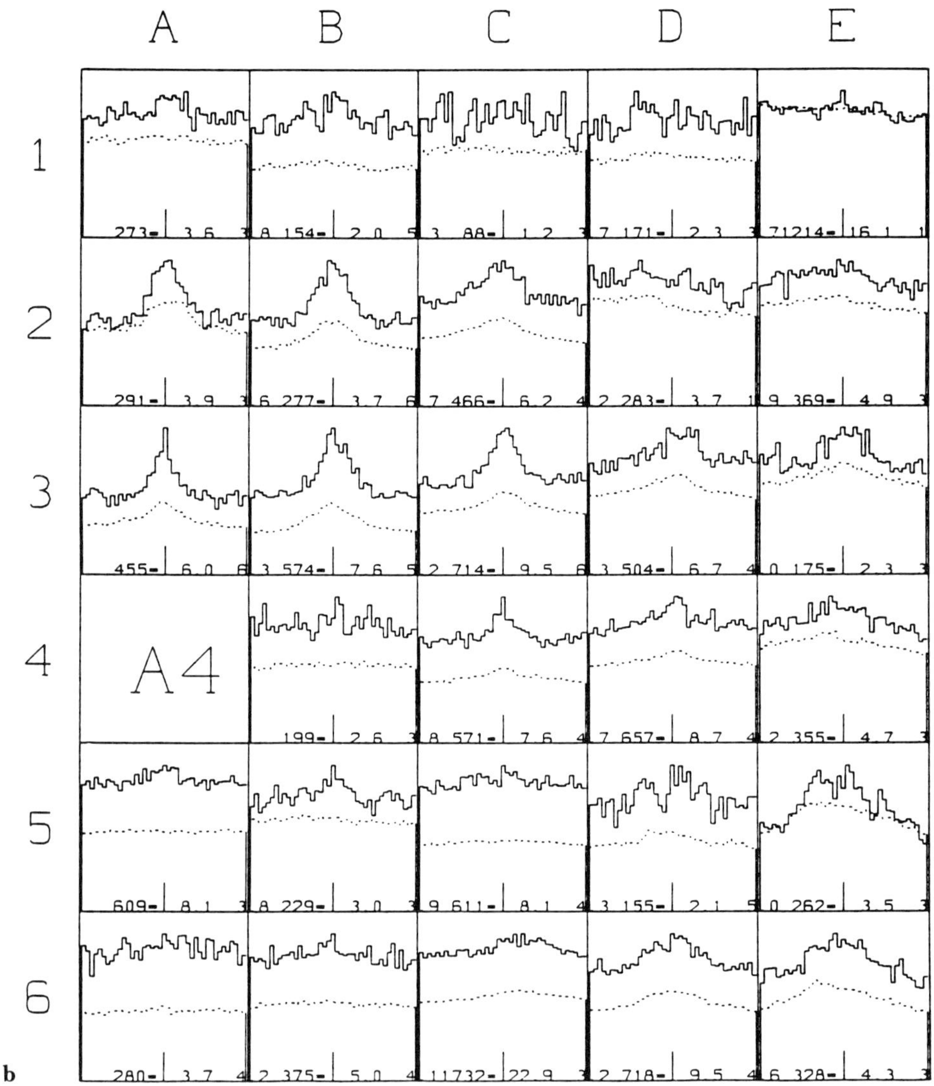

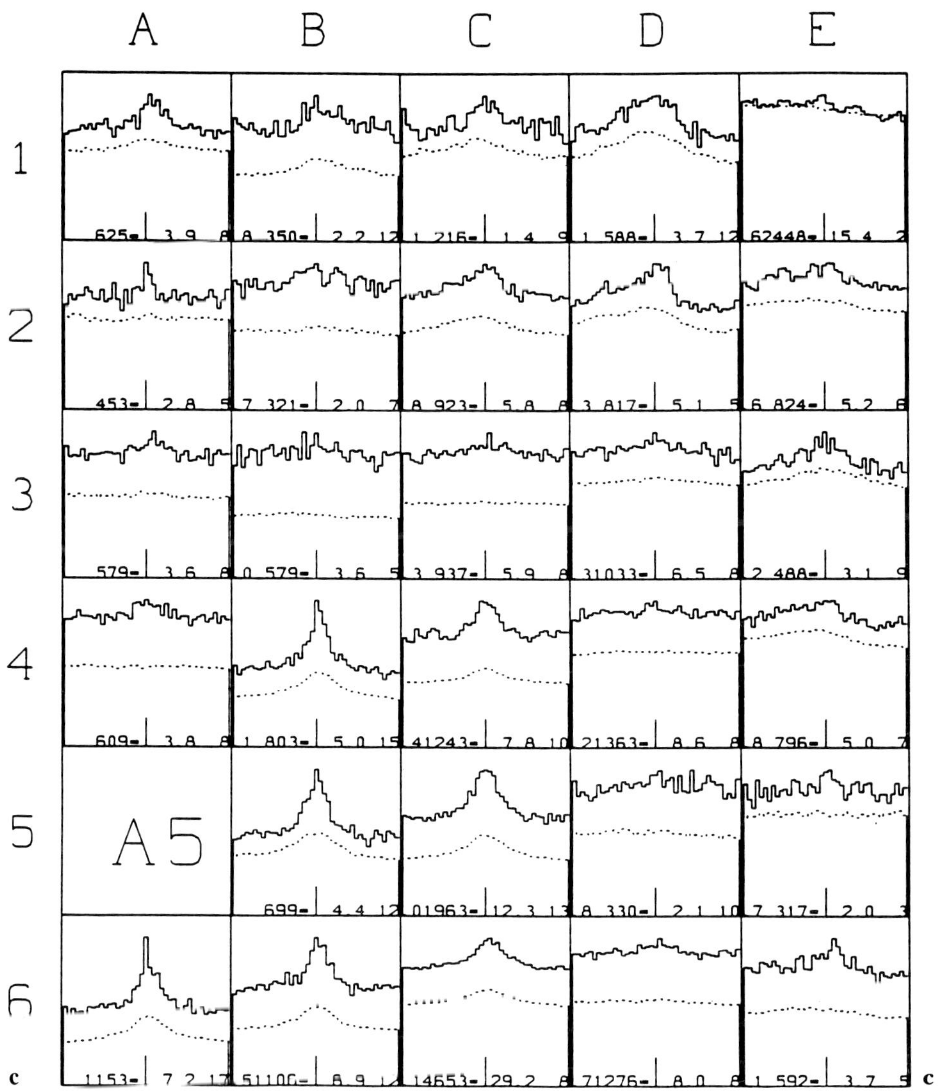

Assuming that the spike rates of A and B were equal before the reductions, the rate of correlated spikes will always be proportional to the product of the two remaining reduced spike rates. A weaker assumption, generating only noise in the relationship, is that the spike rates before and after reduction are not systematically related. This assumes that a rather low spike rate at the outset does not systematically lead, by the process of reduction, to a particularly low resulting rate. The proportionality holds in the visual cortex for the whole range of spike rates encompassing two powers of ten (Fig. 6.3). The point of the argument is that the correlation is generated *before* the spike rates are reduced. The idea to locate the spike sources in the retinae is consistent with the higher spike rates observed there, as compared to those of the cortex.

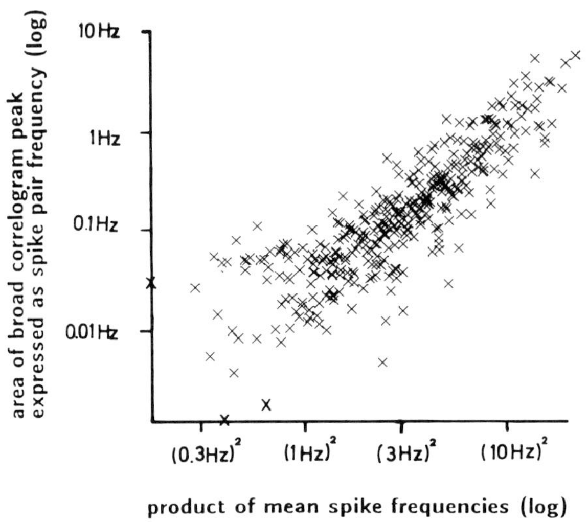

Fig. 6.3. Relationship between the correlogram peak areas and the product of the mean spike rates of the two contributing neurons (double logarithmic plot). The slope of the cloud of points is about 1, indicating a proportionality of the variables

A fourth reason for a retinal origin of the broad correlogram peaks is that many of the peaks occupy eccentric positions. This signifies that on average the spikes of one neuron are delayed or advanced with respect to the correlating spikes of another neuron. The special property of peak displacements is that they are transitive, i.e. the displacement observed between neuron A and C is the sum of that between A and B and the one between B and C (Fig. 6.4). This is not generally true for correlating data. (Note also that more than two electrodes are required to discover that property.) It follows from the transitivity that a scale can be constructed where each neuron occupies a particular position, and the observed peak displacements are proportional to the positional differences on the scale (Krüger and Aiple 1988) Neurons at opposite ends of this "delay scale" can be labelled as "fast" and "slow", respectively. In the infragranular layers so far investigated, spikes from the source of correlation arrive in a fast neuron up to 25 ms earlier than in a slow one. The differences of conduction velocities in the parvo- and magno-cellular subdivisions of the visual afferences immediately come to mind. However, the conduction times measured by cortical stimulation fall short of the above value by a factor of at least five, because the artificial synchronization of large numbers of presynaptic signals under this condition precludes the normal integrative delays of the neurons to become effective. However, with *visual* stimulation latency differences between slow and fast neurons have been found to be comparable to our values (Bolz et al. 1982; Sestokas and Lehnkuhle 1986; Raiguel et al. 1989).

Although solely deduced from spike train correlations, the position on the delay scale is a property of neurons and not of *pairs* of neurons. It correlates with response properties such that the relation to the parvo/magnocellular subdivision of input streams is corroborated: a factor analysis has shown that in the infragranular layers so far analyzed (Krüger, in preparation), one factor explains

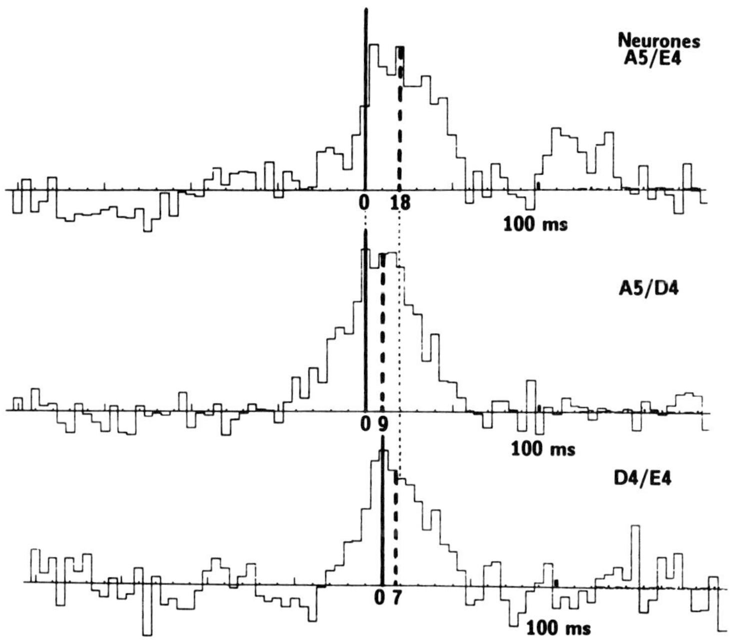

Fig. 6.4. Transitivity of lateral displacements of correlogram peaks. The example shows correlograms from three neurons selected for particularly large displacements. The shift-predictors have been subtracted. For transitivity to hold, the displacement of A5/E4 should be the sum of those of AS/D4 and D4/E4. This is only approximately true in the example

several differences of neuronal properties in the most compact way. This factor correlates positively with the delay scale, direction selectivity, responsiveness to moving visual noise and, most remarkably, with the degree of spike isolation, which is a measure of spike size and thereby a measure of cell soma size. The same factor correlates negatively with color selectivity. Indeed, the factor essentially reflects the degree of dependency on the magnocellular cortical input.

We believe that the picture given so far is sufficiently consistent to make a retinal origin of the broad correlogram peaks most likely.

However, there is one further set of observations that cannot easily be linked to that idea: many correlograms show not only a broad central peak but further lateral undulations (Fig. 6.5a). These are indications of repetitive or oscillatory events. They have frequencies of about 10 Hz which is a much lower value than the about 40 Hz that are ubiquitous in the cat visual cortex (Krüger 1983, see also Chaps. 8 and 9 in the present volume). The oscillations in the monkey cortex are in the range of human α-waves. At some times, or in some monkeys, they appear in many correlograms recorded simultaneously, so that it

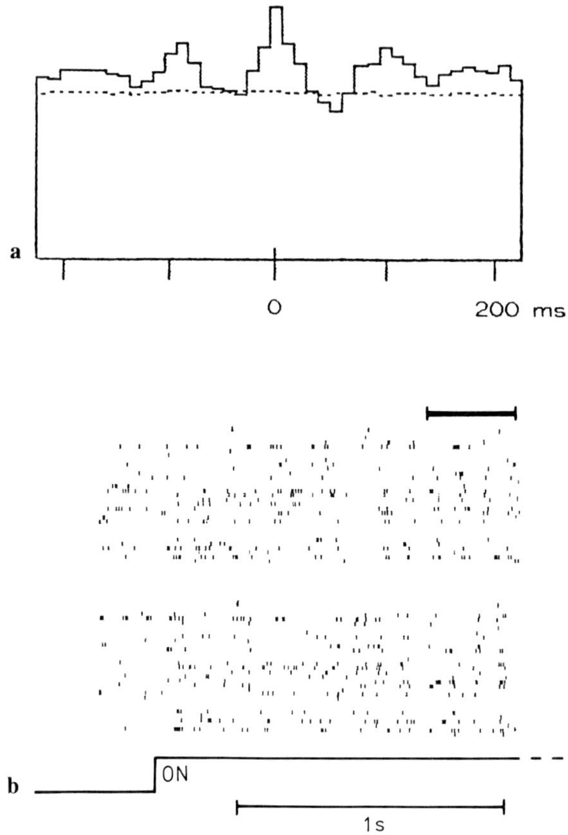

Fig. 6.5. Evidence for oscillatory neuronal activity. **a** Oscillatory correlogram. The bin width is 11 ms. The dashed line is the shift-predictor. The positions of the lateral maxima indicate an oscillation frequency of 10 Hz, similar to that of α-waves. Usually in cases of oscillatory correlations, at least in one of the autocorrelograms (not shown), evidence of oscillations is visible too; **b** 30 simultaneously recorded spike trains showing the responses to a light switched on. The two bands of spikes were evoked by two consecutive repetitions of that stimulus. It can be seen that a slight stimulus-locked oscillation occurs in the first 300 ms of the responses. Later on, a more clear-cut oscillation appears in both repetitions, but in the sections below the black bar it is out of phase with respect to stimulus onset. In the lower trial, a phase shift of half an oscillation period can be seen below the left end of the bar

is likely, just as in man, that they are related to the state of the animal, or that they represent an internal message.

In Fig. 6.5b oscillatory events are depicted in simultaneous recordings obtained with 30 microelectrodes occupying 0.5 square millimetres in cortical layer IVc. The upper band of 30 spike trains shows the spikes evoked by a single presentation of a white large field. Some repetitive firing on most channels can just be recognized at the onset of the responses. A much clearer oscillation, encompassing all electrodes, begins about 0.5 s later. The lower panel shows another presentation of the same stimulus. Again a weak oscillation locked to stimulus onset can be recognized. However, the later oscillation undergoes a spontaneous phase shift in the interval delineated by the horizontal bar above, so that at the right borders the oscillations in the two panels are approximately in counterphase.

It is remarkable that the central peaks of oscillatory correlograms resemble the broad peaks as introduced above, but there is no evidence that the oscillations are generated in the eyes. However, they could be generated in, or transmitted by, the lateral geniculate nucleus (LGN), because this thalamic structure is known to be related to cortical oscillatory behavior (Steriade and Deschênes 1984). It is improbable that the oscillations in the LGN are segregated by layers, i.e. by the eyes. Rather, a phase coupling must be expected. A cursory inspection of many correlograms indeed gave the impression that correlograms between neurons dominated by opposite eyes, if undulations can be seen, have lateral peaks that are relatively larger when compared to the central one. In contrast, correlograms between neurons dominated by the same eye seem to have a relatively larger central peak. If that proves to be correct (the evaluation is in progress), one would tend to suggest two separate mechanisms: a retinal one, generating broad correlogram peaks essentially between neurons dominated by the same eye, and another one independent from and superimposed on the former, that takes place in such a way that there is no distinction by the eyes. The reticular formation, the LGN and the visual cortex itself can play an essential role in the generation and maintenance of the latter process.

If a broad peak is laterally displaced in a correlogram showing oscillations, the undulation peaks usually have about the same lateral shift. It is as if the entire correlogram was rigidly shifted to one side. Two explanations come to mind:

- The oscillations could be an intrinsic property of each piece of cortex that needs to be triggered by correlated spikes. If these spikes arrive with mutual delays at different cortical loci (we have indeed observed a "blob"-like grouping of cell with similar delay values), then different loci would preferentially oscillate with the phase delays set by the trigger impulses. However, it is not clear how the continuing stream of triggering impulses influences the ongoing oscillation.
- The oscillations originate in the brainstem (see Steriade and Deschênes 1984) and enter the visual system at the level of the LGN. They partly use the same pathways to the cortex as the visual signals, and therefore the differential visual delays should manifest themselves correspondingly in the oscillatory signals. The oscillations in the left eye- and right eye-dominated parts of the LGN are likely to be in phase, so that the relatively stronger oscillatory correlations obtained for neurons dominated by opposite eyes can be understood: if there were no systematic phase relationships at all, the correlogram would not show any undulations.

6.3 Correlations in the Range Fractions of Seconds to Minutes ("Covariability")

In the present context, the term "variability" designates fluctuations of spike rates unrelated to controlled parameters in the experiment. Figure 6.6 illustrates the phenomena for two neurons. In both cases, the stimuli were oriented bars.

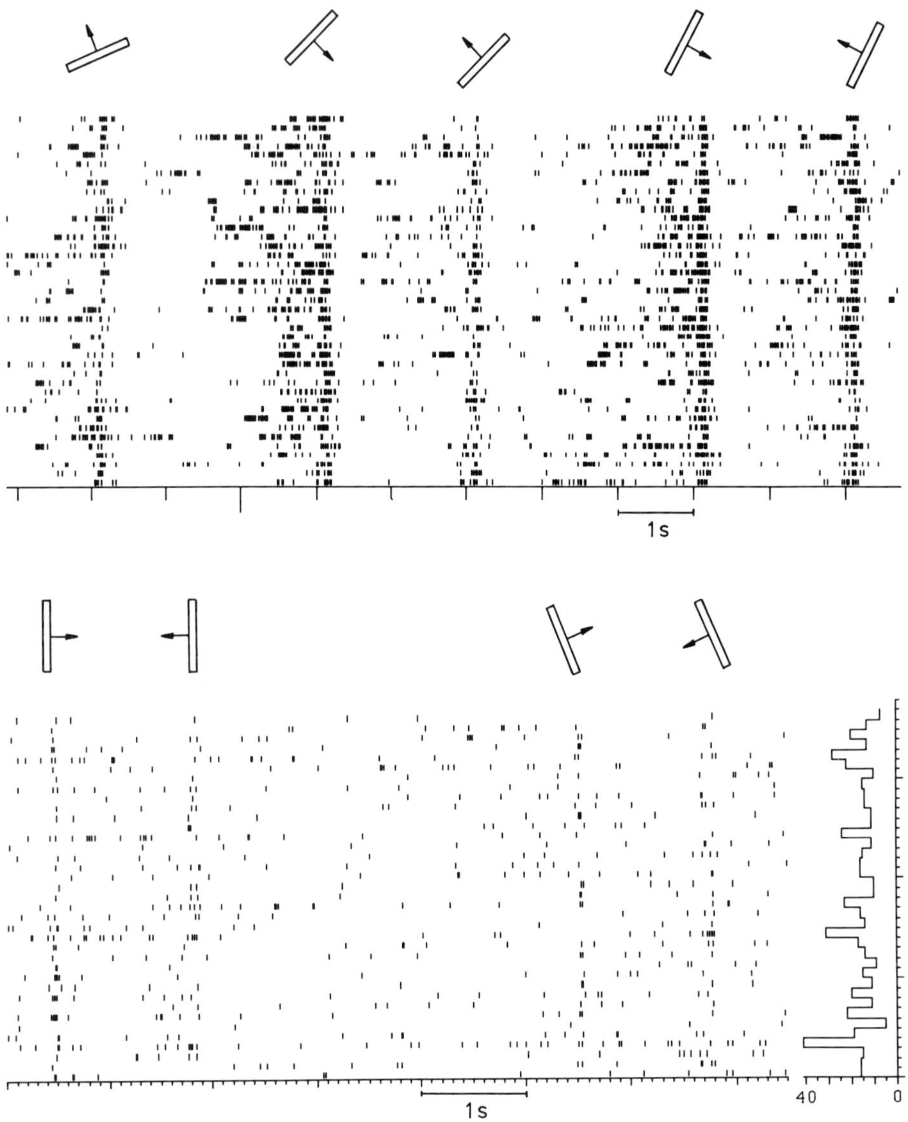

Fig. 6.6. Examples of neuronal variability. The panels show responses of one neuron each to repeated sequences of oriented bars. Upper panel: Variability in the form of spontaneous bursts whose spike frequencies often exceed those of the responses. Lower panel: Variability in the range of seconds. The rotated histogram displays the spike counts in the trials. It can be seen that in this example the counts vary by a factor of up to 6. There is little variability in the times of occurrence of the responses unless the responses disappear altogether

In the upper example, much of the variability occurs in the range of a fraction of a second: some spontaneous variations of the spike rate exceed in magnitude those elicited by the stimuli. In the lower example, the sequences of spikes show variations within seconds, and it can be seen that the spike frequencies of both the responses and the ongoing activity are affected. The histogram on the right (with a vertical abscissa paralleling the repeated trials) displays the spike counts in each trial. It shows the variations occurring within tens of seconds, which in this example amount to a factor of up to about 6. Note that in many stimulus-selective neurons the difference between an optimal and a non-optimal response is not great. The present results were obtained from anaesthetized animals, but in awake monkeys variability has an influence of comparable magnitude on spike rates (Vogels et al. 1989).

If the question ever arises of whether there are interrelations between the variabilities of several neurons within an ensemble, two cases can most easily be imagined: firstly, the variabilities of all neurons could be independent of each other, and secondly, there could be a common variability in *all* neurons, i.e. all spike rates rise and fall together. In the first case, variability could be considered to be similar to noise, and it could be suppressed by averaging over many trials or over many neurons. The second case would be expected if variability depended on a single parameter such as alertness or the state of anaesthesia.

Our actual finding Bach and Krüger 1986 corresponded neither to the first nor to the second case: usually among the 30 neurons recorded simultaneously, there were subgroups of neurons that *co*varied during the entire observation time of up to several hours. A small number of independently varying groups, each with *co*varying members, could be present within the recording area which covered half a square millimetre. Figure 6.7 shows an example of two pairs of covarying neurons. The rotated histograms again display the spike counts during individual trials: the similar shapes of the two upper and the two lower histograms indicate the similarities of the variabilities. At the same time, it can be seen that there is no relation to the similarities of response properties. In Fig. 6.8 the trial counts of all neurons recorded simultaneously are displayed. The four neurons of Fig. 6.7 can be recognized by their electrode labels (A–E: columns of 6 electrodes; 1–6: rows of 5 electrodes). According to a quantitative definition of covariation groups (given by Bach 1982), two groups are distinguished by shading (black and hatched) in Fig. 6.8. It is remarkable that the groups do not only encompass neurons at adjacent electrodes. The arrangement is reminiscent of an underlying anatomical ordering principle, distributing centrifugal input to area 17 in the manner of interleaving eye-dominance columns. (However, a relationship to the actual eye-dominance columns was not found). Whether or not there is a columnar organization, the mere finding of variability groups makes an association of variability with noise highly unlikely. Rather, it is reasonable to assume that all non-visual input to area 17 manifests itself as variability. Our efforts towards chronic multiple electrode recordings are intended, among other things, to elucidate the functions of variability.

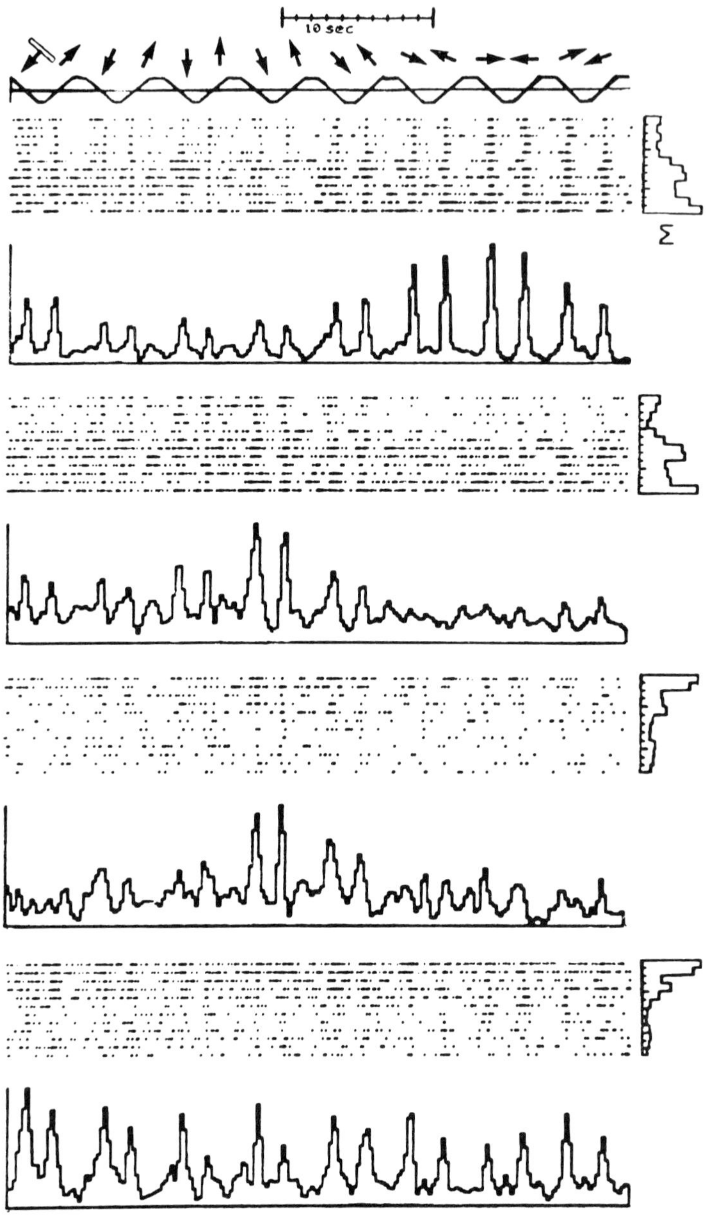

Fig. 6.7. Examples of covariability in layer VI. Stimuli were bright bars moving in 16 different directions. Repeated responses for four simultaneously recorded neurons are shown. Rotated histograms show the spike counts per trial. It can be seen that the similarity of the temporal evolution of variability is unrelated to that of orientation selectivity. Electrodes are (from top to bottom): C1, C3, D3, D6. See also Fig. 6.8

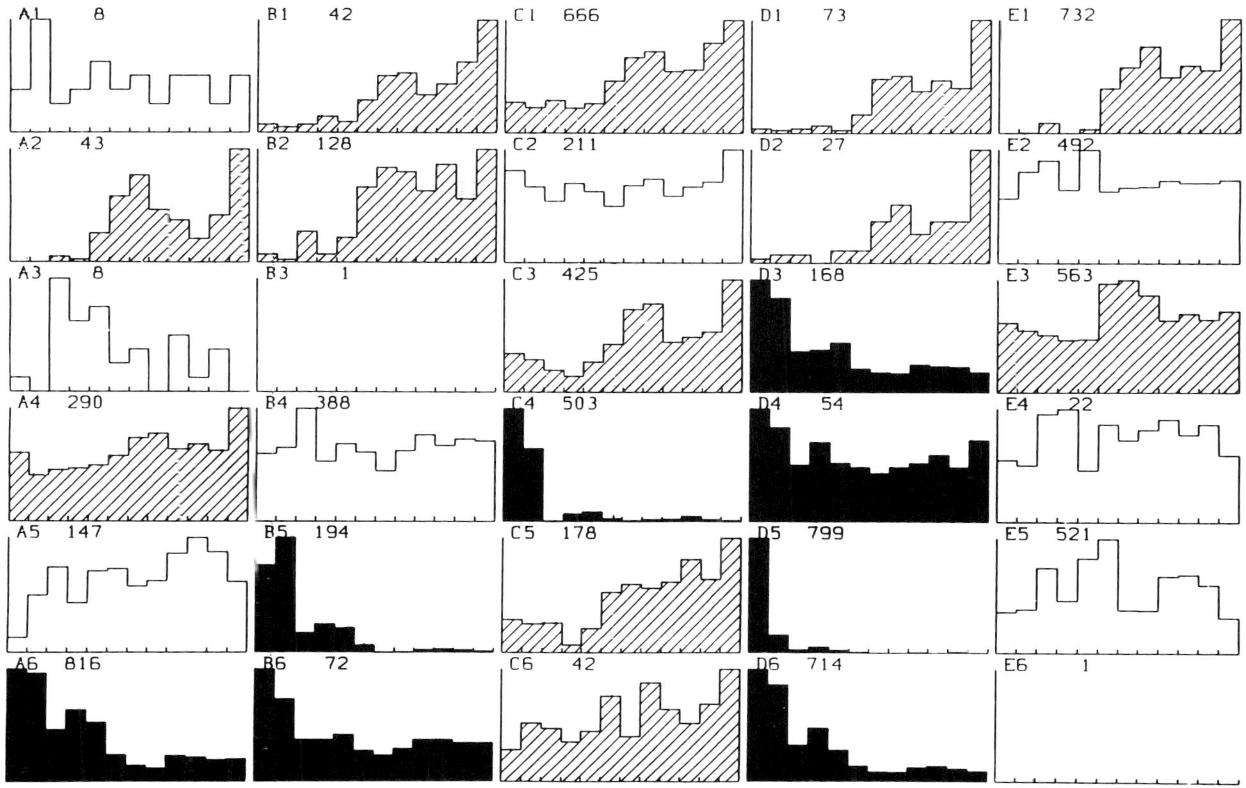

Fig. 6.8. Spatial distribution of covariability in the recording shown in Fig. 6.7. The now upright histograms show the spike counts over trials. Each trial lasted 40 s. The histograms have been classified according to similarity. The procedure has delivered two groups of covarying neurons (black and hatched) and a number of individually varying neurons (white). It can be seen that the covarying neurons are not necessarily neighbors. In this recording, a stable distribution of covariability has been observed for more than four hours

6.4 Operation of the Visual System in the Presence of Correlated Signals

In this section, the present neurophysiological knowledge is used to give a qualitative description of visual processing.

We start from our assertion that the synchronizations giving rise to correlogram peaks about 50 ms wide originate in the retinae. At first it seems odd that signals from different loci in the retina should correlate, since the result would be a loss of independence of messages from neighboring areas of the visual field. However, generally the process of vision consists of a *dissolution* of visuotopy (otherwise a picture would remain a picture), and therefore interactions between signals from different loci *must* be expected. The question is rather to understand how they are organized.

One may think of the retinal output as a multineuronal stream of spikes. The 50 ms correlations signify that some, or perhaps all spikes participate in multineuronal volleys, i.e. preferentially there are spikes in several neurons within intervals of ± 25 ms. However, it must be borne in mind that each spike of a given neuron can form a volley with spikes from a *different* set of neurons. A peak in most autocorrelograms indicates that a similar increase of firing probability next to a given spike is also found in the same neuron, which signifies that a certain tendency to spike bursting is always present. The effect of optical stimulation consists not only of an increase of spike frequencies but also of an increase of *volley* frequency, and a possible time locking of volley occurrences to sharp luminance changes.

As the volleys travel to the brain, they systematically fan out on the time axis, since different neurons differ by conduction velocities and, above all, by temporal integration properties. As has been described above, a spike of a "slow" neuron may be delayed within a volley by up to about 20 ms against the spikes of a "fast" neuron. Having entered the cortex, the volley may bifurcate and propagate to various layers and cerebral regions, introducing additional delays which may be different for each pathway. One can imagine that in this way the visual signal flow remains at least partially segregated into the slower X- and the faster Y-pathways along the visual processing routes from retina to cortical areas beyond the striate cortex, as has been suggested by Livingstone and Hubel (1987). Obviously it could also be that the faster route branches off, so that an additional delay in that branch reunifies the faster and the slower components of the same volley.

Upon all this, neuronal variability is superimposed which introduces huge changes of spike rates, but little variation in the response latencies, as can be seen in the bottom panel of Fig. 6.6. From that, the temporal structures of the volleys can be expected to remain unaltered too.

Thus, three cooperative phenomena determining spike patterns of visual cortical neurons can be distinguished, namely (i) visual stimulation which is cooperative, since any stimulation is shared by several neurons, (ii) volley formation (possibly accompanied by oscillations) and (iii) neuronal variability, including covariability.

What are the purposes of the latter two cooperative mechanisms? An idea that immediately comes to mind is that the volleys, and also the oscillations, could substitute for absent visual stimulation. The need for such a substitute arises during prenatal development, if self-organization principles depending on correlated neuronal activity are put into operation. Similar mechanisms might also be effective in adulthood, and subserve the role of maintenance of the network in a standard state. The volleys share with true stimuli the property that they correlate most strongly at short retinal distances. However, they are monocular. The oscillations could supplement the lacking binocular interactions, although they cannot achieve an alignment of the gaze directions. However, they can provide a means to control the prenatal convergence of afferent fibres on cortical cells: there must be a way to selectively deliver fast visual input via *both* eyes to the same "fast" cortical cell. By virtue of Hebbian (Hebb 1949) reinforcement, pathways that need not necessarily be monosynaptic and that originate in opposite eyes, can be made to terminate preferentially at the same target cell if their oscillatory phases are equal. A convergence of paths with like *visual* delays via the two eyes is thereby obtained.

A further point is related to the mechanism of inhibition: it is known that inhibition is a very inefficient neuronal interaction if the target neuron has a low spike rate (Aersten and Gerstein 1985). This is because a strong continuous inhibition has to be exerted in order to remove with certainty the small number of spikes in the target neuron. The efficiency can be raised if the spikes to be eliminated are more likely to occur at the instances where the inhibiting neurons are active. At first sight it seems odd to *produce* spikes at particular instances with the intention of increasing the efficiency of their subsequent *removal*. However, the formation of volleys occurs at a very early processing stage where little is specified as yet, whereas more specific interactions take place at higher levels. Forming volleys corresponds to setting the stage for efficient interactions of all kinds, but the decision for a peculiar type of interaction (e.g. mutual inhibition of cortical cells with similar orientation preference) is taken later. Since a neuron preferentially reacts to correlated input (see above), in principle an analogous but perhaps weaker mechanism must be effective for excitatory interactions also: it must be easier for one neuron to elicit a spike in another one if both are part of the same volley.

For a further understanding, it may be helpful to first consider the better-known sensory processing of *spatial* parameters: its general aim is to dissolve the spatial relationships given on the sensory surface and to regroup the information into other categories that are adapted to the requirements of further processing, memory storage or motor output. In the visual cortex, a well-known step in that direction consists of the loss of precise visuotopy as one moves from one cell to its close vicinity, and in the appearance of a relatively more accurate order with respect to orientation angles (Hubel and Wiesel 1977). It is noteworthy that the new, more strictly ordered parameter is concerned with a relative analysis: orientation is an indication of where some points are located *relative* to other points. The new order comes in spatial modules, also called hypercolumns, implying that a particular feature is only represented at particular loci separated by regular intervals, and other features are represented in between. Similar modular arrangements are also observed for other parameters, and in other cortical areas (see e.g. Goldman-Rakic 1988). Although the purpose of this type of arrangement is not really understood, for the present discussion it is important to remember that a *spatially* repetitive excitation pattern resulting from the modular structure does not reflect spatially repetitive features of the stimulus, but rather a principle of cortical processing that testifies the dissolution of visuotopy.

For processing of *temporal* parameters, the need to dissolve the input time structure is equally important: if that step is not achieved, the knowledge about an object would not be independent of the sequence of sensory scans required for its exploration. The process of permanent storage of temporal sequences clearly necessitates a complete transformation of explicit time dependency into time-independent distributions of memory elements, and the generation of motor commands is likely to require a complete reorganization of the temporal structures of the causal sensory events.

A partial correspondence between a hypercolumn width of about one millimetre and the correlogram peak width of 50 ms may be suspected. Obviously, the analogy cannot be carried too far: up to now, the spike train correlations between pairs of neurons only show that *some* spikes from one neuron tend to be emitted preferentially together with *some* spikes of another neuron. *Other* (as well as the same) spikes of the first neuron can appear together with spikes of a third neuron. Thus, volleys of spikes can exist continuously at any instant on the time axis, but they are composed of spikes from different neurons. As in the spatial case, the important point is that the volleys do not directly correspond to temporal structures of the stimuli, but rather they reflect a processing principle. If the view of Aertsen and Gerstein (1985, see above) is correct, intracortical inhibition has the effect of cutting out parts of multineuronal volleys: the spikes of certain neurons will be suppressed. More precisely, neurons within a volley will preferentially remove spikes from neurons that are part of an appropriately delayed branch of the same volley. Thus, the *temporal* structure introduced by the volleys is modified as a function of stimulation. The gross time course of the stimulation is still directly represented, with some loss of accuracy, by the sequence of the volleys, whereas time dependencies in short intervals begin to be coded in other terms, e.g. by movement selective neurons.

Neuronal variability is superimposed on the spatiotemporal patterns of neuronal activities and volley correlations. Certainly most of the variability corresponds to real neuronal signals. The earliest entry point of non-visual signals into the visual afferences is at the level of the LGN, which receives many fibres from the brainstem, and from other sources (Sherman and Koch 1986). Area 17 of the monkey is not only a recipient of LGN afferences: these form only 70% of the total input (Doty 1980), whereas the remaining 30% originate in various cortical and subcortical regions. The effect of variability on the spike frequencies of the visual responses is very pronounced, and our finding of covariability in groups of neurons suggests that the visual messages coming from the outer world are systematically modulated by internal signals. Obviously, the significance of variability cannot easily be elucidated under anaesthesia. However, in area 17 of awake animals, a modulation of visual responses by attention has been observed by Boch (1986). The question arises as to how the perceived world can be independent of whether or not one attends to it, if both the stimulus response and the effect of attention manifest themselves as elevations of spike rates of the same neurons. This question is a small part of a more general problem that will be our concern in forthcoming experiments, namely whether one can detect in the brain "entities of meaning" that remain identifiable under a variety of additional influences.

As a first step in this direction we have examined whether the physical constancy of a repeated visual stimulus can be recognized by an analysis of spike trains in the presence of variability, or, to express it differently, whether each one out of several different stimuli can correctly be identified on the basis of the neuronal discharges it evokes. This topic will be treated in the following section.

6.5 Identification of Stimuli from Neuronal Responses

Identical repeated stimuli are known to evoke responses with highly variable spike frequencies. A widely held belief is that only the responses averaged over many similarly reacting neurons carry reliable information (Burns 1968; Shaw et al. 1986). However, from our finding of covariability in groups, we conclude that such ensemble averaging *cannot* easily lead to reproducible responses. From other experiments we conclude that it *should* not, otherwise systematic efferent response modulations (e.g., by attention) would be annihilated.

How then can the identity of two stimuli be signalled? If one looks at the dot displays of multielectrode responses (Fig. 6.9a) to a large number of stimuli (Fig. 6.9b), it comes to mind that the identity of each stimulus might be given by the distribution of black and white within the response column corresponding to that stimulus. One tends to pay little attention to the *amounts* of blackening, but rather to look whether or not a neuron is firing. This contrasts to the prevailing view derived from the analysis of single electrode recordings, namely to search for quantitative links between stimulus parameters and spike frequencies.

Having these partly opposing views in mind, J. D. Becker and I have undertaken to find the best way of recognizing identical stimuli among several other stimuli on the basis of their response patterns. We have analyzed data obtained with 30 microelectrodes from the striate cortex of anaesthetized monkeys. Although other types of stimuli were examined also, we shall describe here the results for bars moving in 16 different directions. Each stimulus, characterized by its movement direction, was applied about 10 times, and a spatiotemporal response pattern averaged over these repetitions ("template") was associated with it. Each stimulus was then applied again, and its *non-averaged* evoked pattern was compared to all 16 templates. The most similar template was sought, hoping that it was the correct one. The aim was to select the procedures and to vary the parameters such that the overall score of correct recognitions became maximal. Behind this was the idea that perhaps the variable response patterns

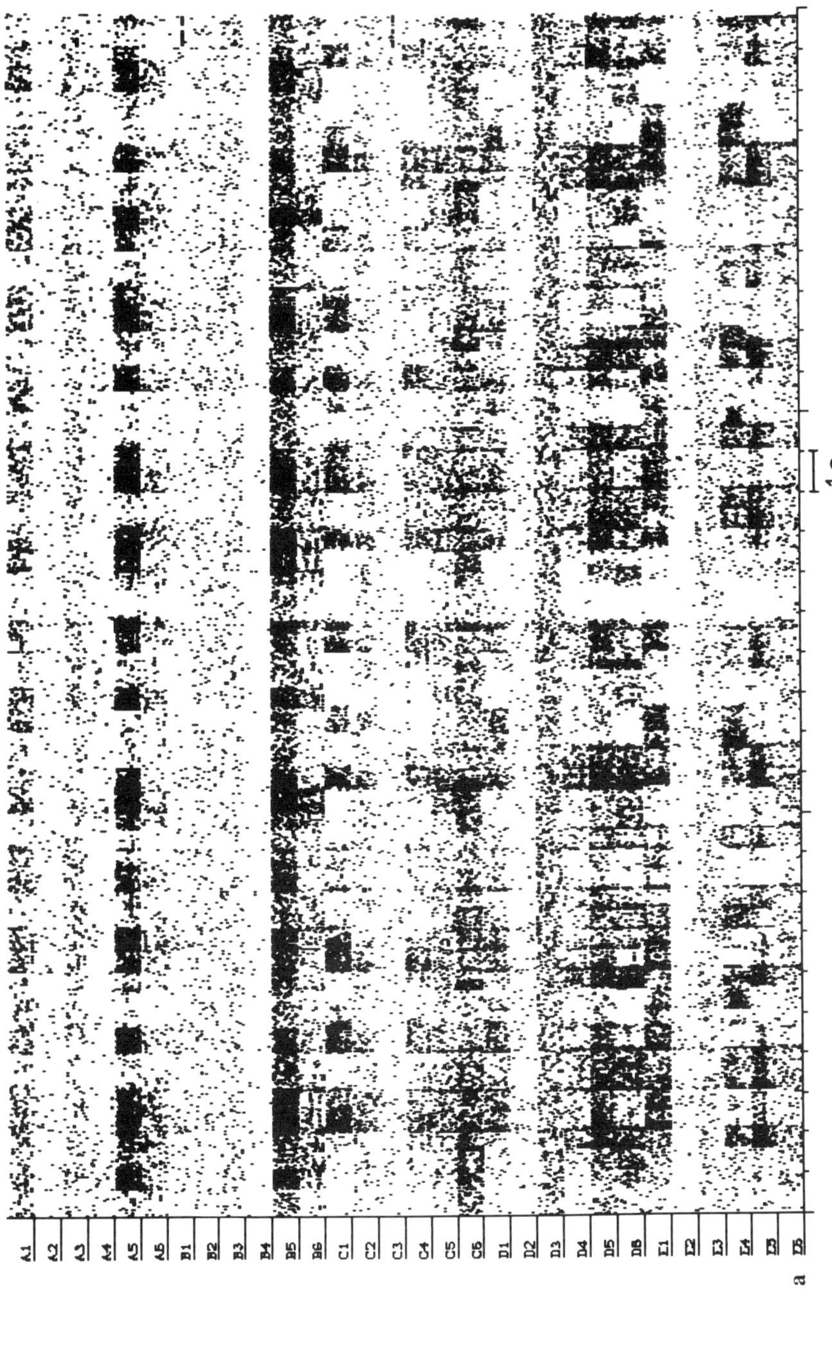
1s
A1
A2
A3
A4
A5
A6
B1
B2
B3
B4
B5
B6
C1
C2
C3
C4
C5
C6
D1
D2
D3
D4
D5
D6
E1
E2
E3
E4
E5
E6

a

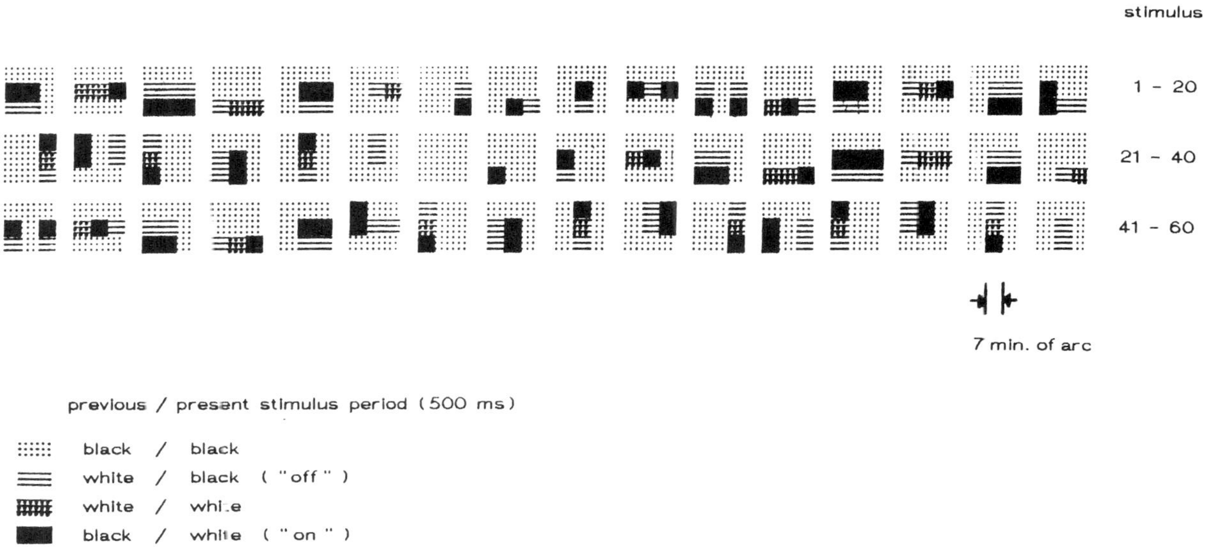

b

Fig. 6.9. Responses to 60 different stimuli recorded by 30 electrodes in foveal layer IVc. **a** Dot displays. Electrodes are indicated on the left. The tic marks on the abscissa indicate the instants of every second stimulus application. **b** stimuli. Bright squares or rectangles on a dark background were abruptly displaced by 7 min of arc (or a shape was changed) within a $0.35 \times 0.35°$ field. For each stimulus the light distributions in the previous and in the present stimulus periods are shown by different shadings. A change occurred every 500 ms

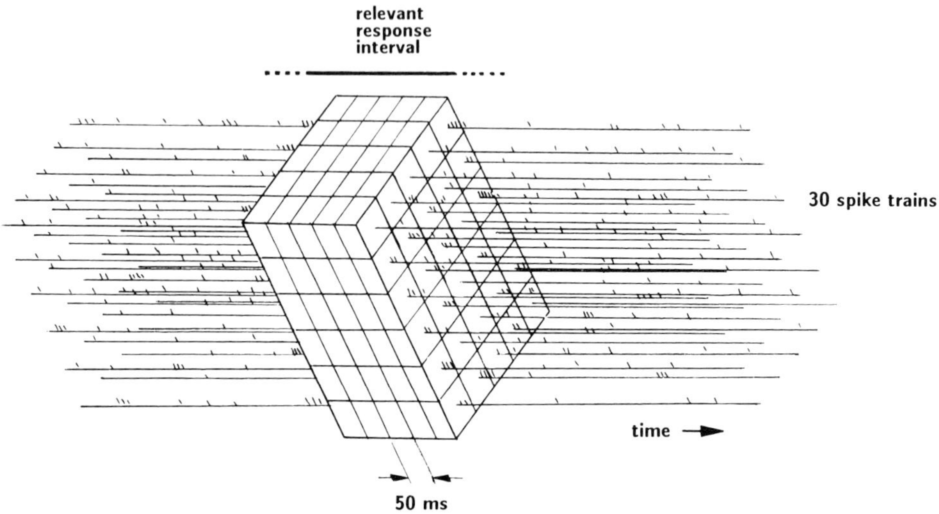

Fig. 6.10. Schematic illustration of a spatiotemporal response pattern. A 5 × 6 array of time axes is shown. The block encompasses the relevant response interval. If the block is subdivided into 5 time slices, as shown, then the spatiotemporal response pattern defining a response consists of 150 small counts. If it is not subdivided, 30 larger counts define a response

could systematically be segregated into a constant part that reflected the stimulus properties, and into a variability component.

We tried to improve the recognition score by variations of the methods of normalization and the definition of similarity, and by taking the time structure into account in various ways.

Time Structure. Firstly, the relevant response intervals were determined and then kept constant. This choice was straightforward and not critical. The intervals ranged between 300 and 700 ms. They were subdivided into a varying number of time slices. Thus, each spatiotemporal response pattern was defined by components whose number was 30 times the number of time slices (Fig. 6.10). Each component was a spike count in a bin (a bin is a small spatiotemporal cube in Fig. 6.10). In the case of orientation selectivity, using only one thick time slice spanning the entire response, interval amounted to using the information as it is contained in the well-known orientation polar plots, with the exception that these plots are usually obtained by averaging over several stimulus expositions. In this case, all temporal structures are ignored and the recognition is based on spike frequency alone. We were surprised at the extremely low performance under this condition even when most neurones were strongly orientation and direction selective. Figure 6.11 shows an example from layer V. The score of correct identifications is shown as a function of the time slice thickness, or resolution, the relevant response interval being 300 ms in this case. It can be seen that there is a broad maximum for resolutions between 20 and 80 ms. The pronounced decrease at the lower resolutions is remarkable; it indicates that little

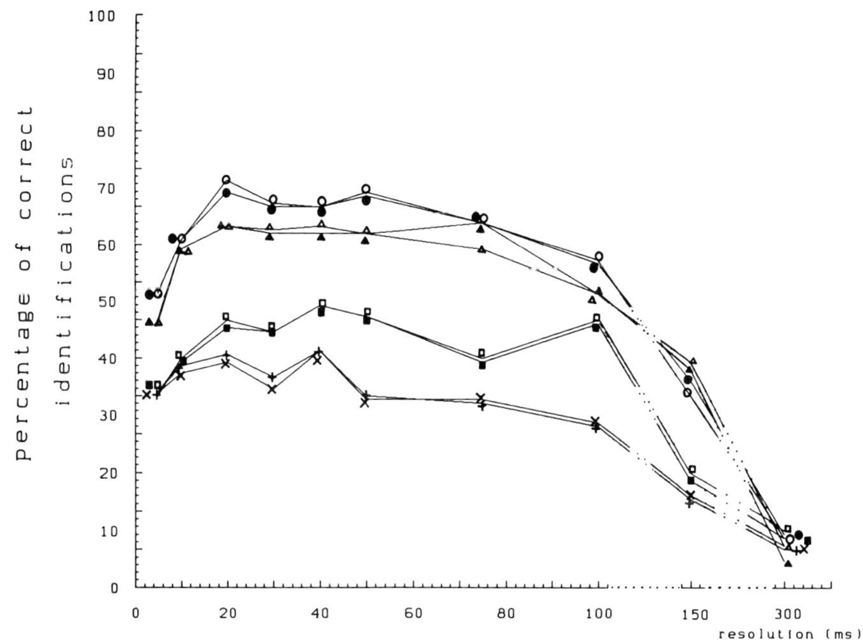

Fig. 6.11. Recognition of stimuli from evoked spatiotemporal spike patterns. Stimuli were bars moving in 16 directions as shown in Fig. 6.7. Ten individual (non-averaged) responses to each stimulus were tested, so that 100% correct identifications correspond to 160 cases. The chance level is at 6.7%. The abscissa indicates the temporal resolution, i.e. the widths of the time slices into which the (fixed) relevant response interval was subdivided. The symbols differentiate between normalizations, done for each electrode separately. Circles: raw spike counts, triangles: spike counts divided by the square root of the average spike rate, squares: spike counts divided by the average spike rate, crosses: spike counts divided by the maximal spike count. Filled symbols and "x", respectively: ongoing spike rate subtracted. This was not done for the curves with open symbols and "+", respectively. The latter measure has no effect. It can be seen that at a resolution of 300 ms the identification rate is at the chance level, but a broad maximum is around 40 ms

information is contained in the spike frequencies. However, the time stucture at an excessive resolution (left side of the figure) again diminishes the score of correctly identified stimuli.

In Fig. 6.12 the difference between the performance at a resolution of 50 ms can be compared to that at a resolution of 300 ms. In the latter case (Fig. 6.12, lower part), no matter which normalization was selected (see below) and which cortical layer was analysed, the recognition scores barely exceeded the chance level. In contrast, Fig. 6.12, upper part, shows a substantial improvement due to the subdivision into 50 ms bins; this, in spite of the fact that the spike counts were larger in the 300 ms case so that one might have expected them to be somewhat more reliable.

It is noteworthy that layer IVc, where the cells are known to lack orientation selectivity (Hubel and Wiesel 1977), does not behave exceptionally within the other layers. This indicates again that orientation selectivity, classically defined as a relationship between stimulus orientations and spike rates, does not seem to

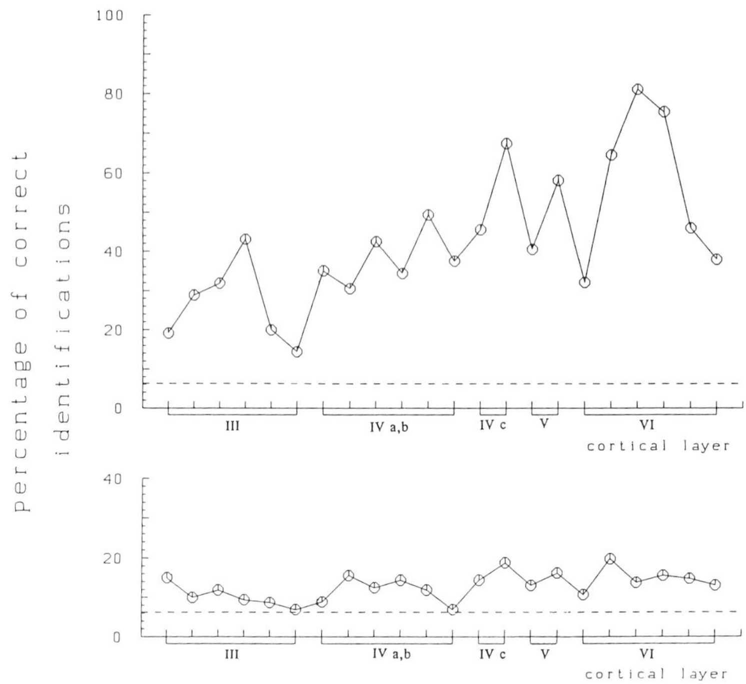

Fig. 6.12. Score of correctly recognized stimuli for different cortical layers. Non-normalized data (see Fig. 6.11). Dashed line: chance level. Top: Each response of each neurone is characterized by a succession of spike counts in 50 ms-bins into which the relevant response interval (lasting 300 to 700 ms) is subdivided. Bottom: Each response of each neurone is characterized by one spike count within the relevant response interval. This corresponds to the assumption of a spike frequency code. Although in layer VI in particular many neurons have very clear cut orientation and direction selectivities, the systematic stimulus dependency of spike rates is not sufficient for a reliable recognition in the presence of neuronal variability. Oriented stimuli can well be identified by non-oriented neurons in layer IVc, if the time structure is taken into account

be relevant for the task of recognizing a slanted line. However, a hypothesis concerning this point will be presented below.

As an additional measure we intentionally degraded the spike-rate information by replacing all bin counts > 1 by 1 but keeping a 50 ms time resolution. This cuts off the maxima in the tuning curves showing spike rates as a function of stimulus angle, so that the optimal angle can no longer be recognized. Even then, the score of correct recognitions fell only to a half but was still clearly superior to the case of 300 ms resolution where the information about the stimulus was drawn from spike rates.

Normalization. Several varieties were applied in conjunction with the option of subtracting the ongoing activity. The reason for doing this was that it is not known whether each spike should have the same weight when the activities of two neurons are compared. There might be one neuron with a zero ongoing firing level that emits precisely one spike as a response to a particular stimulus. Another neuron might have a substantial ongoing activity, but the stimulus responses clearly exceed that level. If raw spike counts are used, the responses of a neuron of the former type have little weight if they appear in calculations together with high activities from other neurons. It seems fairer to use *percentages* of maximal spike frequencies, so that the relative influence of low firing neurons increases. However, to our surprise, the *raw* spike trains yielded the highest recognition rates, although it must be said that the overall effect of varying the normalization was fairly modest (see Figs. 6.11 and 6.12), and subtracting the ongoing activity had practically no effect at all. The reason for the former finding was rather obvious when the individual responses were inspected: when spike rates are transformed to percentages of maximal rates, the result for low rates is that together with the spike rates, the variability is amplified too. This balances the apparent advantage.

Similarity. There is no absolute definition of similarity of two spatiotemporal activity patterns. One can view the responses of 30 neurons, expressed as 30 spike counts, as a vector with 30 components. Once a decision has been reached on the normalization to be used, the 30 components for a given set of responses are fixed. The most obvious idea was that two such vectors were more similar the shorter their difference vector was, relative to the sum of the lengths of the original vectors. However, the performance of this measure, as defined by the score of correct stimulus identifications, was easily excelled by another measure, namely the cosine of the generalized angle (mathematically equivalent to the correlation) between the vectors. This is remarkable because two parallel vectors differing in length are taken to be similar. It means that an increase of the response spike frequencies at all 30 electrodes by a common factor is ignored.

6.6 Signal Contents and Signal Importance

Many researchers have observed that neurons in the central nervous system can be influenced by both sensory input and central messages (see e.g. Fischer and Boch 1989). If the latter are not under experimental control they must necessarily manifest themselves as variability. However, not much attention has been paid to the problem that both types of input superimpose their effects, i.e. they both modulate the spike rates of the same neurons. This makes it difficult to maintain the idea that spike rates code stimulus properties. To circumvent the problem, it was thought to be advisable to maintain the animal in a constant state, so that variability was minimized. This suggestion implies firstly that the nervous system is *disturbed* by the central messages, (why therefore does the nervous system not simply prohibit that input from entering the signal stream?), and secondly, that variability is a poorly structured signal representing only one or a small number of parameters. I strongly oppose to this general attitude which suggests that the brain is essentially a device driven by instantaneous inputs, and without these there are few relevant signals in the brain. Hopefully, the contrary is true.

In accordance with that hope, a different view is presented here. If a stimulus once evokes five spikes and upon repetition eight spikes, then the nervous system does not *want* to re-segregate these discharges into components. Rather, the different spike rates reflect an underlying purpose.

I conclude from the above experiments that the pure stimulus properties are coded not by the spike rates but by the time structures of the spike trains.

What then is the purpose of different spike frequencies? In the above example the "orientation-selective" neurons were unable to signal orientations by their spike rates *not* because they lacked the appropriate orientation dependency, but because of the superimposition of a large variability. Indeed, from the data of Fig. 6.12a one has to conclude that a frequency-based response pattern evoked by a given stimulus is on average similar to *all* patterns tested, irrespective of whether or not they stem from a repetition of the same stimulus. Variability easily produces individual spike rate changes of a factor of 10 within a few seconds, whereas the temporal response structure is much more robust (as can be seen the lower panel of Fig. 6.6) unless response components disappear altogether during instants of powerful suppressive variability. Clearly, at these instants it is the strongest response bursts that are most likely to survive, and therefore they are the most *important* elements to whose temporal information most meaning should be attached. Thus we believe, as do others, that a high frequency burst of an oriented neuron carries the most essential information about the tilt angle. However, the new aspect is that the signal *content* is carried by the temporal response structure, whereas it is the signal *importance* that is represented by the spike rates. There have been earlier attempts to relate the time structures of spike trains to the signals conveyed (Eckhorn and Pöpel 1974, 1975, Optican and Richmond 1987 but no positive role has been attributed to neuronal variability. Here the new variable "importance" is introduced. Its significance is subjective; it depends on expectations, it determines from which of the senses the most relevant information should be extracted or which sensory parameters should be emphasized.

The new coding principle offers the opportunity to modulate the importance of selected components of spatiotemporal excitation patterns by messages from higher brain centers, largely without altering the meaning carried by the patterns. I believe that this is the explanation for the fact that an object looks the same irrespective of whether one directs one's attention to it or not. On the basis of the hypothesis of stimulus coding by spike rates, this is not easy to understand. I also believe that the same principle works at many other places in the brain: it can explain why a sheet of white paper looks homogeneous right to the border, although lateral inhibition in the retina has the effect of a relative enhancement of the responses near the borders. The perception of brightness is not affected because it is the *importances* (the spike rates) of the responses that are modulated. Anyhow, it makes much more sense that borders are distinguished by importance and not by brightness.

The latter example illustrates my view that the suggested coding principle has a more extended validity: in general there is no way to distinguish either between the modulating and the modulated signals, or, by origin, between peripheral and central ones.

In the light of the proposed segregation of meaning and importance, a spike train correlation is a "correlation of importances". Therefore I believe that it is no coincidence that the most prominent spike train correlations and the optimal condition of stimulus recognition both occur at a resolution of 50 ms. However, a more profound understanding of a neural network operating according to the new principle must await further experimental and theoretical work. In our future experiments we shall concentrate on spatiotemporal patterns of central messages in order to establish relationships to behavioral variables (in particular learning), to sensory responses and also to spike train correlations. Using the tools applied in the work described above, the repeated identification of central messages will be attempted. I hope that in this way the nature of higher brain functions can be elucidated.

References

Abeles, M. (1982): "Local cortical circuits". In *Studies of brain function*. Vol 6, ed. by V. Braitenberg (Springer, Berlin, Heidelberg)

Aertsen, A.M.H.J., Gerstein G.L. (1985): Evaluation of neuronal connectivity: Sensitivity of cross correlation. Brain Res. **340**, 341–354

Aiple, F., Krüger, J. (1988): Neuronal synchrony in monkey striate cortex: interocular signal flow and dependency on spike rates. Exp. Brain Res. **72**, 141–149

Bach, M. (1982): Untersuchungen zur Wechselwirkung zwischen Nervenzellen im visuellen Cortex mit Vielfach-Mikroelektroden. Thesis, Freiburg

Bach, M., Krüger, J. (1986): Correlated neuronal variability in the monkey's visual cortex revealed by a multi-microelectrode. Exp. Brain Res. **61**, 451–456

Boch, R. (1986): Behavioral modulation of neuronal activity in monkey striate cortex: excitation in the absence of active central fixation. Exp. Brain Res. **64**, 610–614

Bolz, J., Rosner, G., Wässle, H. (1982): Response latency of brisk-sustained (X) and brisk-transient (Y) cells in the cat retina. J. Physiol. London **328**, 171–190

Braitenberg, V. (1978): "Cortical architectonics: general and areal". In: *Architectonics of the cerebral cortex*, ed by M.A.B. Brazier, H. Petsche (Raven Press, New York) pp. 443–465

Burns, B.D. (1968): *The Uncertain Nervous System* (E. Arnold, London)

Doty, R.W. (1980): Non geniculate afferents to striate cortex in macaques. Exp. Brain Res. **41**, A10–11

Eckhorn, R., Pöpel B. (1975): Rigorous and extended application of information theory to the afferent visual system of the cat. II. Experimental results. Biol. Cybern. **17**, 7–17

Eckhorn, R., Pöpel B. (1975): Rigorous and extended application of information theory to the afferent visual system of the cat. II. Experimental results. Kybernetik **17**, 7–17

Fischer, B., Boch R. (1990): "Cerebral Cortex". In: *Vision and visual dysfunction*, ed. by R. Carpenter *Eye movements* Vol. 9 (Macmillan, London)

Goldman-Rakic P.S. (1988): Topography of cognition: Parallel distributed networks in primate association cortex. Ann. Rev. Neurosci. **11**, 137–156

Hebb, D.O. (1949): The organization of behavior (Wiley, New York)

Hubel, D.H., Wiesel, T.N. (1974): Uniformity of monkey striate cortex: A parallel relationship between field size, scatter, and magnification factor. J. Comp. Neurol. **158**, 295–306

Hubel, D.H., Wiesel, T.N. (1977): Functional architecture of macaque monkey visual cortex. Proc. R. Soc. Lond. B. **198**, 1–59

Krüger, J. (1983): Simultaneous individual recordings from many cerebral neurons: techniques and results. Rev. Physiol. Biochem. Pharmacol. **98**, 177–233

Krüger, J., Aiple, F. (1988): Multi-microelectrode investigation of monkey striate cortex: spike train correlations in the infragranular layers. J. Neurophysiol. **60**, 798–828

Krüger, J., Aiple, F. (1989): The connectivity underlying the orientation selectivity in monkey striate cortex. Brain Res. **477**, 57–65

Livingstone, M.S., Hubel, D.H. (1987): Psychophysical evidence for separate channels for the

perception of form, color, movement and depth. J. Neurosci. **7**, 3416–3468
Mastronarde, D.N. (1983): Correlated firing of cat retinal ganglion cells. I. Spontaneously active inputs to X- and Y-cells. J. Neurophysiol. **49**, 303–324
Optican, L.M., Richmond B. (1987): Temporal encoding of two-dimensional patterns by single units in primate inferior temporal cortex. III. Information-theoretic analysis. J. Neurophysiol. **57**, 162–178
Perkel, D.H., Gerstein, G.L., Moore, G.P. (1967a): Neuronal spike trains and stochastic point processes. I. The single spike train. Biophys. J. **7**, 391–418
Perkel, D.H., Gerstein, G.L., Moore, G.P. (1967b): Neuronal spike trains and stochastic point processes. II. Simultaneous spike trains. Biophys. J. **7**, 419–440
Raiguel, S.E., Lagae, L., Gulyàs, B., Orban, G.A. (1989): Response latencies of visual cells in macaque areas V1, V2 and V5. Brain Res. **493**, 155–159
Sestokas, A.K., Lehmkuhle, S. (1986): Visual response latency of X- and Y-cells in the dorsal lateral geniculate nucleus of the cat. Vision Res. **26**, 1041–1054
Shaw, G.L., Silverman, D.J., Pearson, J.C. (1986): "Trion model of cortical organization: Toward a theory of information processing and memory". In: *Brain Theory*, ed. by G. Palm, A. Aertsen (Springer, Berlin, Heidelberg) pp. 177–191
Sherman, S.M., Koch, C. (1986): The control of retinogeniculate transmission in the mammalian lateral geniculate nucleus. Exp. Brain Res. **63**, 1–20
Steriade, M., Deschênes (1984): The thalamus as a neuronal oscillator. Brain Res. Reviews **8**, 1–63
Vogels, R., Spileers, W., Orban, G.A. (1989): The response variability of striate cortical neurons in the behaving monkey. Exp. Brain Res. **77**, 432–436

7. Connectivity and Functional Organization in the Mammalian Visual Cortex

Daniel Y. Ts'o

Laboratory of Neurobiology, The Rockefeller University, 1230 York Avenue, New York, NY 10021, USA

A remarkable aspect of the early phases of visual processing is the elaboration of receptive field properties along the visual pathway, from the center-surround cells of the retina and lateral geniculate nucleus (LGN), to the orientation and directionally selective simple and complex cells of the mammalian striate cortex. Yet the underlying neuronal circuitry responsible for the construction of new receptive field properties at each stage is poorly understood. Anatomical techniques have revealed several types of neuronal connections that are likely to participate in this process, but the actual functional role of these connections is, in most cases, not known. The present studies focus on the horizontal connections that are intrinsic to the cortex and run laterally, parallel to the surface of the cortex. Anatomical studies have shown that these connections extend for up to several millimeters and the clustered nature of the axonal collaterals suggests a specificity of these connections related to the columnar organization of receptive field properties. The first series of experiments, performed in the cat striate cortex, explore the relationship between receptive field properties of single cells and the participation of these cells in the long-range horizontal connections. These connections were detected and measured physiologically by electrical recording and cross-correlation analysis, a statistical method that provides a sensitive measure of the strength and type of connections. These experiments showed that, among oriented cells, cells having a particular optimal orientation only interacted with other cells having a similar optimal orientation, up to distances of several millimeters in the cortex. A second set of experiments examined, in the primate striate cortex, the functional organization and connectivity of color-selective cells found in the cytochrome oxidase-rich patches (blobs) and the oriented cells found in the interblobs. The connectivity of these two groups of cells was also examined using cross-correlation analysis. It was found that, in the color domain as in the orientation domain, cells having a particular functional specificity, for example, a certain color selectivity, only interacted with other cells having similar functional properties. Taken together, these studies indicate that lateral connections between cells having similar functional properties are a common component of cortical circuitry. Our studies using cross-correlation analysis not only provided information on some of these general rules governing neuronal connectivity, but also revealed specific examples of the neuronal cir-

Springer Series in Synergetics
J. Krüger (Ed.)
Neuronal Cooperativity

cuitry between pairs of cells involved in building new receptive field types in the visual cortex.

7.1 Introduction

Cells found in the visual cortex are characterized by their high degree of selectivity for specific visual stimuli. The selective nature of the visual responses are likely to be linked to the analysis of factors that are important to visual perception, such as color, movement, contour orientation and stereopsis. Through anatomical and physiological studies, it has been revealed that the visual cortex is organized as columns or modules of cells according to these functional properties. However, a detailed understanding of how cells attain their specific response properties and of the underlying neuronal circuitry still eludes us. Most of what we do know about these issues has been achieved through studies of how patterns of connectivity relate to the functional organization of visual cortex.

The columnar structure of the visual cortex and the formulation of the idea of a hypercolumn were important contributions to the concept of a modular organization of the cortex. Cells with like receptive field properties were found to be distributed in vertical columns, from cortical surface to white matter (Hubel and Wiesel 1962). These properties, such as orientation and eye preference, are, in turn, arranged in a periodic fashion such that within a distance of about 800 μm along the cortical surface, a complete set of a particular property, a hypercolumn, is represented (Hubel and Wiesel 1974). Within a few square millimeters, then, is enough cortical machinery to analyze a particular region of the visual field for all orientations and both eyes. The traversal of 2 to 3 millimeters yields receptive fields that are non-overlapping, regardless of eccentricity. Thus it seemed that this stereotyped cortical architecture or module operates relatively independently, locally processing information from a limited portion of the visual field.

Classical anatomical studies using the Golgi procedure supported this view of cortical modularity. These studies showed connections that primarily ran vertically, perpendicular to the cortical surface, from layer to layer (Lorente de No 1933), and thus indicated that connections were restricted to a limited portion of the visual field. Furthermore, the relatively narrow lateral spread of these connections seemed to confirm the independence of the cortical modules. The rich yet specific interlaminar connectivity seen anatomically suggested that this connectivity might play the major role in constructing new receptive field properties.

Horizontal connections. Evidence that horizontal connections in the cortex were also prevalent came first from degeneration studies (Fisken et al. 1975; Creutzfeldt et al. 1977). Small lesions made in the cortex gave rise to degenerating fibers that could be traced for distances of several millimeters, parallel to the cortical surface. Further experiments, using focal extracellular injections of

HRP tracer revealed long-range horizontal connections intrinsic to the cortex that were patchy or clustered in appearance (Rockland and Lund, 1982). At the single-cell level, intracellular injections of HRP allowed a detailed view of the extent and clustered nature of the axonal collaterals of cells that participate in these long-range horizontal connections (Gilbert and Wiesel 1979, 1983). Figure 7.1 shows an example of a filled layer 2 pyramidal cell with extensive clustered horizontal collaterals. The 3D computer reconstruction (Fig. 7.1A–C) reveals more clearly the clustered nature of the collaterals and terminals, and four distinct arms of the axon that project to different neighboring cortical columns. Although the receptive fields of these HRP-filled cells were analyzed, more information is required to determine the functional role of the horizontal connections. It would be important to know, for example, the properties of the target cells, including the type and strength of the synapses made onto the target cells, and the visual field positions and sizes of their receptive fields. Anatomical evidence has suggested that the synapses made onto the target cells are excitatory (Fisken et al. 1975) and the number of terminals on an individual target cell is small (McGuire et al. 1985).

The clustering of the collaterals and terminations have a periodicity that is suggestive of a relationship to the columnar organization of functional properties. In addition, these connections extend over several hypercolumns, and must link cells whose receptive fields are non-overlapping and analyzing different portions of the visual world. It is then expected that the horizontal connections, at least at the longer distances, contribute to functional properties that are not revealed by standard receptive field analysis. These subthreshold properties may require the integration of visual information from large areas of the visual field and thus allow cell responses to be modulated by visual context. In striate cortex, Blakemore and Tobin (1972) observed this type of subthreshold property in a complex cell that had an antagonistic orientation-selective surround. Similar findings were presented by Nelson and Frost (1978) in a more extensive study. In extrastriate visual cortex, cells with direction-selective surrounds have been observed (Allman et al. 1985). However, the relationship between these selective surround properties and the long-range horizontal connections has not been firmly established. One of the primary goals of our studies, therefore, was to explore the functional role of these horizontal connections. In Sect. 7.2 we describe experiments employing cross-correlation analysis to detect and measure interactions between pairs of cells electrophysiologically, and then relate the occurrence of these interactions to the receptive field properties and positions of the two cells under study. These studies, performed in the cat visual cortex, focused on the functional properties of orientation tuning, ocular dominance and directionality. We then used a similar approach to study the receptive field properties, functional organization and connectivity of color selective blob cells in primates, as described in Sect. 7.3. The work described was done in collaboration with Charles Gilbert and Torsten Wiesel.

Cross-correlation methods. Cross-correlation analysis was introduced into neurophysiology as a method of detecting and measuring synaptic connections

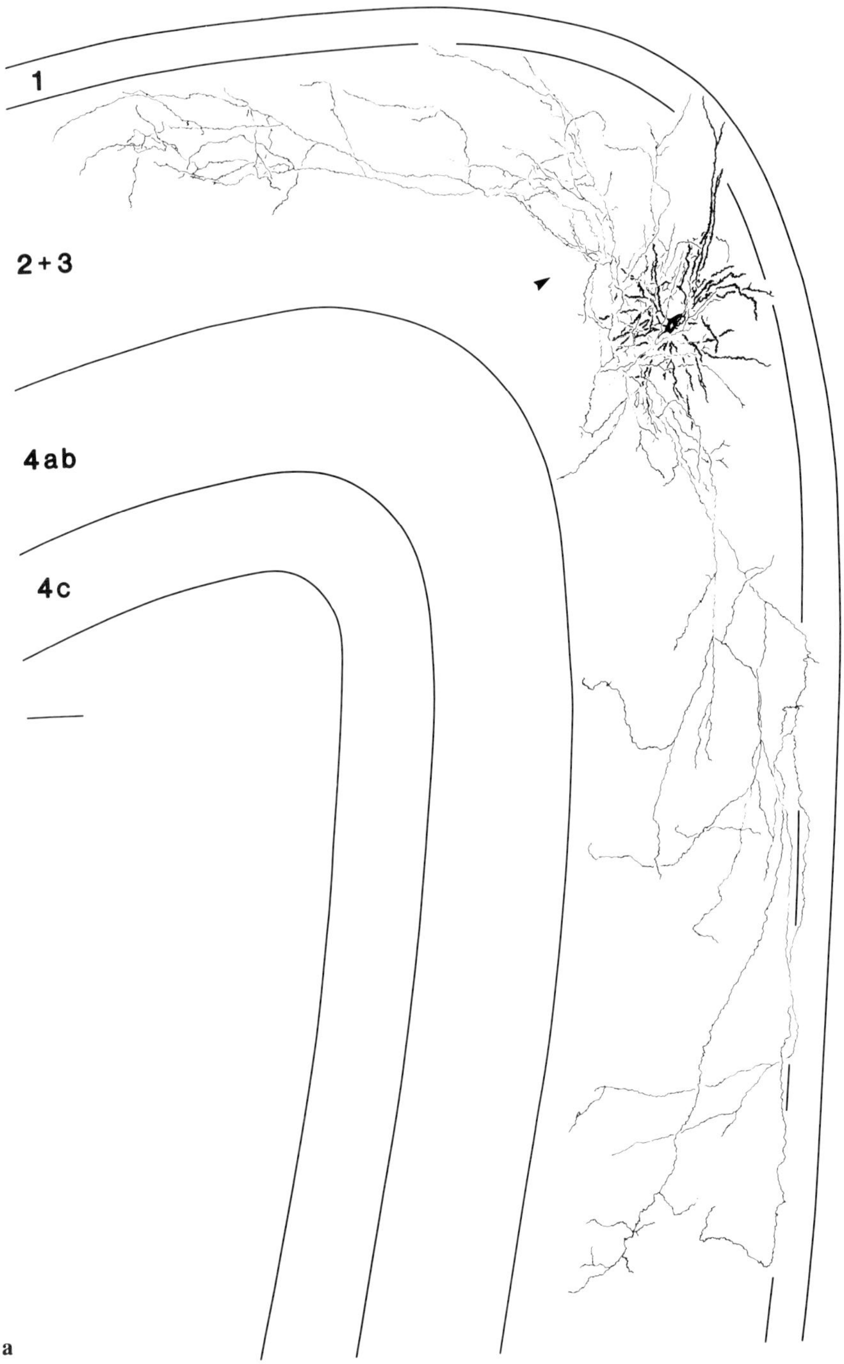

a

Fig. 7.1. a A camera lucida drawing of a pyramidal cell in layer 2 of the cat striate cortex which was injected intracellularly with HRP, after its receptive field properties were determined. This cell was a complex cell, with a 1° × 1° receptive field. *Scale marker* = 100 μm.

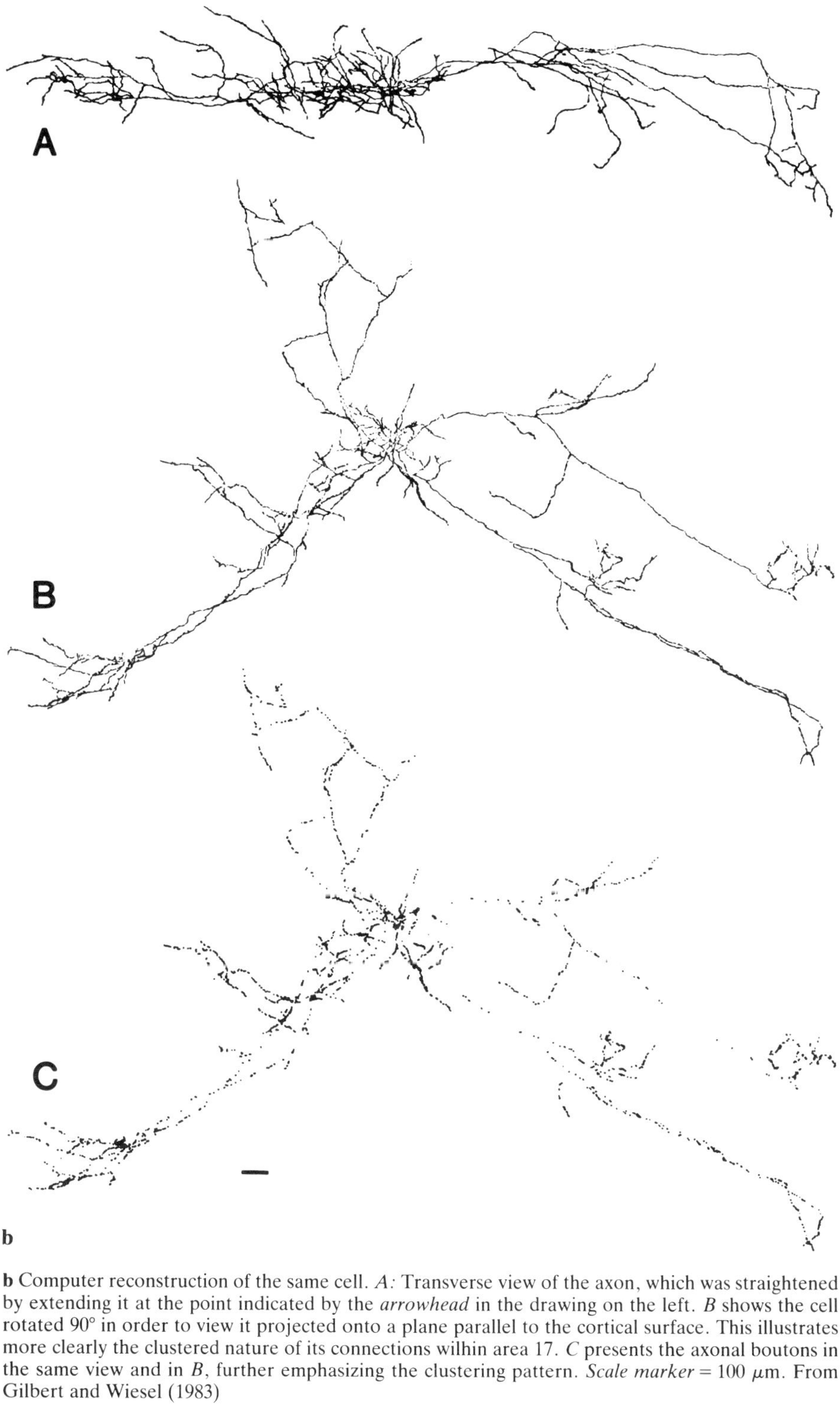

b

b Computer reconstruction of the same cell. *A:* Transverse view of the axon, which was straightened by extending it at the point indicated by the *arrowhead* in the drawing on the left. *B* shows the cell rotated 90° in order to view it projected onto a plane parallel to the cortical surface. This illustrates more clearly the clustered nature of its connections wilhin area 17. *C* presents the axonal boutons in the same view and in *B*, further emphasizing the clustering pattern. *Scale marker* = 100 μm. From Gilbert and Wiesel (1983)

between neurons (Perkel et al. 1967a,b; Moore et al. 1970; Bryant et al. 1972). It is a statistical technique that reveals consistent timing relationships between the firing of pairs of cells. The interpretation of the resultant peaks in the cross-correlogram in terms of particular synaptic arrangements can be ambiguous, but guidelines have been established, based on studies where the actual neuronal circuitry is known and on computer simulations (Moore et al. 1970). We computed raw cross-correlograms by accumulating a histogram of time differences between the occurrence of spikes in the two recorded cells. The left group of bins in the histogram represent the negative times – when the second cell fired prior to the first. The center or zero bin of the histogram accumulates the number of instances when the two cells fired precisely together. The right group of bins, the positive times, account for the various time intervals that the first cell fired prior to the second cell. The temporal resolution of the real-time correlator was 100 μs and cross-correlograms were accumulated with a $\pm$50 ms window. To be able to detect longer lasting synaptic events, we also computed the cross-correlograms for a $\pm$400 ms window. However, none of the interactions observed in our study extended beyond the $\pm$50 ms window, so only the correlograms collected at the short time window were used in subsequent analyses. A minimum of 3000 spikes were collected from each cell to ensure an adequate signal-to-noise ratio in the correlograms. To collect a sufficient number of spikes from each cell in a reasonable amount of time, each cell was stimulated with an optimal light stimulus. In cases where the two cells had quite different receptive field properties and/or position, two different light slits were presented simultaneously to provide an optimal stimulus to each cell. The time required to collect a minimum of 3000 spikes from each cell varied between 10 and 30 minutes. To reduce the noise, resultant cross-correlograms were low-pass filtered.

Previous discussions in the literature have provided guidelines for interpreting cross-correlograms and to infer details of the underlying circuitry (Perkel et al. 1967a,b; Moore et al. 1970; Bryant et al. 1972). Before one can begin to interpret the cross-correlograms, it is important to be certain that any stimulus coordination artifact will not affect the interpretation. To control for the stimulus coordination artifact we first isolated the artifact itself by computing the shuffled correlogram – correlating one spike train with the other spike train shifted in time by one visual stimulus period. This shuffled correlogram is then subtracted from the raw correlogram, thereby removing the component due to stimulus coordination.

The basic patterns of neuronal interaction shown in a cross-correlogram are distinguished by the position of the peaks relative to the zero bin and the shape of the peaks. Simplified examples of neuronal circuitry and their expected resulting cross-correlograms are shown in Fig. 7.2. Common input interactions yield peaks that usually straddle the zero bin and are often symmetrical (Fig. 7.2B). Delayed excitation peaks, attributable to direct monosynaptic input, should be shifted from the zero bin by a time that corresponds to the synaptic latency (Fig. 7.2A). These peaks are generally more asymmetric than a common excitation peak. Their shape is thought to be related to the excitatory postsynaptic potential (EPSP) and synaptic jitter (Kirkwood 1979; Fetz and Gustafsson 1983; Sur-

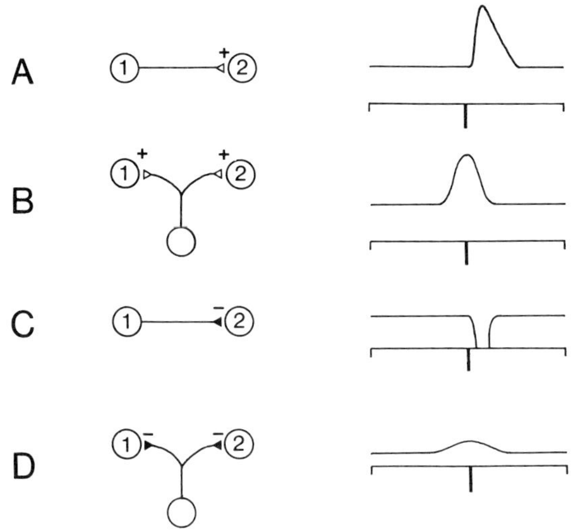

Fig. 7.2. Idealized correlograms from four basic neuronal circuits. *A:* direct excitation from cell 1 to cell 2, producing an asymmetric positive peak, offset to one side of the zero time bin; *B:* common excitation to cells 1 and 2 from a third cell, producing a positive peak, symmetric about zero; *C:* direct inhibition from cell 1 to cell 2, producing a downward trough from baseline, offset to one side of zero; *D:* common inhibition to cells 1 and 2 from a third cell, producing a small, broad, positive peak, symmetric about zero. Adapted from Moore et al. (1970)

meier and Weinberg 1985). Though it may be possible to detect multisynaptic connections in cross-correlograms, it is expected that these peaks would be broader and shallower than peaks due to monosynaptic connections. Delays in peaks may also be due to conduction latencies, but this effect is expected to be small for intrinsic connections. Inhibitory monosynaptic input appears as a valley in the correlogram, with a time shift corresponding to synaptic latency (Fig. 7.2C). Common inhibitory input generates a positive peak, which is expected to be broader and smaller than that seen with common excilation (Fig. 7.2D).

7.2 Horizontal Interactions Among Oriented Cells

We began by studying the relationships between neuronal connectivity and receptive field properties among the cells of cat striate cortex, using cross-correlation analysis. Our first observation was that correlated firing usually depended upon matching receptive field properties, while cell pairs with dissimilar receptive field properties tended to exhibit no correlated firing. An example of this finding is shown in Fig. 7.3. In this instance two electrodes were placed in the superficial layers of cat striate cortex at a separation of 700 μm. Single units were isolated with each electrode, receptive fields characterized and correlograms taken. The first cell had an orientation tuning of 120°, directional preference to the right and an ocular dominance group of 2. The second cell had similar orientation and directional preferences and a similar ocular dominance. Figure 3A is the correlogram obtained from this cell pair. This correlogram has a sharp positive peak that straddled the zero time bin indicating a common excita-

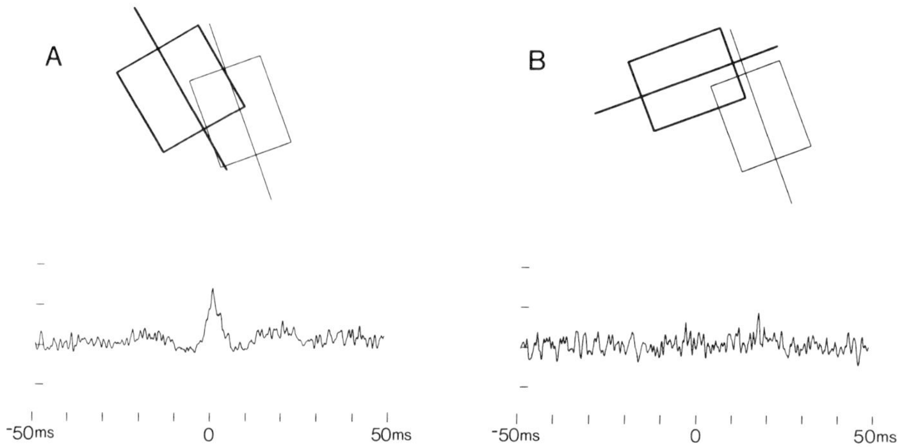

Fig. 7.3. Correlograms obtained from two cell pairs. *A:* the cell pair had similar receptive properties: the first cell had an orientation preference of 120°, directional preference to the right and an ocular dominance group of 2, the second cell had identical orientation and direction preference and an ocular dominance group of 3. *B:* the first cell in *B* was the same as the first cell in *A*. The second cell had different receptive field properties: an orientation preference of 20°, upward directionality and an ocular dominance of 5

tory influence. The second electrode was then moved to isolate a cell with quite different receptive field properties, while the first electrode was kept fixed. This third cell was separated from the first cell by 1 mm and had an orientation preference of 20°, upward directionality and an ocular dominance of group 5. Figure 7.3B is the correlogram obtained between the first and third cells. The correlogram was flat, indicating no observable interaction between these two cells which had differing receptive field properties.

To explore systematically the relationship between the long-range horizontal connections and receptive field properties, we made electrode penetrations that were parallel to the surface of the cortex, down the medial bank of the lateral gyrus (see Fig. 7.4 and compare with Fig. 7.1). This approach enabled us to stay in a particular layer for several millimeters, record from a series of cells with receptive fields displaced from the midline, parallel to the horizontal meridian. Along the penetration we encountered a succession of different orientations and ocular dominance shifts. Another electrode isolated a single unit that was used as a reference cell throughout the experiment. Using this strategy, we were able to study the neuronal interactions between a reference site in a single cortical column and other cells in other columns at varying horizontal distances and with a range of receptive field properties. To limit our study to the horizontal in contrast to interlaminar connectivity, we restricted our penetrations to layers 2 + 3.

The dependence of correlated firing on matching receptive field properties as seen in the single cell pair (Fig. 7.3) was also observed over the entire length of our medial bank penetrations, covering several millimeters. Figure 4 shows an example of such a penetration. The reference cell had an orientation of 60°. The

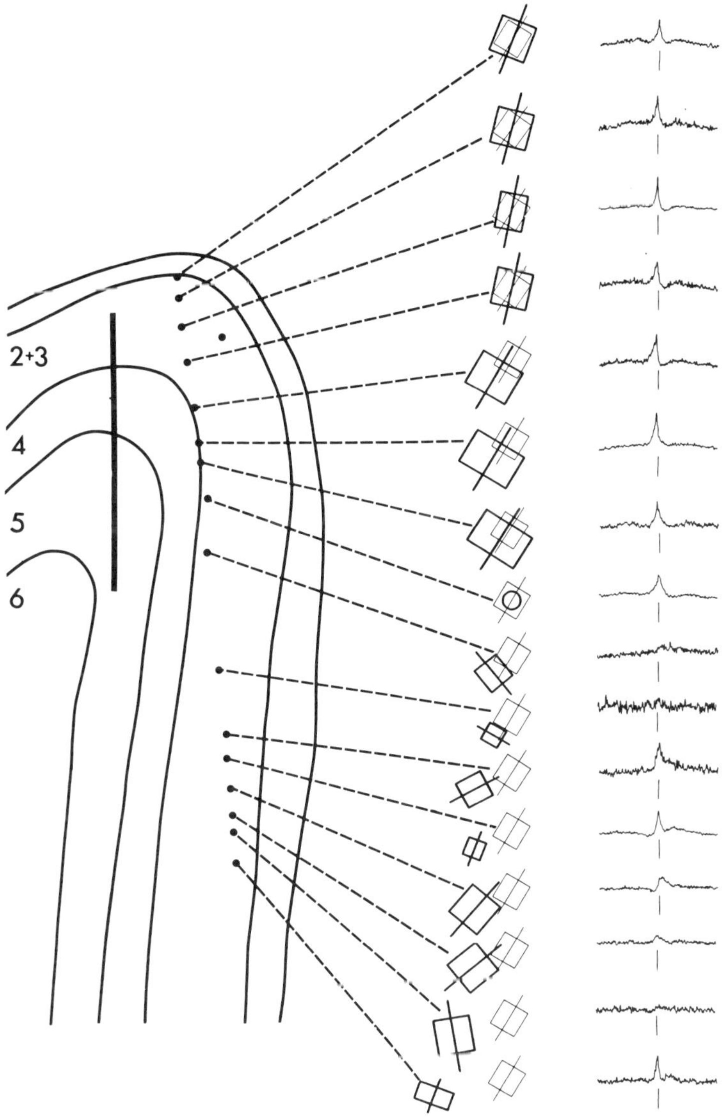

Fig. 7.4. Medial bank penetration with correlograms between a reference cell and various test cells. The reference recording site is shown as a single dot (●), and had an orientation preference of 60°. Its receptive field map and position is shown in each pair with light lines. The size of the reference cell receptive field was 1° × 1°. Each test cell recorded in the second electrode has its receptive field shown in bold lines. The shuffled, adjusted, noise-filtered and normalized correlogram obtained from each test cell (with respect to the reference cell) is shown at the far right. The cell displayed with a circular receptive field had no detectable orientation preference. The correlograms were collected with a ±50ms window and the bottom vertical tick mark midway between each correlogram denotes the zero time bin. Scale bar equal 1mm

cells initially encountered by the second electrode had orientation preferences that were similar to the reference cell. For example, the first cell isolated had an orientation of 70°. The resultant cross-correlogram was sharply peaked. Throughout the penetration, peaked correlograms were only found with groups of cells having similar orientation preferences, interposed by regions of cells that had flat correlograms. The correlograms had robust peaks for orientation differences up to about 30°. The waxing and waning of correlated firing extended for the entire 2mm length of the electrode penetration. Towards the end of the penetration we still encountered peaked correlograms, even though the receptive fields of the cells did not overlap. The last pair of receptive fields was separated by 3° (center to center).

Figure 7.5 is another example of cross-correlograms from a penetration of a few millimeters. As in the previous penetration, peaked correlograms were found in regions where the two cells had similar receptive field properties and flat correlograms were observed when the cells had properties that differed. Figure 7.5 also shows that the strength of the correlogram peaks tended to diminish as the distance between the electrodes increased. The shrinking size of the peaks with distance, though not as apparent in Fig. 7.4, was a common observation in these penetrations. Typically the interactions observed at 1 mm were less than one half the strength of those observed more proximal to the reference site (less than 500 μm).

Of the peaked correlograms obtained, 29% showed peaks characteristic of monosynaptic interactions (cf. fifth in Fig. 7.4). Another 20% had peaks with an ambiguous interpretation, possibly a combination of monosynaptic and common input interactions (cf. sixth in Fig. 7.4). The remaining 51% appeared to be due to common input interactions alone (cf. first in Fig. 7.4).

To compile our results on the relationship between the cross-correlograms and orientation specificity, we first computed an index of the strength of the interaction between the cell pairs, as derived from the correlogram. This strength index was calculated for the 115 pairs of cells and grouped into 4 bins. Figure 7.6 shows representative correlograms that fell in each of the four strength groups. Group 0 correlograms were flat, group 1 were either uncertain or weak, group 2 were of moderate strength and group 3 were strong. The cell pairs were also grouped according to the difference in orientation preference between the two cells. Histograms were plotted for each of five orientation difference groups showing the relative proportion of cell pairs in each correlation strength bin (Fig. 7.7). For cell pairs with similar orientation tuning, e.g. 0 to 30 degrees difference, over 75% of the cell pairs yielded correlograms in groups 2 and 3. For cell pairs with very different orientation tuning, 60 to 90 degrees difference, all (100%) of the correlogram strengths fell within groups 0 and 1. Some (less than 25%) cell pairs which had similar orientation preference did not exhibit significant interactions. These findings on the relationship between matching orientation preference and strength of correlation yield a x^2 significance level of $p << 0.01$. As would be expected from the columnar organization of the cortex, nearby cells (e.g. within a few hundred μm) had similar receptive field properties, which might lead to confusion between a dependence

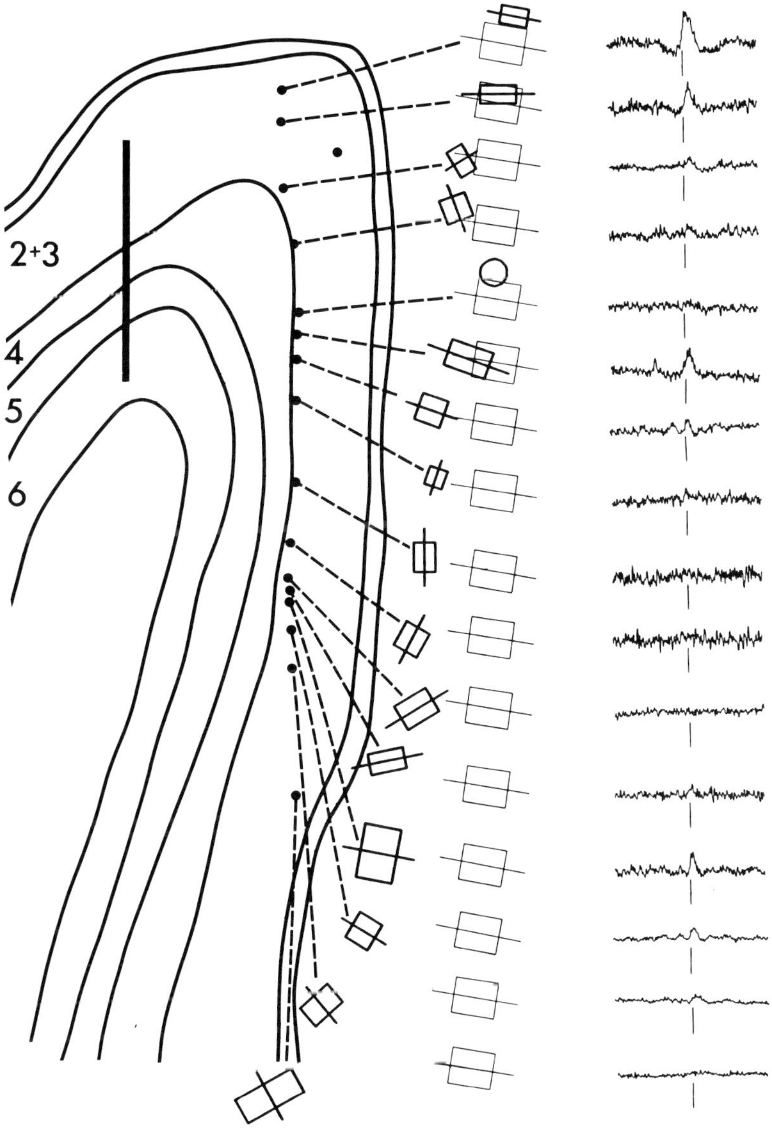

Fig. 7.5. Medial bank penetration similar to Fig. 4. Note tendency for the diminishing strength of the peaked correlograms with increase in distances between cells along the medial bank penetration and the reference cell. Also note the large difference in receptive field positions between the recording sites furthest apart

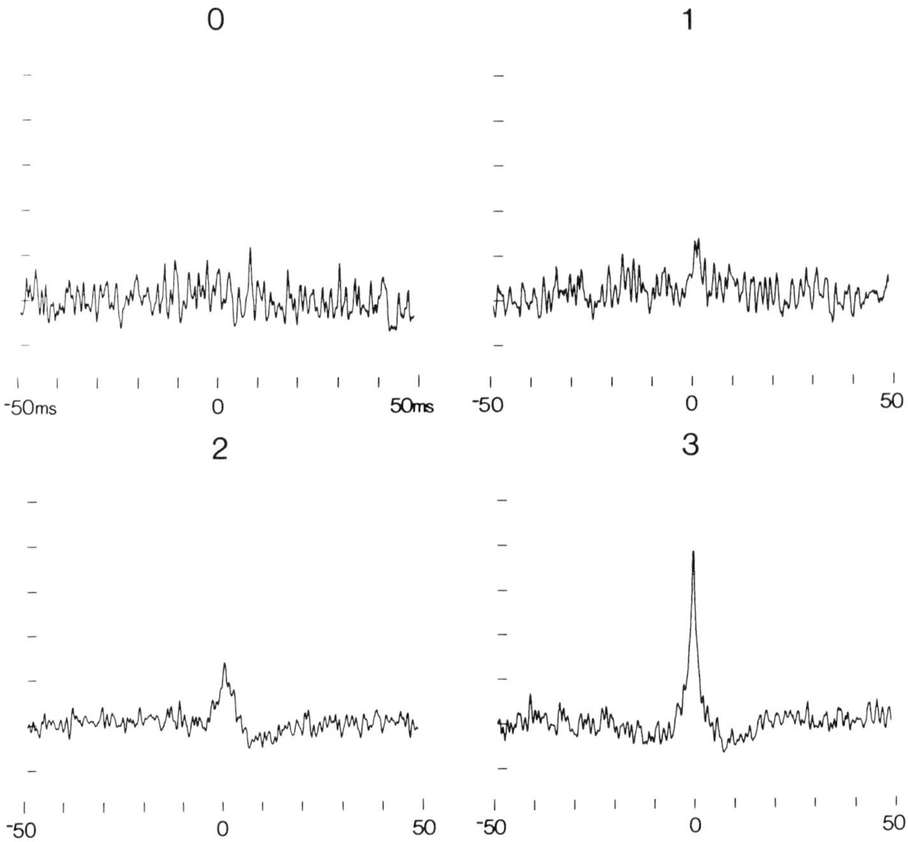

Fig. 7.6. Correlograms from cell pairs representative of each of four correlation strength groups. Group 0 correlograms were flat. Group 1 correlograms were almost flat. Group 2 and 3 correlograms were moderate and strong, respectively. Placement of a cell pair in its correlation strength group was performed by a computer program that estimated the energy in the peak

on interelectrode distance and orientation preference. However, the relationship between orientation preference and correlation strength was observed irrespective of interelectrode distance.

Dependency of interactions on other properties. Other receptive field features were examined for their relationship to the strength of correlation. We found that there was a tendency for little or no correlation when the two cells in the pair had opposite ocular dominances, and a tendency towards a moderate or strong correlation when cells had matching ocular dominance. However, strength of correlation was much more closely related to matching orientation preference than matching ocular dominance. For directionality, on the other hand, no consistent difference was found in the strength of correlation between cells having matched or opposing directional preference. Also, we saw no significant relationship between the strength of correlation and end-inhibition, edge

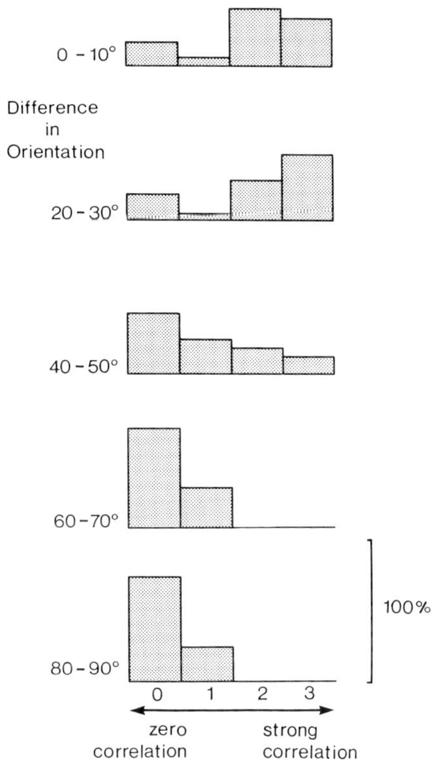

Fig. 7.7. Histograms displaying percentage of total cell pairs falling in each of four correlation strength groups for various differences in orientation. The difference in orientation preference between the cells in each pair was computed along with its correlation strength group. Total sample: 115 cell pairs

preferences (leading or trailing) or response to stationary flashing stimuli (ON versus OFF). However, since our resultant sample size was small, a weak relationship between the strength of correlation and the receptive field properties measured might be undetectable. Furthermore, our analysis depended on consistent relationships within a population of cells. The absence of a consistent relationship in the population does not preclude a specific relationship between any two single cells.

Thus, we found that cell pairs in cat striate cortex exhibited interactions only when their orientation preferences were similar, regardless of the horizontal distance of up to several millimeters that separated the two cells. This arrangement is shown schematically in Fig. 7.8. These results provide a possible physiological counterpart to the anatomical studies of the horizontal connections. The observed interactions were exclusively excitatory, and extended over several millimeters including cells with fields separated by several degrees. These findings confirmed predictions that the longest horizontal projections involve cells and columns with non-overlapping receptive fields and like-orientation specificity (Gilbert and Wiesel 1981, 1983; Mitchison and Crick 1982). Furthermore, we believe that the distribution of correlated firing and its relationship to columnar organization of receptive field properties corresponds to the clustered nature of the horizontal connections seen anatomically (Gilbert and

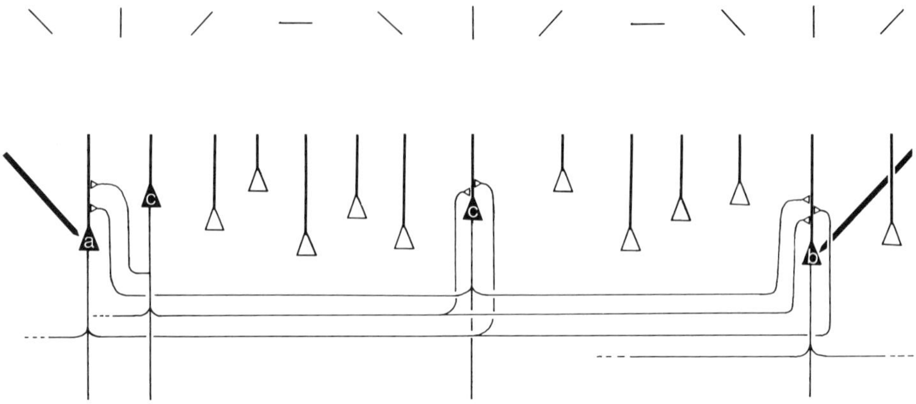

Fig. 7.8. Schematic representing results of long-range horizontal connections between columns of cells with similar receptive field properties. Recording from the two cells *a* and *b*, having similar orientation preferences (shown at the top); cell *a* is seen to provide a direct input to cell *b*. In addition, the pair receives common input from several *c* cells in cortical columns with like orientation specificity and intervening columns of differing specificity. The common input connections outnumber the direct inputs from *a* to *b*

Wiesel 1979, 1983). Another study comparing 2-DG orientation maps with the distribution of transported fluorescent beads injected in a known orientation column, also found that connections existed between columns of like rather than orthogonal orientation tuning (Gilbert and Wiesel 1989).

Distance and the strength of interactions. Another measure of the strength of interaction has been termed the "contribution" (Levick et al. 1972). This index is defined as the percentage of the total number of spikes collected from a cell that are contained in the cross-correlogram peak. The contributions represented by our correlograms ranged from approximately 0.5 to 20% of the cells' total firing, and the average was 6%. At the shorter distances studied (500 μm and less) the strength of interactions found are in general agreement with the previous studies of interlaminar cross-correlations by Toyama (1981a,b). Michalski et al. (1983) also found strong direct input and common input interactions among cells in close proximity. However, at longer distances we observed that the strength of the interactions between columns of like specificity dropped as the horizontal distance increased and became significantly weaker. This observation is consistent with the decrease in the extent of arborization and number of synaptic boutons in the distant clusters of HRP-injected cells (Gilbert and Wiesel 1983). Although the observed horizontal interactions, shared and direct, between two given cells may be relatively weak, each cell is likely to receive input from many other cells distributed over a wide territory. This organization would increase the importance of the horizontal connections to a cell.

The excitatory nature of horizontal interactions. The interactions observed were excitatory in nature, though there has been speculation concerning the existence of inhibitory horizontal connections in the cortex. It has been shown that the cross-correlation technique may be incapable of detecting weak inhibitory interactions (Aertsen and Gerstein 1985), and in fact we rarely saw direct inhibitory interactions at the interelectrode distances studied. At shorter distances, Toyama (1981a,b) and Michalski et al. (1983) did see inhibitory interactions with the cross-correlation technique. Of the common input correlograms which have broad positive peaks, it is possible that some could be due to common inhibitory input (Moore et al. 1970). Nevertheless, the longest range horizontal connections are probably made by pyramidal cells, which are presumably excitatory. Presumptive inhibitory cells, such as the smooth stellate cells, have axons that tend to be more restricted in their horizontal extent than the distances over which we have observed correlated firing. Basket cells, also presumed to be inhibitory, do exhibit axons lengths of the order of one millimeter (Somogyi et al. 1983). However, since we observed correlated firing at separations of 3mm, the underlying projections should be in that range, and we believe that at least the majority of the interactions observed were excitatory.

There is anatomical evidence suggesting that horizontal projections are excitatory. One piece of evidence comes from experiments examining degenerating terminals of horizontal connections seen after a narrow lesion spanning all the cortical layers (Fisken et al. 1975; Creutzfeldt et al. 1977). These terminals formed asymmetric type I synapses, which are thought to be excitatory. More recently, McGuire et al. (1985) studied the terminals of long-range horizontal projections of an HRP-injected cell in the monkey visual cortex and found that, in addition to forming asymmetric synapses, a majority of boutons ended on spines, presumably on spiny dendrites. Since spiny cells are thought to be excitatory, these findings support the idea that horizontal connections produce excitatory influences.

The prevalence of common input and its possible source. Many of the cross-correlograms had peaks which were highly asymmetric but contained significant counts on both sides of the zero bin, possibly due to a combination of common input and monosynaptic excitatory interactions. Michalski et al. (1983) also found common input interactions to be more prevalent than direct interactions. The large proportion of common input correlation disturbed us at first, but upon further reflection seemed quite reasonable because of the pattern of connections observed with intracellular HRP injections. An individual cell gives off clusters of collaterals near its own dendritic field and at distant sites (Fig. 8). Consequently, one would expect that a given cell receives input from a group of cells that is highly interconnected. This pattern of connections would lead to common input together with direct excitation. If a large number of cells are involved in this network, the contribution of common input to the correlated firing would be relatively high. It is then expected that the strength of input shared by two cells with like specificity would exceed the strength of any direct connection they may have between each other. Correlated firing due to common input should then be more readily detectable than monosynaptic interactions.

Since we were interested in horizontal intrinsic cortical connections, we needed to be certain that most or all of the correlated firing found came from within the cortex. Correlograms with a monosynaptic component clearly represent intrinsic connections. However, the prevalence of common input correlation introduces the possibility that the common input might not be cortical but geniculate in origin. This explanation is unlikely since the presence of correlated firing is shown to be dependent on matched orientation preferences. Since geniculate cells are strictly monocular and have center-surround receptive fields, common geniculate input would not be expected to provide input to cells of a specific orientation preference, but rather to cells that have the appropriate ocular dominance. In addition, reconstructions of geniculate axonal arbors have shown them to innervate a full ocular dominance column, making it unlikely that they would only contact cells of a single orientation (Hubel and Wiesel 1972; Ferster and LeVay 1978; Gilbert and Wiesel 1983; Blasdel and Lund 1983). Another piece of evidence is that on a number of occasions we found the shape of the correlogram between a particular cell pair to be dependent on the orientation of the stimulus, indicating that the source of common input to the cells was itself oriented, and thus unlikely to be geniculate in origin. Other possible sources of common input are the recurrent projections from extrastriate cortex, which also have a clustered distribution (Wong-Riley 1979a; Gilbert and Kelly 1975). It is possible that both cortico-cortical connections and intrinsic connections contribute to the interactions between columns of like specificity.

7.3 Functional Organization and Connectivity in the Blob and Interblob Regions of Primate Striate Cortex

In addition to demonstrating "like talks to like" connectivity in the orientation domain in cat striate cortex, we were interested in examining the patterns of connectivity in the primate striate cortex. As in the cat, intracellular HRP injections have demonstrated the existence of clustered horizontal connections in monkey striate cortex (Fig. 9, McGuire et al. 1985). Livingstone and Hubel (1984b) had also shown in the monkey a specificity of connections, anatomically through the use of extracellular focal HRP injections. The cytochrome oxidase-rich blob regions were exclusively connected to other blob regions and the interblob regions were only connected to other interblob regions. We sought to explore the connectivity of blob regions and interblob regions physiologically using cross-correlation analysis to investigate the relalionship between the patterns of connectivity and the functional properties of the cells involved.

The blob color system of striate cortex. An important feature of the modular organization of the monkey striate cortex was revealed by the finding of a patchy labeling of striate cortex to a stain developed for the mitochrondrial enzyme cytochrome oxidase (Wong-Riley 1979b; Horton and Hubel 1981). High concentrations of this enzyme are thought to indicate long-term elevated metabolic activity. The regions of dense cytochrome oxidase reactivity, in tangential sections of macaque striate cortex, are centered on ocular dominance columns, and

Fig. 7.9. A camera lucida drawing of a pyramidal cell in layer 3 of the monkey striate cortex that was injected intracellularly with HRP after its receptive field properties were determined (it was an oriented complex cell, McGuire et al. 1985). *Scale marker* = 100m. Note the long-range horizontal projection of the cell axon. Its clustered terminations are suggestive of a relationship with the cortical columns

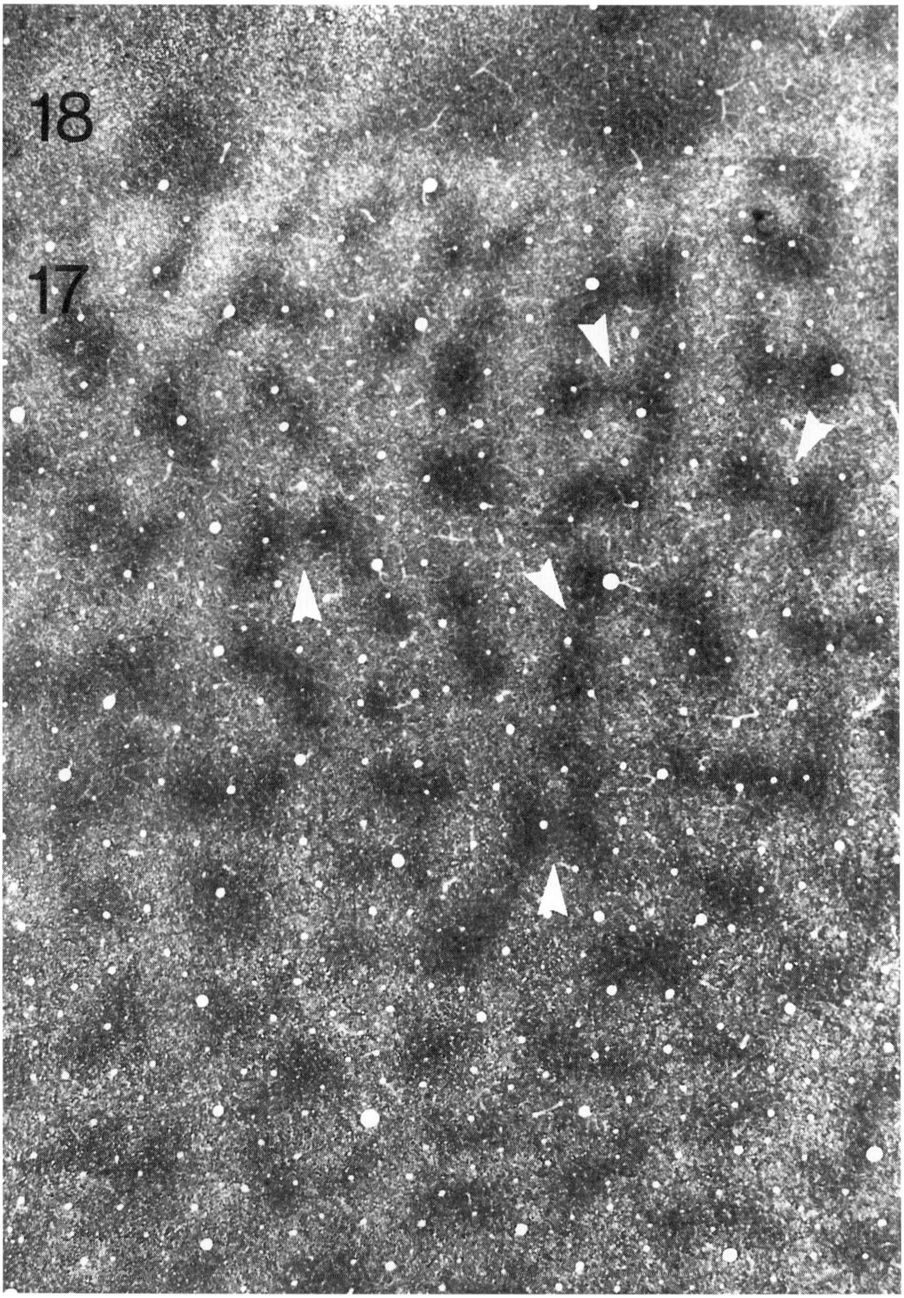

Fig. 7.10. Tangential section through the superficial layers of macaque striate cortex, including the area 17/18 border. The section has been stained for cytochrome oxidase and shows the patchy staining pattern of the blobs in area 17 and the banded or striped pattern of staining in area 18. Note the bands of staining between individual blobs, both along the ocular dominance columns (perpendicular to the 17/18 border) and across ocular dominance columns

laid out as a matrix of oval patches or blobs of approximately 150 × 200 μm each. These blobs exist primarily in layers 2 and 3, although they are more faintly seen in layers 4B, 5 and 6. Figure 7.10 shows a tangential section of monkey visual cortex stained for cytochrome oxidase. Livingstone and Hubel (1984a) found that the blobs contained cells with receptive fields that were monocular, unoriented and often color selective. Cells outside the blobs in the inter-blob regions, were non-color selective and had orientation tuning. The blob cells could be categorized into groups or types that were similar to types found in the lateral geniculate nucleus (LGN), the thalamic relay from which the striate cortex received its visual input (Wiesel and Hubel 1966). In addition, they found a class of blob cells that did not correspond to any cell type found in the LGN. These cells were described as double-color opponent, having spatial opponent fields with color opponency in both center and surround. They reported finding double-color opponent cells in large numbers (44% of the nonfoveal blob cells). Since the blobs apparently contained the majority of color-selective cells in the superficial layers of the striate cortex, the blob regions are likely to play a major role in the processing of color information in primate striate cortex.

7.3.1 Receptive Field Properties of Blob Cells

We began our studies by re-examining the receptive fields of blob cells and then investigating the functional organization of the color cells. We made multiple vertical penetrations of a single electrode in the superficial layers of striate cortex to locate regions of cells with unoriented receptive fields. Subsequent histological reconstructions of the electrode tracks and lesions in sections stained for cytochrome oxidase were matched with our recording protocols to determine which penetrations were in blob regions. We found that cells in blobs had receptive fields that were poorly oriented or completely unoriented (class C or D of Livingstone and Hubel 1984a) and were strongly monocular (groups 1, 2, 6 or 7 of Hubel and Wiesel 1962). The spatial organization and color selectivity of the receptive fields of blob cells could be divided into categories or types, most of which resembled those found in the LGN (Wiesel and Hubel 1966). The various types of cells are illustrated in Fig. 7.11 (see Ts'o and Gilbert 1988). In addition to cells showing red/green opponency, we also encountered cells with blue/yellow opponency. Blue/yellow opponent and red/green opponent cells were found with Type I, II and modified Type II receptive fields.

Although Livingstone and Hubel (1984a) reported finding a preponderance of double-color opponent cells in the blobs, we could not confirm that finding. There were a only few (1%) blob cells that could be classified as true double color opponent. However we did find another cell type to be quite common, the modified Type II cell. The similarities between our modified Type II cells and double-opponent cells suggests that the modified Type II cells may have been mistaken for double opponent cells in other studies. For example, both show opponent color responses to small spots and are suppressed by large spots of any color. However, the modified Type II cell, unlike a true double-opponent cell,

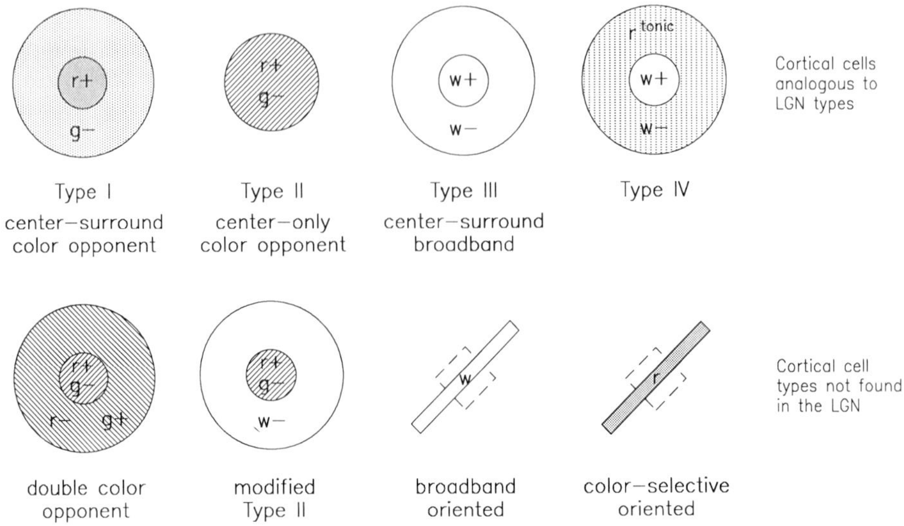

Fig. 7.11. Receptive field types of striate cortex. The oriented cell types included simple and complex fields. Our studies focused on the cells found in the superficial layers of striate cortex, in the blob and interblob regions. In addition to color cells with red vs. green opponency, cells with blue vs. yellow opponency were found, as were cells of both center sign, OFF (−) and ON (+)

does not respond well to isoluminant color contrast and, in the surround, does not exhibit opponent color responses of opposite sign from those of the center (Daw 1968). Since occurrence of true double-opponent cells seems quite rare and the stimulus used in earlier studies could not distinguish the double opponent from modified Type II cells, and because of the significant functional difference between the two types, we believe the modified Type II classification is more appropriate.

In summary, the principal difference found between receptive fields of LGN cells and those of unoriented blob cells was seen in the modified Type II cell. These cells combine color opponency in the center with broadband spatial opponency, and thus may provide an integration of information concerning color contrast with information concerning luminance contrast or spatial positioning. Since standard Type II cells have relatively larger receptive fields without surround suppression, they would give poor spatial resolution. The surround suppression exhibited by the modified Type II seems similar to other surround suppression, such as that of the Type III cells, except that the antagonism is both to the ON and OFF responses in the center.

7.3.2 Functional Organization of Color Opponency

An important feature of the distribution of cells in the blobs was that cells of the same color opponency were clustered. Within individual vertical electrode penetrations (Fig. 7.12) and in groups of neighboring penetrations within the

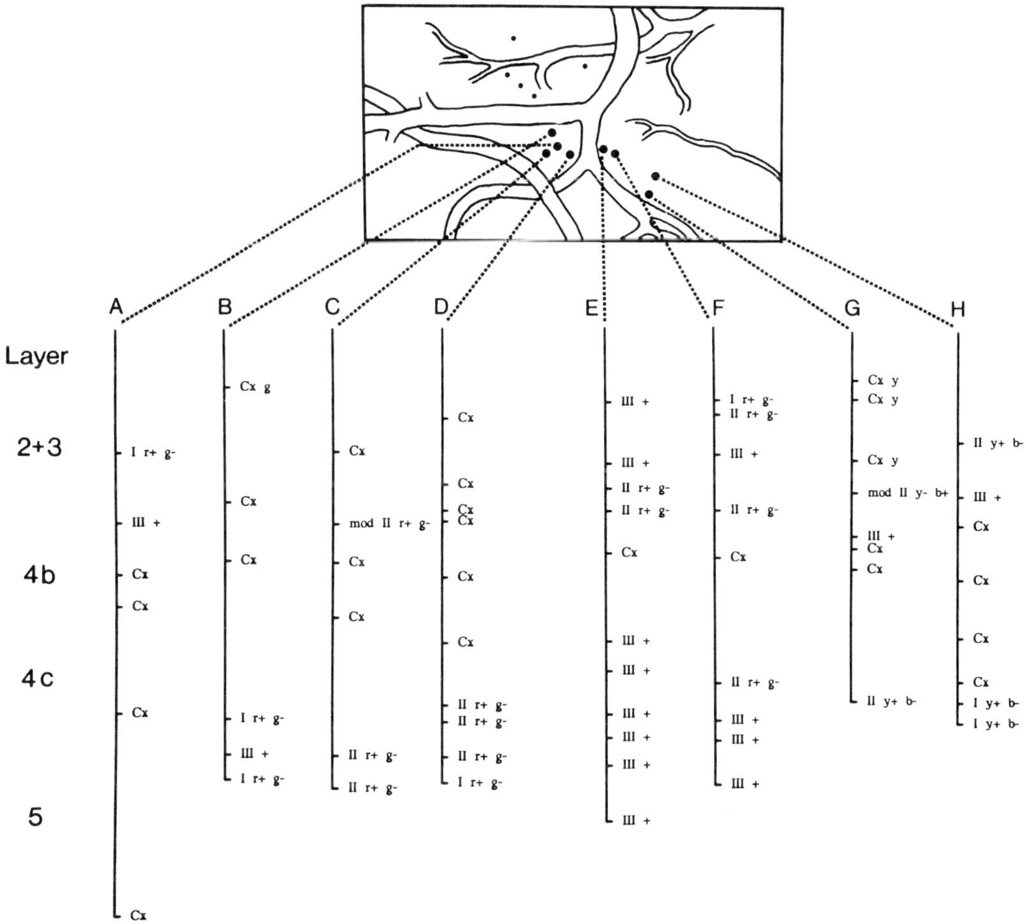

Fig. 7.12. Electrode penetrations in blob regions, extending to layer *4C* and corresponding cortical surface map showing the surface vasculature (top of figure). Smaller black dots on surface maps indicate positions of other electrode penetrations not detailed. Approximate cortical layer shown on left, derived from histological reconstruction of electrolytic lesions. Adjacent to each penetration record, the types of cells found are indicated: *Cx* ≡ complex oriented, *mod II* ≡ modified Type II. For Type I cells, the first color (*r* ≡ red, *g* ≡ green, *b* ≡ blue, *y* ≡ yellow) indicates the center color and the second color, the surround color. Penetrations *A, B, C, D, E, F* were through or near blob regions containing cells of red/green color opponency, with like color opponency found in layer *4C*. An adjacent blob region, *G, H,* had cells with blue/yellow color opponency in layers *2 + 3* and *4C*. Note the region of complex cells in layer *4B*

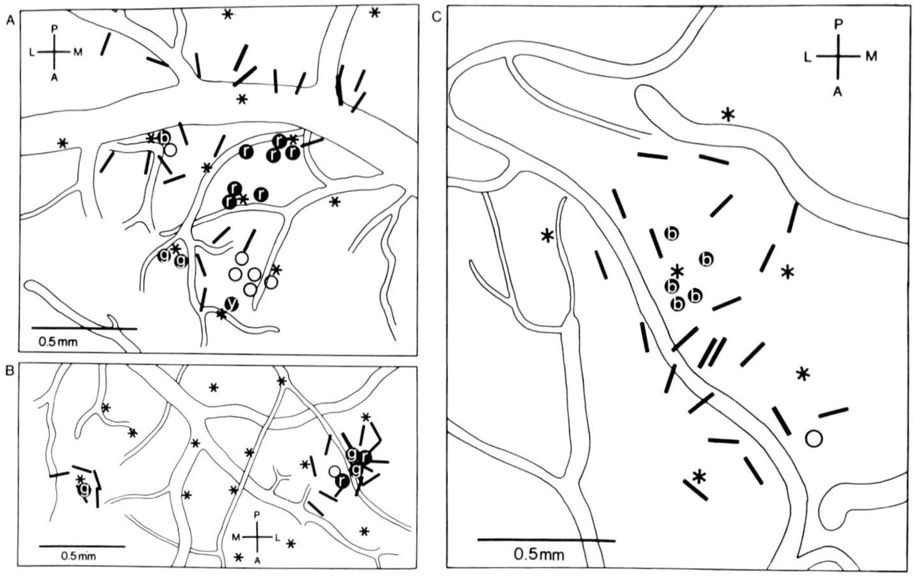

Fig. 7.13. Functional maps of a region of striate cortex derived from multiple-electrode penetrations and recordings, suggesting that individual blobs are dedicated to a single color opponency system. Electrode placement initially guided by a photograph of cortical surface and vasculature and later confirmed with electrolytic lesions and subsequent cytochrome oxidase histology. The asterisks mark the approximate location of cytochrome oxidase-rich blob regions. The black bars denote penetrations encountering cells with oriented fields, the open circles – only unoriented non-color selective Type III cells, the labeled circles – unoriented color selective cells having an ON response to the corresponding color. Thus, *r* circles indicate penetrations where red-ON green-OFF unoriented cells were found and *g* circles mark sites where green-ON red-OFF cells where found (*b* = blue and *y* = yellow. A color reproduction of this figure may be requested from the author). Generally, Type III cells were found in most penetrations, in addition to color selective cells. **A** shows sites containing unoriented cells with red/green or blue/yellow color opponency with no intermixing of color opponency in individual blobs. **B** shows two blobs of red/green color opponency. Each blob was examined with multiple penetrations whose placement extended to the surrounding oriented interblob regions. No cells exhibiting blue/yellow color opponency were fround. Note the red and green oriented cells found intermixed with the unoriented red/green blob cells on the right. **C** shows a blob containing only blue/yellow color opponency. No red/green color opponency was found

same blob (Fig. 7.13A), all the color selective cells had the same opponency, either blue/yellow or red/green. These findings led us to speculate that the entirety of a single blob might be devoted to one color opponency system. To further test the idea that blobs are dedicated to one color opponency, we made systematic maps of portions of cortex in and around single blobs with closely spaced multiple penetrations (Fig. 7.13B,C). We began by locating a blob, determining the color opponency represented and then completely surrounded the blob with penetrations until we found transitions between the blob and the oriented interblob regions. In Figs. 7.13, 7.14, the labeled dots on the map represent the ON color of the response of color-selective cells found at that recording site. A shift in center color sign (e.g. red-ON to red-OFF) was rarely seen in an

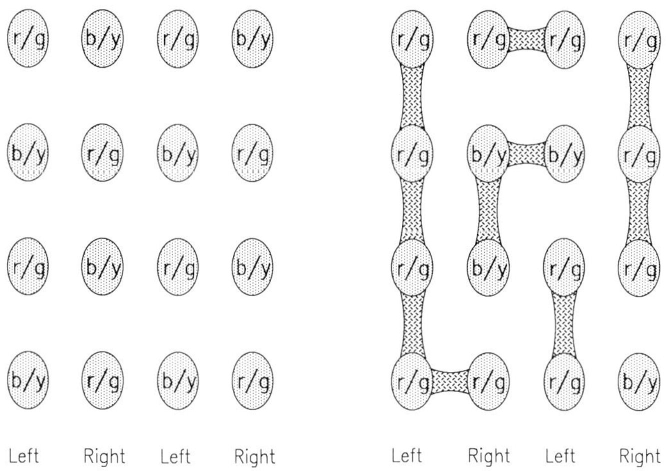

Fig. 7.14. Two possible arrangements for the red/green and blue/yellow blobs. *Left:* a checkerboard arrangement, similar to that proposed by Dow and Vautin (1987). Our results do not support such an arrangement, but are more consistent with the scheme shown on the *right*. There was a 3 to 1 ratio of red/green to blue/yellow blobs and the blue/yellow blobs were often clustered together. The diagram also depicts the cytochrome oxidase-rich bridges, connecting neighboring blobs and running both along and across the ocular dominance columns

individual penetration. In the example shown in Fig. 7.13B, only red/green color opponency was found. Of the blob regions recorded from, we determined 75% to be red/green and 25% to be blue/yellow, a 3 to 1 ratio. It is interesting to note that colored-oriented cells, usually with color selectivity related to the blob cells, were sometimes found near the non-oriented blob cells, at times even in the same vertical penetration (Fig. 7.13B). Figure 7.13C shows an example where the same strategy was used for a blue/yellow blob. Only blue/yellow cells were found. These findings suggest that individual blobs are color-opponency specific. However, different parts of one blob may contain cells with either center color (e.g. red or green) and/or sign (ON or OFF). Dow and Vautin (1987) have suggested a related, but somewhat different organization of the color-selective cells, with centers of red or blue wavelength selectivity surrounded by regions sensitive to the middle wavelengths. A possibility that we have not ruled out is whether certain blobs contain only Type III broadband cells. An example of a series of penetrations that suggest this possibility is shown in Fig. 7.13A, where adjacent penetrations only yielded Type III cells.

We then asked if the red/green and blue/yellow blobs were evenly or randomly distributed and what their relative abundance was. By mapping a larger area of cortex and examining the subsequent cytochrome oxidase histochemistry, we reconstructed the layout of the red/green and blue/yellow blobs for that portion of cortex. However, no general rule governing the layout of red/green and blue/yellow blobs, such as a simple alternation or checkerboard, was apparent. Thus, our evidence does not support the checkerboard model presented by Dow and Vautin (1987). There is evidence, however, that the blue/yellow blobs are often adjacent to other blue/yellow blobs, providing a clustered organization on a larger scale. Further functional mapping over an even greater cortical area will be required to uncover any systematic arrangement of these two classes of blobs. Figure 7.14 illustrates our current view of the arrangement of the color-opponency-specific blobs. A checkerboard model is also shown for comparison.

Is the color specificity of blobs maintained from the superficial layers to the deeper cortical layers? This question is important in considering the patterns of connectivity leading to the receptive field properties of blob cells. Anatomical evidence suggests that layer 4C projects to the superficial layers (Lund and Boothe 1975; Fitzpatrick et al. 1985) and may contribute input to the blob cells in those layers (Michael 1986). Thus we were curious as to the color specificity of the layer 4C cells underlying blobs of known color specificity. We made extended vertical penetrations, first entering a blob and determining the color opponency represented, and then advancing to layer 4C and determining the color opponency of cells in that layer. Figure 7.12 shows examples of such penetrations. In penetrations in red/green blobs, underlying layer 4C cells that were color opponent were also red/green. Penetrations into blue/yellow blobs subsequently found blue/yellow cells in layer 4C also. In viewing Fig. 7.12, the penetrations in a given blob should be considered together in determining the relationship between the color opponency of the superficial layer blob cells and the underlying 4C cells, even though certain individual penetrations strayed into neighboring interblob regions containing non-color-selective complex cells.

Our finding that cells in layer 4C often had identical color specificity to the cells in the overlying blobs is consistent with the projection from layer 4C to layer 3. In addition to their input from layer 4, blob cells receive a direct projection from the interlaminar layers of the LGN (Fitzpatrick at al 1983). The properties of the cells within the interlaminar layers have not been described, and the relative roles of the direct thalamic and layer 4C input to the blobs are not yet understood. The correspondence of the color opponency type between layer 2 + 3 and layer 4C supports the notion of a color column. However, it is not known whether the cells underneath the blobs in layers 5 and 6 share the color properties of the blobs.

7.3.3 Connectivity of Blob and Non-Blob Cells

Though the functional architecture of the color system is an important element in understanding how the cortex analyzes visual information, the main thrust of our work has been to understand how the functional properties of cells are transformed at progressive stages along the visual pathway, and to understand the mechanism of this transformation at the level of connectivity. Our findings concerning the relationships between receptive field properties and the overall organization of blob cells and their connectivily, provide some insight into this transformation within the submodality of color.

For the sake of analysis, one can classify the intrinsic connections of the cortex into those connections that run vertically across the cortical layers, and those connections that run laterally or horizontally parallel to the cortical surface. This classification is likely to have a functional significance since vertical connections in the visual cortex run between cells that subserve the same visual field position and have many receptive field properties in common. In contrast, the horizontal connections join different columns of cells that, in the short range, have differing receptive field properties, and in the long range deal with different positions in the visual field. The blob color system participates in connections running in both domains. Therefore, we were not only interested in the columnar or vertical organization of color opponency and receptive field structure, but also the horizontal connectivity of the blob regions.

Connectivity of the blob regions. Since the blob regions represent a major component of the functional and cytoarchitectonic organization of the monkey striate cortex, it is natural to ask how the blobs might relate or participate in horizontal connections. As mentioned above, the anatomical tracing studies of Livingstone and Hubel (1984b) suggested a segregation of the intrinsic connections between the blob color system and the oriented, non-color-selective interblob regions. We wanted to study this pattern of connectivity physiologically, and to further explore the functional role of the connections.

To further understand the functional significance of the interconnectivity of the blob color system, we combined the cross-correlation technique with receptive field analysis of the blob cells. In addition to the issues of the segregation of connections and its relationship to functional organization, we were interested in the neuronal circuitry underlying the various receptive field types found in the blobs and interblob regions. Our general strategy in the cross-correlation analysis was to record from pairs of electrodes spaced 50μ to 1 mm apart in the superficial layers, and to isolate single units in each recording channel. This enabled us to correlate the firing between cells with various combinations of receptive field types, columnar specificity and varying cortical distances, and also among blob and interblob cells.

Horizontal interactions among interblob regions. We found, among the oriented cells in the interblob regions, that peak correlograms were obtained only between cells with matched orientation preference. These peaked correlograms were positive and indicated common excitatory input, monosynaptic excitation or a combination of the two. To further examine the horizontal extent of these interactions, we mapped a portion of cortex using multiple vertical penetrations and determined the orientation preference and ocular dominance at each site. A single unit was then isolated at a chosen reference site and a series of cross-correlations collected between this reference cell and cells located at the other test sites (Fig. 7.15). Among the various horizontal distances tested, peak correlograms were only obtained when the orientation preference of the test cell matched that of the reference cell.

Therefore our findings, using cross-correlation analysis, of correlated firing between pairs of oriented interblob cells with matched orientation preference are similar to those found for cat striate cortex (Ts'o et al. 1986). As in the cat, excitatory correlograms were obtained over a range of horizontal distances, sug-

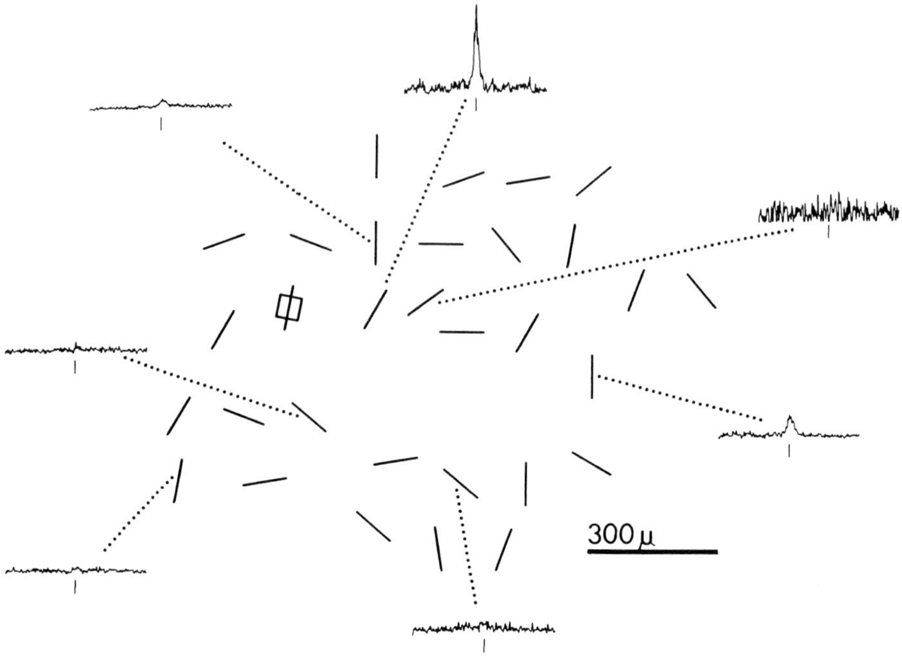

Fig. 7.15. Cross-correlations between a reference site and other selected test sites in the interblob regions. The orientation map of a region of striate cortex, derived from multiple electrode penetrations is shown in the black bars. The reference site had an orientation preference of 70°. Correlograms were normalized by the baseline correlation. Peaked correlograms, with varying electrode separations, occurred only if the orientation preference of the test site was close to 70°. The recording site at the lower left corner had an ocular dominance different from the reference site

gesting a correspondence to the anatomically demonstrated clustered long-range horizontal connections. Since these interactions were dependent on matched orientation preference and not on matched eye preference alone, it is likely that the underlying connections are intrinsic to the cortex and not due to common input from the LGN.

Horizontal interactions between blob and interblob regions. We also used cross-correlation analysis to study the pattern of connections among blob and interblob cells. Figure 7.16 shows an example of a peaked cross-correlogram between two oriented interblob cells with matched orientation preference (cells A and B1). We then moved one of the electrodes to a blob region with unoriented, monocular fields and isolated a Type III blob cell. The resultant cross-correlation was flat (cells A and B2). We later confirmed the location of the Type III cell in a blob region, through histological reconstruction. In this example, an oriented interblob cell showed interactions with another interblob cell with matched orientation preference, but showed no interactions with a neighboring Type III blob cell. We also have recorded from pairs of blob cells and obtained peak correlograms, and then moved one electrode to an interblob region – these cor-

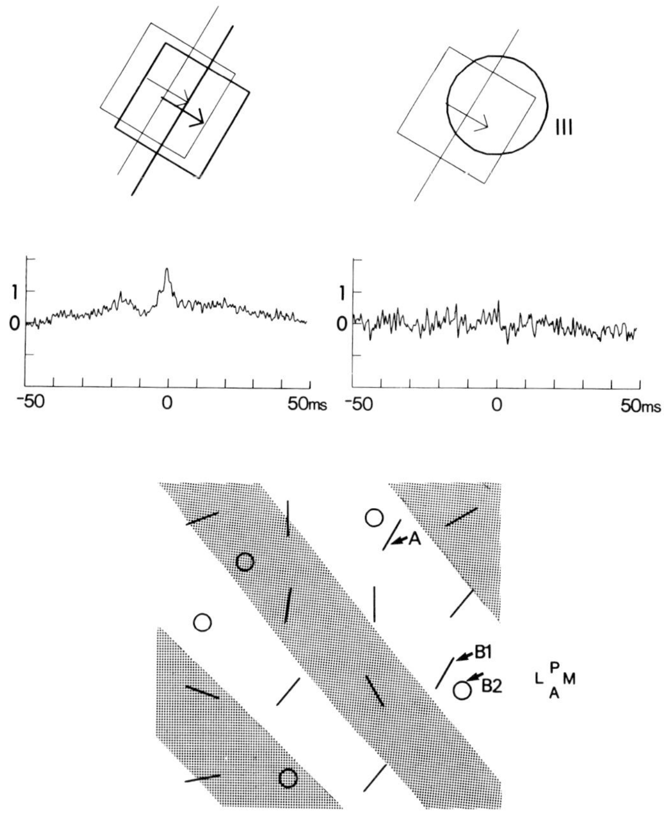

Fig. 7.16. Correlograms obtained from two cell pairs (above). Functional map (below) showing orientation preferences at several sites (marked by black bars), ocular dominance columns derived from the recordings (shown in stippled bands) and locations of several blob regions (indicated by open circles). First cell pair (left) were both oriented, interblob cells with similar receptive field properties: orientation preference of 60°, directional preference to the right and ocular dominance of 3. One electrode remained fixed at site *A*, while the second electrode moved from site *B1* to site *B2*, in a blob region and isolated a Type III blob cell. The resultant correlogram is shown on the right

relograms were invariably flat. Together, these findings suggest that the connections between blob–blob regions are segregated from those between interblob–interblob regions.

This physiological segregation of connections between blob–blob and interblob–interblob is consistent with the anatomical demonstration of the segregation of connections shown by focal extracellular injections of HRP (Livingstone and Hubel 1984b). An additional set of cross-correlations between color-selective oriented cells and color-selective blob cells revealed that only when the color selectivity of the pair was matched did peaked correlograms occur. The blob cells that interacted with the orientation and color-selective complex cells had either Type I or modified Type II fields. Thus, although in general the unoriented cells in the blob regions do not interact with the oriented, non-color-

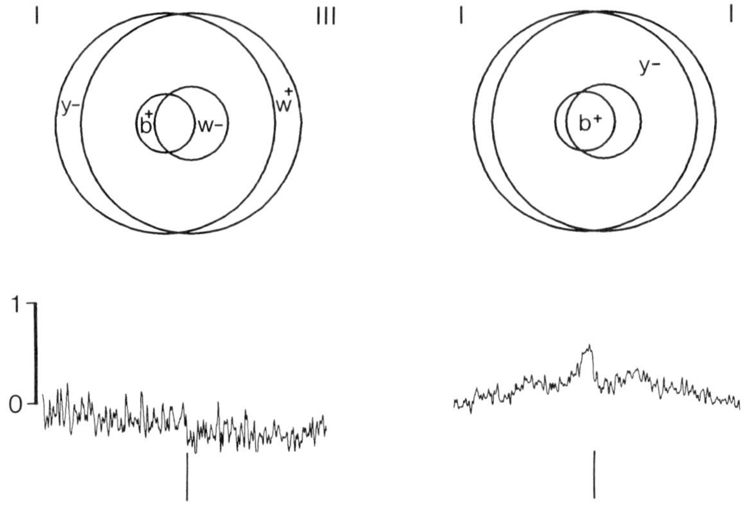

Fig. 7.17. Left: Blue-ON yellow-OFF Type I to Type III OFF-center correlogram. Right: Blue-ON yellow-OFF Type I to Blue-ON yellow-OFF Type I cell. The shape and pattern of the peak indicates an excilatory monosynaptic connection from one Type I to the other, $N_a = 4378$, $N_b = 19370$, Contribution 13.2%, Effectiveness 3.0%

selective interblob cells, there does seem to be an interaction between the color-selcctive oriented cells and the unoriented color-selective blob cells.

The segregation of blob and interblob connections, taken together with the partitioning of color-specific unoriented cells into the blobs, supports the idea of a separation of form and color into different cell populations. However, the very existence of color-specific oriented cells is indicative of a convergence of form and color information at the single-cell level. Direct demonstration of interactions contributing to this convergence comes from our finding of peaked correlograms between color-selective unoriented and color-selective complex cells. These interactions represent further evidence that color and form are not completely segregated. While the anatomical substrate of such interactions is not known, one possibility is that there exist connections between blob and interblob cells which were not visible in the studies using extracellular HRP injection.

Interactions between blob regions. Our results show that blob cells tended to interact with other blob cells of the same receptive field type and color opponency. For example, in one pair of penetrations we first recorded from a blue-ON yellow-OFF Type I cell in one electrode, and an OFF-center broadband Type III cell in the other electrode (Fig. 7.17). We obtained a nearly flat correlogram, with perhaps a trace of inhibition. Moving the second electrode, we then recorded from another Type I blob cell similar to the Type I cell at the first electrode–blue-ON yellow-OFF, and obtained a correlogram with an excitatory monosynaptic pattern (asymmetric peak, and shifted entirely to one side of the zero bin). This interaction represents a monosynaptic connection between Type

I blue/yellow blob cells. While the reference electrode continued to record from one Type I blue/yellow cell, the second electrode was advanced further. Peaked correlograms were obtained only when other Type I blue/yellow cells were encountered. The correlograms reflected both monosynaptic and common input interactions. Peaked correlograms were not found between red/green and blue/yellow color-opponent cells. We observed this pattern of interaction both between pairs of cells in the same blob and in two different blobs.

Thus we found a specificity of blob connections for receptive field type (e.g. I, II, or III) and color selectivity. This pattern of connections in which "like talks to like" mirrors the specificity of the interblob connections for columns of matched orientation preference. Interactions between blob cells both within a single blob and between two different blobs were observed (as long as the two blobs were dedicated to the same color opponency). We obtained both common input and monosynaptic excitatory correlograms. While the monosynaptic excitatory correlograms are probably due to intrinsic connections in the cortex, it is less easy to be certain of the source of the common input correlograms. It is possible that a portion of the interactions between two blob cells with similar receptive field properties is due to shared input arising either from layer 4C or the LGN. Further studies are required to clarify this issue.

Our results also contributed to understanding the connections underlying the generation of specific receptive field properties. For example, we wished to determine the sources of intrinsic input to the modified Type II cell. A cross correlogram was obtained between a standard red-ON, green-OFF Type II and a red-ON, green-OFF modified Type II cell (Fig. 7.18, left). The correlogram indicates an excitatory monosynaptic connection from the standard Type II to the modified Type II cell. Such connections may contribute to the construction of the modified Type II receptive field in the cortex. One possible scheme uses a standard Type II field to provide the center response and derive the surround antagonism from the summing inputs from Type I, Type II or Type III cells, or a mixture of cells of each type having suitably positioned fields. Such models, however, at least in their simplest form, are not entirely satisfactory since they do not adequately explain the properties of the surround antagonism of the modified Type II cell.

As mentioned above, we also found peaked correlograms between modified Type II cells and color-selective oriented complex cells with matching color specificity (Fig. 7.18, right). In this example, the red-ON green-OFF modified Type II cell may have contributed to the green trailing edge response of the green-selective complex cell.

Since our cross-correlation studies revealed a further segregation of the blob–blob connections based on like color opponency, we wished to establish an anatomical correlate of this pattern of connections. We made a focal injection of HRP into a blue/yellow blob dominated by the contralateral eye. Subsequent histological processing and HRP tracer reconstruction revealed that a blob next to the injection site was heavily labeled. This labeled blob was also a blue/yellow blob dominated by the contralateral eye. Another neighboring blue/yellow blob with the same eye dominance was somewhat more weakly labeled. In contrast,

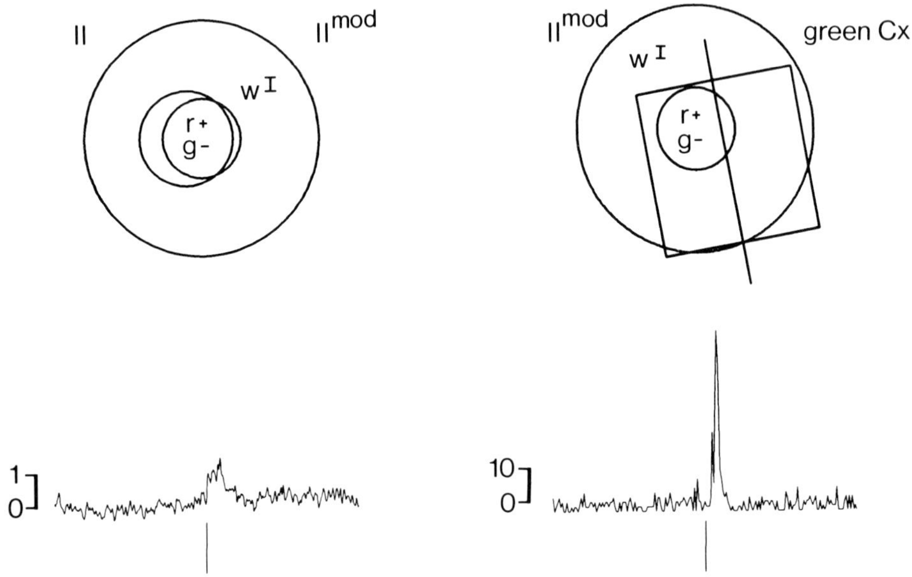

Fig. 7.18. Left: Correlogram from a red-ON green-OFF Type II cell to a modified red-ON green-OFF Type II cell. The correlogram indicates an excitatory monosynaptic connection from the Type II to the modified Type II cell. Correlation window: ±50ms, $N_a = 8988$, $N_b = 28610$, Contribution 1.3%, Effectiveness 4%. Right: Correlogram from a red-ON green-OFF modified Type II cell to a oriented, complex cell that preferred a green moving bar or edge stimulus. The correlogram indicates an excitatory monosynaptic connection from the modified Type II to the green complex cell. The correlogram is shown with a narrower correlation window (±5ms) and smaller scaling factor to display the peak in better detail, $N_a = 1092$, $N_b = 2080$, contribution 11.7%, effectiveness 6.2%

neither a neighboring red/green blob with the same eye dominance as the injection site, nor a neighboring blue/yellow blob with the opposite eye dominance were labeled. No interblob regions were labeled either. These results support the notion that blob-blob connections are segregated by color opponency. Thus, we found a strong correspondence between the specificity of blob–blob connections seen physiologically with the cross-correlation technique, and the pattern of blob–blob connections seen anatomically, via HRP tracing.

7.4 Conclusions

Our studies, using cross-correlation analysis, have revealed specific functional connections that extend horizontally among oriented cells and color-selective cells. These connections are likely to correspond to the anatomically demonstrated clustered horizontal connections. The segregation of neuronal connections in a "like talks to like" fashion was a recurring finding, and is apparently a common feature of cortical circuitry and cortical processing strategies. It is possible that various classes of horizontal connections subserve different functions. We suggest that closer range connections, including some of those found between blob cells and those found between oriented cells, may be involved in the building of new receptive field types and properties, such as the sharpening of orientation tuning and the properties of the modified Type II and color-selective oriented cells. On the other hand, the long-range horizontal connections between cells of like functional properties may contribute to subthreshold effects or other receptive field properties not revealed by classical receptive field analysis (Allman et al. 1985). These properties may require visual information integrated over a large area of the visual field and thus allow a cell's responses to be influenced by context.

References

Aertsen, A., Gerstein, G.L. (1985): Evaluation of neuronal connectivity: sensitivity of cross-correlation. Brain Res. **340**, 341–354

Allman, J., Miezin, F., McGuinness, E. (1985): Stimulus specific responses from beyond the classical receptive field: Neurophysiological mechanisms for local-global comparisons in visual neurons. Ann. Rev. Neurosci. **8**, 407–430

Blakemore, C., Tobin, E. (1972): Lateral inhibition between orientation detectors in the cat's visual cortex. Exp. Brain Res. **15**, 439–440

Blasdel, G.G., Lund, J.S. (1983): Termination of afferent axons in macaque striate cortex. J. Neurosci. **3**, 1389–1413

Bryant, H.L. Jr., Ruiz Marcos, A., Segundo, J.P. (1972): Correlations of neuronal spike discharges produced by monosynaptic connections and by common inputs. J. Neurophysiol. **35**, 864–878

Creutzfeldt, O.D., Garey, L.J., Kuroda, R., Wolff, J.R. (1977): The distribution of degenerating axons after small lesions in the intact and isolated visual cortex of the cat. Exp. Brain Res. **27**, 419–440

Daw, N.W. (1968): Colour-coded ganglion cells in the goldfish retina: extension of their receptive fields by means of new stimuli. J. Physiol. **197**, 567–592

Dow, B.M., Vautin, R.G. (1987): Horizontal segregation of color information in the middle layers of foveal striate cortex. J. Neurophysiol. **57**, 712–739

Ferster, D., LeVay, S. (1978): The axonal arborization of lateral geniculate neurons in the striate cortex of the cat. J. Comp. Neurol. **182**, 923–944

Fetz, E.E., Gustafsson, B. (1983): Relation between shapes of post-synaptic potentials and changes in firing probability of cat motoneurones. J. Physiol. **341**, 387–410

Fisken, R.A., Garey, L.J., Powell, T.P.S. (1975): The intrinsic, association and commissural connections of area 17 of the visual cortex. Phil. Trans. Roy. Soc. **272**, 487–536

Fitzpatrick, D., Itoh, K., Diamond, I.T. (1983): The laminar organization of the lateral geniculate body and the striate cortex in the squirrel monkey (*Saimiri sciureus*). J. Neurosci. **3**, 673–702

Fitzpatrick, D., Lund, J.S., Blasdel, G.G. (1985): Intrinsic connections of Macaque striate cortex: afferent and efferent connections of layer 4C. J. Neurosci. **5**, 3329–3349

Gilbert, C.D., Kelly, J.P. (1975): The projections of cells in different layers of the cat's visual cortex. J. Comp. Neurol. **163**, 81–106

Gilbert, C.D., Wiesel, T.N. (1979): Morphology and intracortical projections of functionally identified neurons in cat visual cortex. Nature **280**, 120–125

Gilbert, C.D., Wiesel, T.N. (1981): Laminar specialization and intracortical connections in cat primary visual cortex. In *The Organization of the Cerebral Cortex*, ed. F.O. Schmitt, F.G. Worden, G.A. Adelman, S.G. Dennis (MIT, Cambridge, Massachusetts) 163–191

Gilbert, C.D., Wiesel, T.N. (1983): Clustered intrinsic connections in cat visual cortex. J. Neurosci. **3**, 1116–1133

Gilbert, C.D., Wiesel, T.N. (1989): Columnar specificity of intrinsic horizontal and corticocortical connections in cat visual cortex. J. Neurosci. **9**, 2432–2442

Horton, J.C., Hubel, D.H. (1981): A regular patchy distribution of cytochrome-oxidase staining in primary visual cortex of the macaque monkey. Nature **292**, 762–764

Hubel, D.H., Wiesel, T.N. (1962): Receptive fields, binocular interaction and functional architecture in the cat's visual cortex. J. Physiol. **160**, 106–154

Hubel, D.H., Wiesel, T.N. (1972): Laminar and columnar distribution of geniculo-cortical fibers in the macaque monkey. J. Comp. Neurol. **146**, 421–450

Hubel, D.H., Wiesel, T.N. (1974): Uniformity of monkey striate cortex. A parallel relationship between field size, scatter, and magnification factor. J. Comp. Neurol. **158**, 295–306

Kirkwood, P.A. (1979): On the use and interpretation of cross-correlation measurements in the mammalian central nervous system. J. Neurosci. Meth. 1, 107–132

Levick, W.R., Cleland, B.G., Dubin, M.W. (1972): Lateral geniculate neurons of cat: Retinal inputs and physiology. Invest. Ophthalmol. **11**, 302–311

Livingstone, M.S., Hubel, D.H. (1984a): Anatomy and physiology of a color system in the primate visual cortex. J. Neurosci. **4**, 309–356

Livingstone, M.S., Hubel, D.H. (1984b): Specificity of intrinsic connections in primate primary visual cortex. J. Neurosci. **4**, 2830–2835

Lorente de No, R. (1933): Studies on the structure of the cerebral cortex. J. Psych. u. Neur. **45**, 382–438

Lund, J.S., Boothe, R.G. (1975): Interlaminar connections and pyramidal neuron organization of the visual cortex, area 17, of the Macaque monkey. J. Comp. Neurol. **159**, 305–344

McGuire, B.A., Gilbert, C.D., Wiesel, T.N. (1985): Ultrastructural characterization of long-range clustered horizontal connections in monkey striate cortex. Soc. Neurosci. Abst. **11**, 11.

Michael, C.R. (1986): Functional and morphological identification of double and single opponent color cells in layer IVCb of the monkey's striate cortex. Soc. Neurosci. Abst. **12**, 1497.

Michalski, A., Gerstein, G.L., Czarkowska, J., Tarnecki, R. (1983): Interactions between cat striate cortex neurons. Exp. Brain Res. **51**, 97–107

Mitchison, G., Crick, F. (1982): Long axons within the striate cortex: their distribution, orientation and patterns of connection. Proc. Nat. Acad. Sci. USA **79**, 3661–3665

Moore, G.P., Perkel, D.H., Segundo, J.P. (1970): Statistical signs of synaptic interaction in neurons. Biophys. J. **10**, 876–900

Nelson, J.I., Frost, B. (1978): Orientation selective inhibition from beyond the classic visual receptive field. Brain Res. **139**, 359–365

Perkel, D. H., Gerstein, G. L., Moore, G.P. (1967a): Neuronal spike trains and stochastic point processes. I. The single spike train. Biophys. J. **7**, 391–418

Perkel, D.H., Gerstein, G.L., Moore, G.P. (1967b): Neuronal spike trains and stochastic point processes. II. Simultaneous spike trains. Biophys. J. **7**, 419–440

Rockland, K.S., Lund, J.S. (1982): Widespread periodic intrinsic connections in the tree shrew visual cortex. Science **215**, 1532–1534

Somogyi, P., Kisvarday, Z.F., Martin, K.A.C., Whitteridge, D. (1983): Synaptic connections of morphologically identified and physiologically characterized large basket cells in the striate cortex of cat. Neurosci. **10**, 261–294

Surmeier, D.J., Weinberg, R.J. (1985): The relationship between cross-correlation measures and underlying synaptic events. Brain Res. **331**, 180–184

Toyama, K., Kimura, M., Tanaka, K. (1981a): Cross-correlation analysis of interneuronal connectivity in cat visual cortex. J. Neurophysiol. **46**, 191–201

Toyama, K., Kimura, M., Tanaka, K. (1981b): Organization of cat visual cortex as investigated by cross-correlation techniques. J. Neurophys. **46**, 202–214

Ts'o, D.Y., Gilbert, C.D., Wiesel, T.N. (1986): Relationships between horizontal interactions and functional architecture in cat striate cortex as revealed by cross-correlation analysis. J. Neurosci. **6**, 1160–1170

Ts'o, D.Y., Gilbert, C.D. (1988): The organization of chromatic and spatial interactions in the primate striate cortex. J. Neurosci. **8**, 1712–1727

Wiesel, T.N., Hubel, D.H. (1966): Spatial and chromatic interactions in the lateral geniculate body of the rhesus monkey. J. Neurophysiol. **29**, 1115–1156

Wong-Riley, M.T.T. (1979a): Columnar cortico-cortical interconnections within the visual system of the squirrel and macaque monkeys. Brain Research **162**, 201–217

Wong-Riley, M.T.T. (1979b): Changes in the visual system of monocularly sutured or enucleated cats demonstrable with cytochrome oxidase histochemistry. Brain Research **171**, 11–28

8. The Formation of Cooperative Cell Assemblies in the Visual Cortex

W. Singer

Max-Planck-Institute for Brain Research, Deutschordenstrasse 46, D-6000 Frankfurt a.M. 71

The retinal image of a visual scene consists of a two-dimensional continuous distribution of grey levels. In order to identify particular figures or objects it needs to be determined which of the local luminance gradients result from particular objects and which are generated from the embedding background. Some grouping must be performed in order to associate these luminance distributions with the contours of a single object, to segregate signals from objects with overlapping contours from each other and from the signals generated by the background. These operations are commonly addressed as scene segmentation or figure-ground segregation. Because most of them are usually carried out subconsciously and do not require directing selective attention to particular features of the scene, these operations are called "preattentive visual processes" or "early visual processes" (for reviews and examples see Julesz 1971; Marr 1976; Treisman 1986; Ramachandran 1988).

8.1 Coherence of Features as Criterion for Scene Segmentation

The first step in scene segmentation consists of the extraction of certain features from the retinal image. Local luminance gradients are categorized according to their contrast, orientation, length and spatial frequency content, and when the gradients are not stationary, according to direction and speed of motion. In animals endowed with color vision, the spectral composition of the images is added as a further feature dimension, and animals capable of stereopsis use the interocular disparity of contrast borders to evaluate their relative distance in space.

The next step apparently consists of a grouping operation whereby relationships are established between these spatially distributed features. The principle criteria for grouping seem to be firstly, spatial contiguity and secondly, coherence in particular feature domains. The visual system interprets luminance gradients as originating from the same object or figure if they are closely spaced, and especially if they are continuous. The same is true for spatially separate luminance gradients if they share similarities within one of the feature domains

Springer Series in Synergetics
J. Krüger (Ed.)
Neuronal Cooperativity

listed above. In the simplest case, objects are distinguished from other objects and background if they have a continuous outline. When an object is partly obscured by another object, its spatially separate luminance gradients may still be interpreted as coming from one-and-the-same object if they are sufficiently similar within one or more feature domains. Thus, if the spectral composition of the parts of the object or their retinal disparity (distance from the observer) are the same, they will still be interpreted as belonging to a single object.

Colinearity is a particularly powerful criterion according to which interrupted contrast borders are interpreted as belonging to the same object. Discontinuous but colinear contrast borders tend to be interpreted as the interrupted outline of the same figure rather than as independent borders.

Another very powerful criterion is coherence of motion. Spatially distributed contrast borders are interpreted as belonging to the same object if they move with the same speed in the same direction.

The following thought experiment exemplifies this. If one produces on a television display a cloud of dots of identical size and luminance which move with the same speed but in randomized directions, and then suddenly makes those dots which happen to be located on the thought outlines of a triangle move in the same direction, one perceives these dots as the outline of a moving shape, in this case the triangle. Apparently, the visual system interprets those dots which share a common feature, in this case the same movement vector, as belonging together, and segregates the assembly of these coherently moving dots from the cloud of randomly moving dots. The former are interpreted as components of a figure, and the latter as components of the background.

These psychophysical data predict that the visual system is capable of establishing relations between the responses of spatially distributed feature detectors, and of identifying the responses to those features which have some coherent properties. An important constraint of this evaluation process is that topological information must be strictly preserved. After the responses resulting from figures are isolated, the figures need to be identified, which requires that the spatial relations of their contours are accessible. A mechanism is needed, therefore, which allows one to establish relations between spatially distributed feature detectors without losing positional information.

8.1.1 Representation of Coherent Features by Assemblies of Coherently Active Neurones

One solution for the detection of coherent features in a scene is to let feature detectors interact with one another and to provide a mode of interaction such that responses of detectors encoding coherent features become distinguishable. This can be achieved by selectively coupling neurones with *similar* feature preferences through reciprocal excitatory connections.

The effect of such connections is that they selectively enhance the responses of feature detectors which respond to similar features in a scene. Thus, if an image contains a figure that is distinguished by the presence of similar features, the corresponding feature detectors in visual cortex would become activated, and, due to mutual facilitation, their responses would be enhanced and become stronger and perhaps also more sustained than responses to non-related elements of the image. Neurones responding to coherent features would then become distinguishable from neurones which encode non-coherent features because their responses are enhanced.

However, if cell assemblies encoding coherent properties of the image are solely distinguished by the increased amplitudes of their responses, some problems arise when the scene contains more than one coherent figure or when the background has itself some coherent properties such as a regular texture. In this case it would become impossible to identify which of the many neurones with enhanced responses actually belong to the assemblies coding for the various figures and for the background. Individual figures would no longer be distinguishable.

This problem has been addressed as the "superposition problem" by v.d. Malsburg and Schneider (1986). These authors have postulated that assemblies of neurones coding for some global and coherent properties of figures ought to be defined by temporal coherence of the responses of their constituting elements, rather than by enhanced amplitudes (see also Crick 1984). They have proposed that cell assemblies coding for different figures should be activated in alternation. Recent simulation experiments have confirmed that this approach can, in principle, solve the superposition problem (v.d. Malsburg and Bienenstock 1986; Bienenstock and v.d. Malsburg 1987).

8.1.2 Constraints on Connectivity

The postulate that feature detectors with particular preferences need to interact selectively with each other implies an extremely complex network of connections between spatially distributed neurones in the visual cortex. In order to extract the property of colinearity, for example, selective connections must be implemented between neurones which have the same orientation preference and whose receptive fields are aligned colinearly. Because of the columnar organization of striate cortex, such neurones are distributed in spatially separate columns. Thus, it is not possible to simply connect nearest neighbors. Since the arrangement of iso-orientation domains in the striate cortex is not very regular, and in some species even shows marked anisotropies, it is also not possible to define a simple rule for these selective connections, such as that only neurones which are a fixed distance apart should be interconnected (Fig. 8.1). The same holds true for connections between neurones which code for the same direction of motion or for the same color, etc. This raises the problem of how such selective connections can develop.

One way to establish specific connections is selection according to functional criteria. This is particularly attractive in the present case since the feature detectors which must be selectively coupled can be identified with high probability on the basis of their responses. If, for example, an object moves across the visual

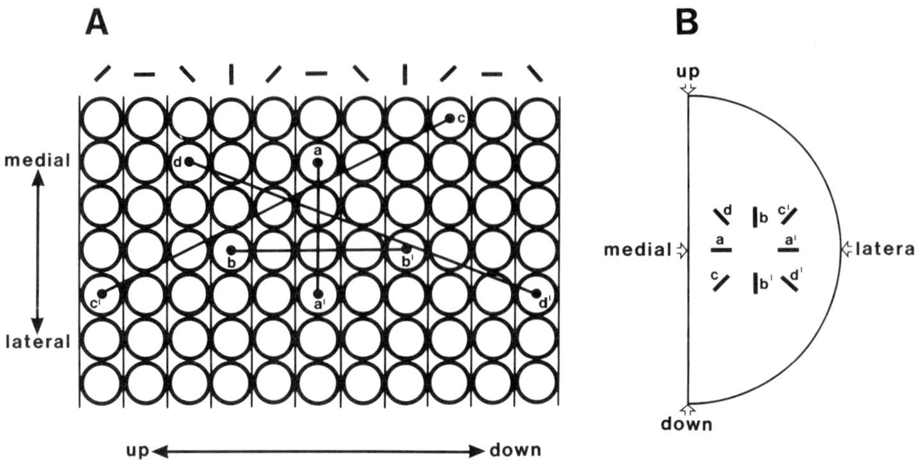

Fig. 8.1. Schematic diagram showing the representation of retinal coordinates and isoorientation domains in striate cortex. **A** Top view of the cortical sheet. Isoorientation domains are assumed to be arranged in parallel stripes which is an idealization of the conditions found in the cat visual cortex. **B** Representation of the contralateral visual field which is mapped onto the cortical sheet in **A**. The oriented contours a to d are assumed to activate the corresponding feature detectors at the appropriate locations, as indicated in **A**. The connections in **A** link cell clusters which are activated by contours which have the same orientation and are aligned colinearly, as indicated in **B**. Note that each of the combinations requires trajectories of connections of different length and direction. From (Singer, 1990)

field, movement detectors with similar preferences for speed and direction will become activated simultaneously by the moving contours of the object. The same holds true for the assembly of orientation detectors whose receptive fields are aligned colinearly if a straight contrast border is present, or for detectors of a particular disparity if contours are present at a certain distance. It would thus be sufficient to provide initially redundant sets of connections which link in an unselective way the various subpopulations of feature detectors, and then to selectively stabilize connections between neurones which are often activated simultaneously. Such a process would be advantageous for two reasons. Firstly, it would greatly economize on the genetic instructions. Secondly, it would ensure that selective coupling is established preferentially between feature detectors whose combined properties are matched by actual and frequent feature constellations present in the physical environment.

In the following paragraphs experimental evidence will be reviewed that bears on these postulates. Developmental results will be summarized first, which demonstrate that there are mechanisms which allow for activity-dependent selective stabilization of connections in the striate cortex. The development of binocular connections will be used as a model to illustrate the principles, and then it will be examined to which extent these selection processes can be generalized to the development of connections between spatially distributed feature detectors. Subsequently, data will be reviewed from adult animals which indicate that the responses of feature detectors in striate cortex have a distinct temporal structure. It will further be demonstrated that spatially distributed feature detectors with related response properties interact selectively with each other and engage in coherent activity if patterns are presented which contain the corresponding coherent features.

8.2 The Development of Binocular Correspondence as an Example for Experience Dependent Selection of Neuronal Connections

Higher mammals and humans, who have frontally positioned eyes with overlapping visual fields, can compute the distance of objects from the differences between the images in the two eyes. The basis for this function are neurones in the visual cortex which possess two receptive fields, one in each eye, that are tuned to the same features and whose relative positions on the two retinae are precisely defined. Thus, during development the afferents arriving from each eye must be arranged in a highly selective way so that only those pairs of afferents which originate from retinal loci with similar interocular disparities converge onto individual cortical target cells. In principle this requires the establishment of a precise match between two receptor surfaces; i.e., both the neighborhood relations of ganglion cells within the same retina and their relative interocular disparities have to be represented in a single map at the level of striate cortex. In analogy to other developmental processes in the brain, one might consider specific recognition molecules as a solution for the specification problem. However, in this case there is a principle limit to the degree of selectivity that can be achieved with chemical markers: there is no way to predict with any great precision which retinal loci will actually be corresponding in the mature visual system. Retinal correspondence depends on parameters such as the size of the eyeballs, the position of the eyeballs in the orbit and the interocular distance. Clearly, these parameters are strongly influenced by epigenetic factors. Moreover, they change while the skull is growing. In principle, it follows that positional markers alone, even if they were quantitatively sufficient, cannot suffice to determine with the required precision the pattern of interocular connections. An elegant possibility exists, however, to identify fibers as coming from retinal loci of similar disparity by evaluating their responses to contours. When a two-dimensional pattern is fixated with both eyes, its contours stimulate corresponding retinal loci. When the target is a three-dimensional object, its outlines also stimulate retinal loci with different disparities, whereby any point of the object produces responses at sites which have similar disparities. Thus, responses of afferents from retinal loci with similar disparity are elicited by the same parts of the object and, therefore, are likely to be more similar than those of afferents coming from non-corresponding loci. What is required, then, is a developmental mechanism capable of selecting those retinal afferents for convergence on common cortical target cells which convey correlated activation patterns. The mechanism postulated here is thus very similar to the one that would be required for the selective association of feature detectors.

As Wiesel and Hubel have demonstrated in their pioneering studies (Wiesel and Hubel 1963, 1965a,b; Hubel and Wiesel 1965), the connections between the two eyes and the cells in the striate cortex are malleable during early postnatal development and subject to use-dependent modifications. The crucial role of retinal signals in these developmental processes becomes particularly obvious when visual experience is manipulated. For example, if one eye is deprived of contour vision while the other is allowed to view normally, the large majority of cortical cells rapidly lose the ability to respond to the deprived eye (Wiesel and Hubel 1965a; for review see Frégnac and Imbert 1984).

An important step in the analysis of the mechanisms underlying these use-dependent modifications of ocular dominance has been the discovery that the deprivation-induced changes in ocularity do not solely depend on the level of activity in the afferents from the two eyes (Cynader and Mitchell 1977; Singer et al. 1977; Wilson et al. 1977; Greuel et al. 1987). Other critical variables include the state of activation of the postsynaptic neurone and, in particular, the degree of temporal correlation between pre- and post-synaptic activation (Rauschecker and Singer 1979; Rauschecker and Singer 1981; Frégnac et al. 1988; Greuel et al. 1988). The use-dependent modifications of excitatory transmission actually seem to follow rules which closely resemble those postulated by Hebb (1949) and Stent (1973) for adaptive neuronal connections (Rauschecker and Singer 1981). As summarized in Fig. 8.2, the direction of the change – increase or decrease of efficacy – depends on the correlation between pre- and post-synaptic activation. Connections stabilize if the probability is high that the presynaptic afferents and the postsynaptic cell are active in temporal contiguity and they destabilize when the postsynaptic target is strongly activated while the presynaptic terminal is silent. These rules, when applied to circuits where two or more afferent pathways converge onto a common postsynaptic target cell, have the effect of selectively stabilizing and hence associating pathways that convey correlated activity. For such afferents, the probability is high that they are active in temporal contiguity with the common target cell and therefore they consolidate. Likewise, these modification rules lead to competition between converging pathways if they convey uncorrelated activity. In that case one subset of afferents is always inactive while the other is driving the postsynaptic cell, and vice versa. Hence the active subset increases its gain while the other weakens, and these conditions alternate. Eventually, the subset of afferents which has the highest probability of being active in temporal contiguity with the postsynaptic target cell will 'win'.

Thus, according to these modification rules the converging pathways arriving from the two eyes will only remain connected to a common cortical cell if their activities are sufficiently correlated. Direct proof for this has been provided by Stryker and Harris (1986) who stimulated the optic nerves electrically and showed that only synchronous activation led to maintenance of binocular connections. If the responses from the two eyes are too asynchronous, the corresponding afferents compete with one another. In that case only one pathway consolidates and the other becomes repressed. The maximal interval of asynchrony still compatible with the maintenance of binocular connections was found to be in the order of 200 to 400 ms (Altmann et al. 1987).

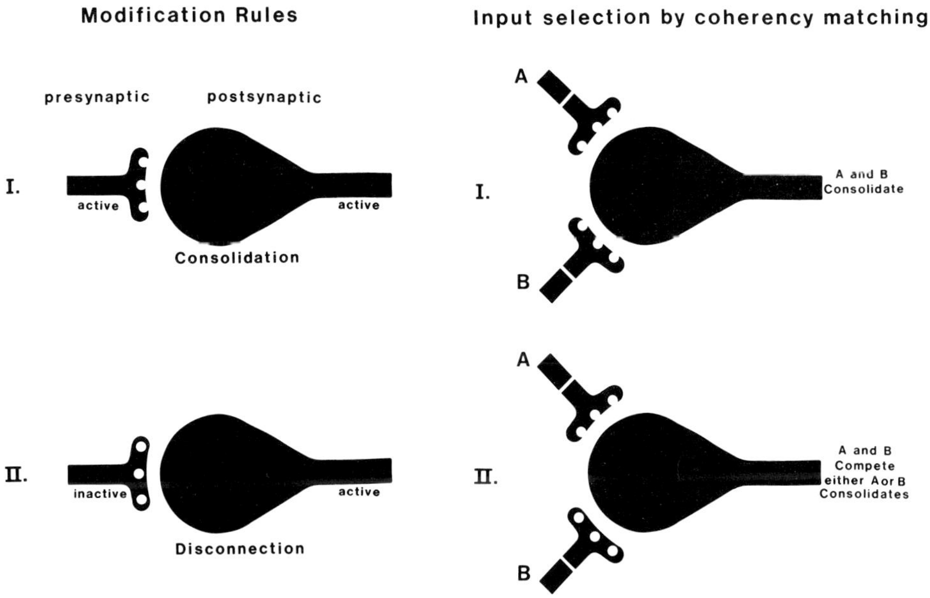

Fig. 8.2. Schematic representation of the rules which describe activity-dependent modifications of synaptic connections in the developing visual cortex. As the rules in the left column indicate, connections consolidate if the probability is high that pre- and post-synaptic elements are active in temporal contiguity (rule 1), while connections destabilize if the probability is high that the presynaptic terminal is inactive at the same time as the postsynaptic target is activated (rule 2). When applied to conditions where two inputs converge on the same target (right column) these local rules lead to selective stabilization of converging inputs that convey correlated activity (condition I) while they lead to competition between converging inputs if these convey non-correlated activity (condition II). In this latter case one input will consolidate at the expense of the other. From (Singer, 1990)

8.3 The Control of Local Circuit Modifications by Global Gating Systems

The data reviewed above indicate that there are developmental mechanisms which are, in principle, capable of selectively stabilizing pathways conveying coherent activity. However, this selection can only be successful in optimizing binocular connectivity if it is restricted to moments when the animal actually fixates a non-ambiguous target with both eyes. Pruning must not take place when the visual axes of the two eyes are not properly aligned. If the images on the two retinae are too different and cannot be fused, all signals from the two eyes, including those originating from retinal loci with similar disparity, are uncorrelated. If selection occurred under such conditions, all afferents from the two eyes would compete with one another and the consequence would be complete disruption of binocular connections. The same would be the case if the spontaneously produced bursts of activity that occur for example in the geniculate afferents during certain sleep stages were capable of inducing changes in circuitry. Moreover, prior to selection, the two eyes must be aligned to assure an optimal match between the images in the two eyes. Thus, the direction of gaze must be adjusted until the activity patterns arriving from the two eyes are maximally correlated within the coarsely prespecified retinotopic representation. Only once these adjustments have been made can selection occur. In order to achieve its goal, selection must therefore be gated by non-retinal control systems which enable use-dependent modifications only when conditions are appropriate.

In agreement with this postulate it has been found that a number of non-retinal afferents to the striate cortex play a crucial role in gating ocular dominance plasticity. If these projections are inactivated or destroyed, retinal signals are no longer capable of inducing changes in binocular connectivity. The following projections have been identified so far as having a permissive role in ocular dominance plasticity: the proprioceptive afferents from extraocular muscles (Buisseret and Singer 1983), the noradrenergic afferents from locus coeruleus (Kasamatsu and Pettigrew 1979; Bear and Singer 1986) and the cholinergic projection from the basal forebrain (Bear and Singer 1986). Results of lesion studies (Singer 1982) have further suggested that retinal signals only influence the development of cortical functions when the animal uses them for the control of behavior. Correspondingly, retinal signals do not lead to changes of cortical functions when the kittens are paralyzed and/or anaesthetized while exposed to visual patterns (Buisseret et al. 1978; Freeman and Bonds 1979; Singer 1979; Singer and Rauschecker 1982).

8.4 Involvement of a Molecular Coincidence Detector in Use-Dependent Plasticity

The results reported above suggest that use-dependent modifications of synaptic transmission require a certain amount of cooperativity between retinal input and internally generated gating signals. This is supported by a number of studies which indicate that the process mediating competitive disconnection has a threshold which is reached only when retinal signals are coincident with additional facilitatory input (Singer and Rauschecker 1982; Greuel et al. 1988; Frégnac et al. 1988). Recently, a molecular mechanism could be identified that is likely to be responsible for this threshold. It was found that the activation of the N-methyl-D-aspartate (NMDA) receptor is a necessary prerequisite for the activity-dependent disconnection of the deprived pathway (Kleinschmidt et al. 1987; Gu et al. 1989).

A special property of this receptor-gated channel is that it is also voltage dependent. It becomes permeable for Ca^{++} ions only when the receptor is occupied by the synaptic transmitter *and* when the membrane of the postsynaptic neurone is sufficiently depolarized (Mayer et al. 1984). Thus, this synaptic mechanism is ideally suited to "evaluate" contingency of pre- and post-synaptic activation. Furthermore, this mechanism accounts for the evidence that the activation threshold for the induction of long-term modifications is high and reached only if there is sufficient cooperativity between converging inputs. The fact that this channel, when activated, becomes permeable for Ca^{++} ions makes it particularly well adapted for the mediation of changes in synaptic transmission, because Ca^{++} ions serve as second messengers for the initiation of a variety of biochemical processes in nerve cells. Another interesting aspect is that the NMDA-receptor mechanism is also involved in use-dependent synaptic plasticity in cortical structures of the mature nervous system such as the hippocampus (Collingridge and Bliss 1987) and the visual cortex (Artola and Singer 1987). Together with recent data that are reviewed elsewhere (Singer 1989), this suggests that the use-dependent modifications of synaptic transmission which presumably mediate learning in the adult brain depend on similar mechanisms as the experience-dependent self-organization of neuronal connectivity during development.

8.5 Cooperativity Between Distributed Neuronal Responses as Criterion for Local Circuit Modifications

The data reviewed thus far allow us to define some of the constraints of use-dependent modifications of neuronal connectivity. It can be inferred from the requirement that the activation threshold of the NMDA mechanism be reached in order to obtain competitive disconnections, that a critical level of postsynaptic depolarization must be reached. The probability of reaching this threshold increases to the extent that excitatory and inhibitory inputs to the neurone under consideration are activated and silenced, respectively. Since nearby cortical neurones are tightly coupled to each other through excitatory connections, this implies that those stimuli will be most effective in inducing modifications which best match the resonance properties of local arrays of excitatory cortical neurones. Since some of the excitatory input to individual cortical neurones is recruited also from remote cortical and subcortical centers, it can furthermore be expected that the efficacy of stimuli to induce modifications of cortical circuitry will increase to the extent that the stimuli match not only the response properties of local neurone clusters in the striate cortex, but also conform with the resonance properties of more distributed neuronal assemblies. Thus, part of the criteria which determine whether a particular activation pattern actually reaches the modification threshold reside in the distributed response properties of the interconnected nerve nets. The "now print" signal is, at least in part, derived from the global response properties of the processing circuits themselves. However, even if these local and global requirements for cooperativity are fulfilled, modifications may not occur unless the permissive modulatory systems –

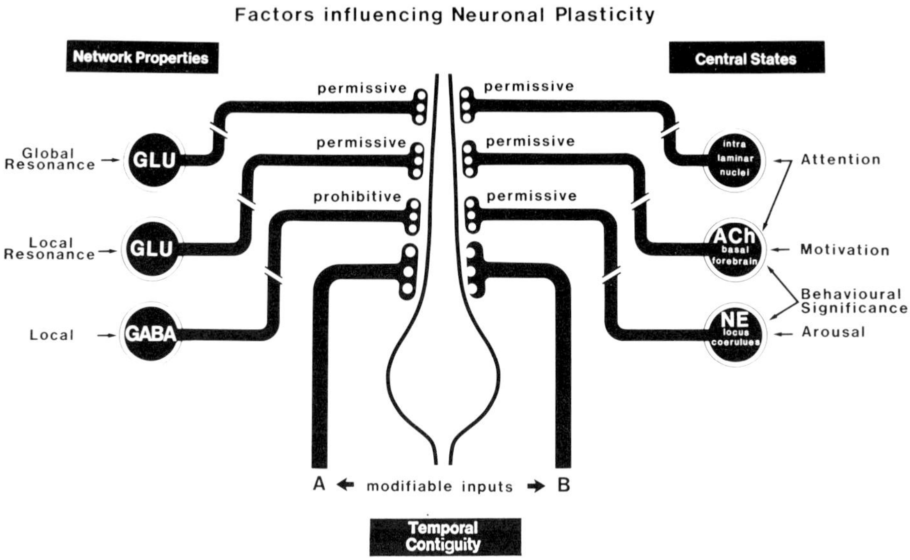

Fig. 8.3. Schematic representation of factors influencing use-dependent modifications of neuronal circuits in the developing visual cortex. The sign of the change in modifiable connections A and B depends on the temporal correlation of activity conveyed by the modifiable inputs. Whether a change occurs in response to this activity depends on the state of activation of additional synaptic inputs to the same neuron. These comprise feedback loops originating within the same cortical column, the same cortical area and probably also remote cortical and subcortical structures. Additional gating inputs are derived from the more globally organized modulatory systems which originate in central core structures of the brain. For further details, see text. From (Singer 1989)

the noradrenergic, the cholinergic and probably also other central-core projections – are in the appropriate state of activation. These systems in turn receive input from numerous other centers in the brain, both visual and non-visual. They can therefore be expected to influence local adaptive changes as a function of a very global evaluation of the adequacy of the visual signals actually processed. (For a summary see Fig. 8.3).

As reasoned above, such a multilevel control of local modifications is indispensable in our particular case of map-matching where fixed relations must be established between the neuronal representations of two mobile sensory surfaces. However, a global control of plasticity by distributed gating is probably not confined to this special case of use-dependent modifications of synaptic efficacy. Because of the distributed organization of the central nervous system and the high degree of parallelism of its processes, local modifications of synaptic efficacy will always have a widespread influence on numerous other cell groups. Therefore, to ensure that local changes lead to ordered states at a global scale, these changes must be made dependent upon a global evaluation of their functional consequences.

8.6 Activity-Dependent Self-Organization at Higher Levels of Cortical Processing

The mechanisms mediating use-dependent modifications of ocular dominance possess many of the features that have been postulated above for the development of assemblies of selectively coupled feature detectors. In this paragraph it will be examined whether there is evidence for similar self-organization processes at the level of feature representation. It has been known for a long time that one of the prominent features of cortical organization is the presence of an extremely dense network of far-reaching connections which are tangential to the cortical lamination (Szentagothai 1973). Electronmicroscopic (McGuire et al. 1985, LeVay 1988) and electrophysiological (Luhmann et al. 1990b) data indicate that these connections are excitatory, originate predominantly from pyramidal cells and terminate preferentially on the apical dentrites of other pyramidal cells. These pathways are thus capable of mediating interactions between cortical neurones that are nonadjacent and located in different columns. Furthermore there are indications that these connections are selective, linking in a reciprocal way neurone clusters that tend to be spaced periodically (Rockland and Lund 1983) and that share certain functional properties such as the same orientation preference and/or the same eye dominance (T'so et al. 1986; Gilbert and Wiesel, 1989; but also see Matsubara et al. 1985).

Developmental studies in the cat have shown that these tangential connections essentially appear postnatally, pass through a phase of exuberant proliferation during which they are particularly numerous and far-reaching, and subsequently become pruned (Price and Blakemore 1985a,b). This pruning occurs at a time when visual signals are readily available and appears to be influenced by retinal activity. If visual experience is unrestricted, subpopulations of these pathways are stabilized; if vision is prevented by dark rearing or binocular deprivation, only a rudimentary network of horizontal connections is maintained (Luhmann et al. 1990a).

The anatomical indications that excitatory tangential connections are initially exuberant and imprecise, and assume their selectivity through pruning, is supported by physiological data. In kittens the receptive fields of individual cortical neurones were found to be significantly larger than in adult cats. Moreover, in kittens about 20% of the cells had additional ectopic receptive fields that were excitatory, had similar orientation preferences as the dominant receptive field and could be located as far as 20° away from the center of the conventional receptive field. These ectopic fields occurred mainly in cells located in supragranular layers and, with one exception, were not found in adult cats (Luhmann et al. 1990c; see also Singer and Tretter 1976). This age-dependent decrease in receptive field size, the laminar distribution of cells having an ectopic receptive field, and the numerical reduction of such cells with age, correlate well with the organization and postnatal pruning of tangential projections, suggesting a causal relation between ectopic receptive fields and exuberant tangential connections.

The development of the network of intrinsic tangential connections thus resembles in a number of aspects that of the connections between the eyes and their target structures in the visual cortex. Both continue to develop postnatally and achieve topological selectivity through a pruning process. For the connections from the eyes to the visual cortex, there is direct electrophysiological evidence that this pruning is guided by visual experience (see above). For the intracortical connections this proof is still lacking, but there is indirect evidence that this may also be the case.

Support for this possibility comes from experiments with kittens whose visual experience had been restricted to vertically oriented gratings of constant spatial frequency. Subsequent electrophysiological analysis revealed that a large majority of cortical neurones had developed a preference for vertical contours. Moreover, a substantial fraction of neurones in supragranular layers had, in addition to the conventional receptive field, one or two ectopic fields. These had also a preference for vertically oriented contours, and the angular separation of these ectopic fields from the conventional field corresponded to the spatial frequency of the experienced grating. The angular separation of the stripes of the grating was 10°. This excludes that the ectopic excitatory input was mediated by far-reaching axon collaterals of thalamic afferents because these do not span such long distances (Humphrey et al. 1985a,b; Freund et al. 1985).The fact that neurones with such unconventional receptive fields were located preferentially in supragranular layers, where tangential connections are particularly far-reaching and numerous, is compatible with the interpretation that the ectopic receptive fields are generated by tangential intracortical connections.

The striking match between the spacing of the ectopic fields and the spacing of the bars of the grating suggests, furthermore, that the connections responsible for the generation of the ectopic fields have been selected as a function of experience. The effect of viewing a vertical grating is that those cortical columns which prefer vertically oriented contours and are spaced at a distance that corresponds to the angular separation of the stripes of the grating become coherently and selectively activated. Connections between such columns thus have a high probability of being activated in temporal contiguity with their respective target cells. According to the rules that have been identified for the pruning of thalamocortical connections, this should lead to selective stabilization of these connections, and hence to enhanced coupling of these columns. If this coupling is strong enough to produce neuronal responses – and in the kitten this seems to be the case (see also Luhmann et al. 1990c) – ectopic fields with the reported features are to be expected. The same argument applies in the case in which the ectopic fields are not mediated by intra-areal tangential connections but by reciprocal connections with cells in other visual areas.

In conclusion, there is at present no direct evidence that the excitatory connections between spatially distributed feature detectors self-organize according to the same rules as the thalamocortical connections, but there are indications suggesting that this may be the case.

The result of such self-organization would be that those intracortical connections which extend between neurone clusters whose activation patterns show some statistical correlation, are selectively stabilized. For neurone clusters too remote from each other to share a common input from retino-cortical afferents, the degree of correlated activation does not depend upon particular neighborhood relations, but is determined essentially by coherences between particular features of the visual scene (see above). Selective stabilization of tangential intrinsic connections would thus generate a non-topographically organized map which matches the coherent properties of "feature constellations" in physical reality. The effect is similar to that of use-dependent pruning of binocular connections. The selective connections between the two eyes and common cortical target cells have the effect that the cortical cells become "detectors" of a particular interocular disparity. As an analogy, selective connections between distributed clusters of feature detectors with related preferences have the effect that these, as an assembly, become a detector of coherent constellations of features.

Thus, by simple iteration of the very same processes of self-organization which, at peripheral levels of the visual system increase the precision of topographic maps, it is possible to generate non-topographic maps which represent relations in feature space. Once such maps are established, they can be used to detect coherences among features of visual scenes.

It is of particular importance and should be emphasized here that such maps of "conceptual" relations in feature space can actually coexist in the same cortical area as the retinotopically organized maps which represent the topological relations between features. It is thus not necessary to sacrifice retinotopy in order to decode coherences between parts of a figure which are separated on the topographic map. As discussed above, preservation of retinotopic relations is a necessary prerequisite for a network that is capable of performing the segmentation processes required at the level of preattentive vision.

8.7 Detection and Representation of Coherent Features by Synchronously Oscillating Cell Assemblies

In this last paragraph, experimental evidence is reviewed which supports the postulate that cell assemblies should be distinguished by coherent activity of the constituting neurones in order to overcome the "superposition problem" inherent in ensemble coding. Recently we have obtained evidence which suggests that cell assemblies coding for coherent features in visual scenes may not simply be distinguished by the fact that the constituting neurones are particularly active. Rather it appears as if such cell assemblies are characterized by synchronization of oscillatory responses.

We have discovered that a large fraction of neurones in the cat striate cortex engage in oscillatory activity in a frequency range of 40 to 60 Hz when activated with light stimuli to which the neurones are tuned (Gray and Singer 1987; Gray and Singer 1989). This phenomenon is illustrated in Fig. 8.4. Units close enough to be recorded with a single electrode, if responsive to the same stimulus, always

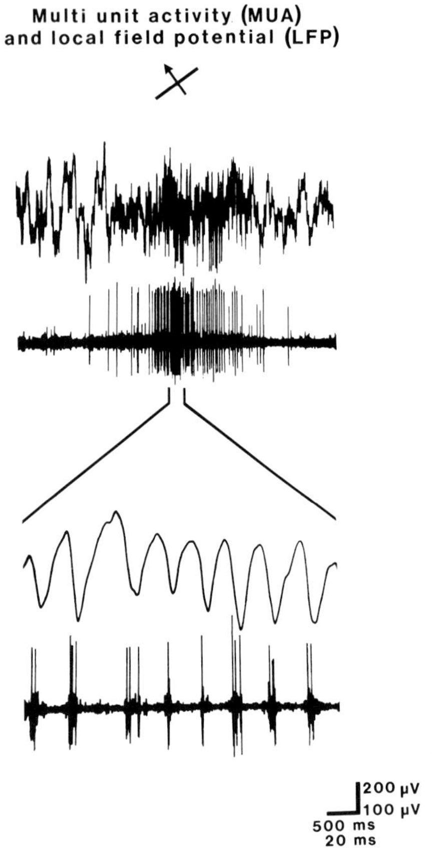

Fig. 8.4. Multiunit activity (MUA) and local field potential (LFP) responses recorded from area 17 in an adult cat to the presentation of an optimally oriented light bar moving across the receptive field. Oscilloscope records of a single trial showing the response to the preferred direction of movement. In the upper two traces, at a slow time scale, the onset of the neuronal response is associated with an increase in high-frequency activity in the LFP. The lower two traces display the activity at the peak of the response at an expanded time scale. Note the presence of rhythmic oscillations in the LFP and MUA (35–45 Hz) that are correlated in phase with the peak negativity of the LFP. Upper and lower voltage scales are for the LFP and MUA, respectively. From (Gray and Singer 1989)

synchronize their respective oscillatory responses. In most instances oscillatory responses are also in phase for neurones aligned along the vertical axis of a cortical column.

Of particular interest in the present context is the finding that the oscillatory responses can also synchronize over considerable distances across spatially separate columns (Gray et al. 1989) and even between cortical areas (Eckhorn et al. 1988). Thus far, three parameters have been identified which determine the degree of synchrony within area 17: the distance between the units, the similarity of their orientation preference and the coherence of the stimulus itself. When neurones are less than 2 mm apart, in which case their receptive fields (RFs) are usually overlapping, they always synchronize their oscillatory responses when they show the same orientation preference, and they often synchronize even if their orientation preferences differ, as long as these differences are sufficiently small to allow activation of both neurone clusters with a single stimulus. At larger distances, when the receptive fields are no longer overlapping, cell clusters tend to synchronize their oscillatory responses only when they have similar orientation preferences and/or are activated by stimuli that have the same orientation and that move in the same direction. In such cases correlation breaks down when the two stimuli pass in opposite directions over the two receptive fields, and reaches its maximum when both neurone clusters are activated with a single continuous stimulus (Gray et al. 1989).

This phenomenon is illustrated in Fig. 8.5. In each of the three stimulation conditions the auto-correlations show a periodic modulation, indicating that the local responses are oscillatory. However, when the two stimuli move in opposite directions, where they are perceived as two independent contours, the cross-correlation function is flat, implying that the respective oscillatory responses have no consistent phase relation. When the two stimuli move in the same direction, where they are perceived as related, the cross-correlogram shows a periodic modulation, indicating that the respective oscillatory responses are in phase. Synchronization improves further when the two stimuli are replaced by a single bar of light. This can be inferred from the deep modulation of the oscillatory cross correlogram in Fig. 8.5C.

Interestingly, if cross-correlograms are periodically modulated, they are always centered around zero-phase angle. This indicates that oscillations tend to synchronize in phase if they synchronize at all.

Important and still unresolved questions are how oscillatory responses are generated and how synchrony is achieved across spatially separate columns. An answer to the first question may be provided by the recent finding of R. Llinas that a subpopulation of neurones in the visual cortex is endowed with pacemaker currents which support oscillations in the 40 Hz range (Llinas, private communication). The substrate of long-range synchronization remains to be identified. Clearly, the tangential intracortical connections are good candidates but reciprocal connections between different cortical areas need to be considered too. In conclusion, synchrony in the oscillatory responses of spatially separate neurone clusters signals coherence inherent in the stimuli that give rise to the responses. Particularly good synchronization is achieved among clusters of cells activated by colinear and continuous contours. However, cell clusters also synchronize their respective responses if they are activated by spatially distributed pattern elements that share particular features, such as the same orientation or the same direction of motion.

Thus, the assembly of detectors which respond to coherent features of a scene becomes distinguished from detectors responding to non-coherent features by the fact that the oscillatory responses of the cells in the assembly coding for coherent features are synchronous and in-phase. This allows identification of responses originating from the contours of a coherent figure without losing the information on the location of the respective features. The activity of a particular neurone signals when and where a particular feature is present in the receptive field, while the phase relation of the oscillation with respect to other responses signals relations with other features in the image. Since figures or objects in a scene are defined by coherent features, such a mechanism is ideally suited for the segmentation of visual scenes into figures and background.

Expressing coherence by synchrony among oscillatory responses also resolves the superposition problem which arises when more than one figure is pres-

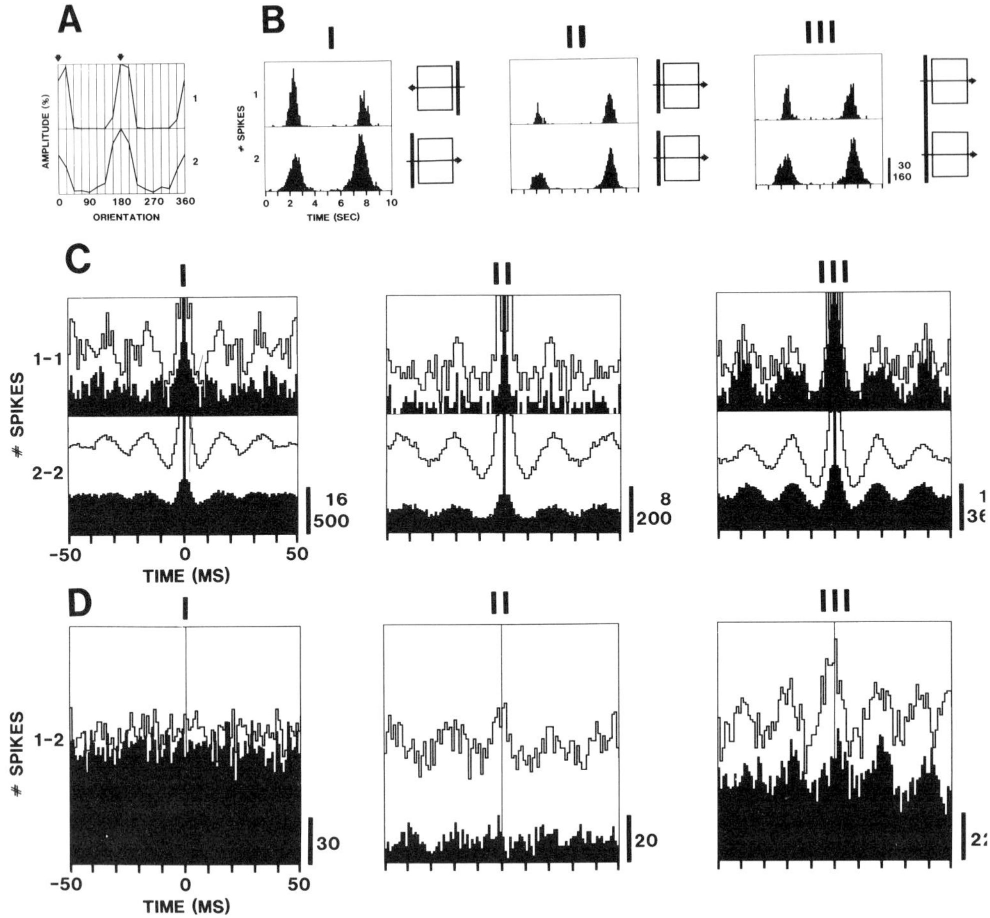

Fig. 8.5. Long-range oscillatory correlations reflect global stimulus properties. **A** Orientation tuning curves of neuronal responses recorded from two electrodes (1, 2) separated by 7 mm show a preference for vertical light bars (0 and 180°) at both recording sites. **B** Post-stimulus time histograms of the neuronal responses recorded at each site for each of three different stimulus conditions: (I) two light bars moved in opposite directions, (II) two light bars moved in the same direction and (III) one long light bar moved across both receptive fields. A schematic diagram of the receptive field locations and the stimulus configuration used is displayed to the right of each post-stimulus time histogram. **C**, **D** Auto-correlograms (**C**,1-1, 2-2) and cross correlograms (**D**,1-2) computed for the neuronal responses at both sites (1 and 2 in **A** and **B**) for each of the three stimulus conditions (I, II, III) displayed in **B**. For each pair of correlograms except the two displayed in **C** (I, 1-1) and **D** (I), the second direction of stimulus movement is shown with unfilled bars. Numbers on the vertical calibration correspond to the number of coincident events (spikes) in the respective auto- and cross-correlograms. From (Gray et al. 1989)

ent in a scene or when the background also has some coherent properties. If an assembly is defined by the synchrony of the oscillatory responses of its constituting elements, more than one coherently active assembly of cells can be represented simultaneously in the same cortical sheet without becoming confounded. It would be sufficient that the various assemblies oscillate independently of each other.

In order to segregate neuronal responses related to a particular figure from those related to other figures or to the background, a sampling process is required that is sensitive for temporal coincidence and capable of selecting those responses which are synchronous or coherent. Such a sampling device at higher processing levels could consist of an assembly of distributed cells if these are able to generate oscillatory responses and to establish fixed-phase relations with the oscillating assemblies at lower levels. The higher-order assemblies would thus serve as "detectors" of synchronously active lower-order assemblies. This suggests the attractive possibility that the basic principles of cortical organization are similar at different processing levels.

References

Altmann, L., Luhmann, H.J., Greuel, J.M., Singer, W. (1987): Functional and neuronal binocularity in kittens raised with rapidly alternating monocular occlusion. J. Neurophysiol **58**, 965–980

Artola, A., Singer, W. (1987): Long-term potentiation and NMDA receptors in rat visual cortex. Nature **330**, 649–652

Bear, M.F., Singer, W. (1986): Modulation of visual cortical plasticity by acetylcholine and noradrenaline. Nature **320**, 172–176

Bienenstock, E., von der Malsburg, C. (1987): A neural network for invariant pattern recognition. Europhys. Lett. **4**, 121–126

Buisseret, P., Singer, W. (1983) Proprioceptive signals from extraocular muscles gate experience dependent modifications of receptive fields in the kittens visual cortex. Exp. Brain Res. **51**, 443–450

Buisseret, P., Gary-Bobo, E., Imbert, M.(1978): Ocular motility and recovery of orientational properties of visual cortical neurones in dark-reared kittens. Nature **272**, 816–817

Collingridge, G.L., Bliss, T.V.P. (1987): NMDA-receptors – their role in long-term potentiation. TINS **10**, 288–293

Crick, F. (1984): Function of the thalamic reticular complex: The searchlight hypothesis. Proc. Natl. Acad. Sci. USA **81**, 4586–4590

Cynader, M., Mitchell, D.E. (1977): Monocular astigmatism effects on kitten visual cortex development. Nature **270**, 177–178

Eckhorn, R., Bauer, R. Jordan, W., Brosch, M., Kruse, W., Munk, M., Reitböck, H.J. (1988): Coherent oscillations: A mechanism for feature linking in the visual cortex? Biol. Cybern. **60**, 121–130

Freeman, R.D., Bonds, A.B. (1979): Cortical plasticity in monocularly deprived immobilized kittens depends on eye movement. Science **206**, 1093–1095

Frégnac, Y., Imbert, M.(1984): Development of neuronal selectivity in primary visual cortex of cat. Physiol. Rev. **64**, 325–434

Frégnac, Y., Shulz, D., Thorpe, S., Bienenstock, E. (1988): A cellular analogue of visual cortical plasticity. Nature **333**, 367–370

Freund, T.F., Martin, K.A.C., Whitteridge, D. (1985): Innervation of cat visual areas 17 and 18 by physiologically identified X- and Y-type thalamic afferents. I. Arborization patterns and quantitative distribution of postsynaptic elements. J. Comp. Neurol. **242**, 263–274

Gilbert, C.D., Wiesel, T.N. (1989): Columnar specificity of intrinsic horizontal and corticocortical connections in cat visual cortex. J. Neurosci. **9**, 2432–2442

Gray, C.M., Singer, W. (1987): Stimulus-dependent neuronal oscillations in the cat visual cortex area 17. IBRO Abstr. Neurosci. Lett. Suppl. **22**, 1301

Gray, C.M, Singer, W. (1989): Stimulus-specific neuronal oscillations in orientation columns of cat visual cortex. Proc. Natl. Acad. Sci. USA **86**, 1698–1702

Gray, C.M, König, P., Engel, A.K., Singer, W. (1989): Oscillatory responses in cat visual cortex exhibit inter-columnar synchronization which reflects global stimulus properties. Nature **388**, 334–337

Greuel, J.M., Luhmann, H.J., Singer, W. (1987): Evidence for a threshold in experience-dependent long-term changes of kitten visual cortex. Dev. Brain Res. **34**, 141–149

Greuel, J.M., Luhmann, H.J., Singer, W.(1988): Pharmacological induction of use-dependent receptive field modifications in the visual cortex. Science **242**, 74–77

Gu, Q., Bear, MF., Singer, W. (1989): Blockade of NMDA-receptors prevents ocularity changes in kitten visual cortex after reversed monocular deprivation. Dev. Brain Res. **47**, 281–288

Hebb, D.O. (1949): *The organization of behavior*. (Wiley, New York)

Hubel, D.H., Wiesel, T.N. (1965): Binocular interaction in striate cortex of kittens reared with artificial squint. J. Neurophysiol. **28**, 1041–1059

Humphrey, A.L., Sur, M., Uhlrich, D.J., Sherman, S.M. (1985a): Projection patterns of individual Y- and Y-cell axons from the lateral geniculate nucleus to cortical area 17 in the cat. J. Comp. Neurol. **233**, 159–189

Humphrey, A.L., Sur, M., Uhlrich, D.J., Sherman, S.M. (1985b): Termination patterns of individual X- and Y-cell axons in the visual cortex of the cat: Projections to area 18, to the 17/18 border region, and to both areas 17 and 18. J. Comp. Neurol. **233**, 190–212

Julesz, B. (1971): *Foundations of cyclopean perception*. (University of Chicago Press, Chicago)

Kasamatsu, T., Pettigrew, J.D. (1979): Preservation of binocularity after monocular deprivation in the striate cortex of kittens treated with 6-hydroxydopamine. J. Comp. Neurol. **185**, 139–162

Kleinschmidt, A., Bear, M.F., Singer, W. (1987): Blockade of "NMDA" receptors disrupts experience-dependent plasticity of kitten striate cortex. Science **238**, 355–358

LeVay, S. (1988): Patchy intrinsic projections in visual cortex, area 18 of the cat: morphological and immunocytochemical evidence for an excitatory function. J. Comp. Neurol. **269**, 265–274

Luhmann, H.J., Singer, W, Martinez-Millan, L. (1990a): Horizontal interactions in cat striate cortex: I. Anatomical substrate and postnatal development. Eur. J. Neurosci. **2**, 344–357

Luhmann, H.J., Greuel, J.M., Singer, W. (1990b): Horizontal interactions in cat striate cortex: II. A current source-density analysis. Eur. J. Neurosci. **2**, 358–368

Luhmann, H.J., Greuel, J.M., Singer, W. (1990c): Horizontal interactions in cat striate cortex. III. Ectopic receptive fields and transient exuberancy of tangential connections. Eur. J. Neurosci, **2**, 369–377

Marr, D. (1976): Early processing of visual information. J Phil Transactions Roy. Soc. Lond. B **275**, 483–524

Matsubara, J., Cynader, M., Swindale, N.V., Stryker, M.P. (1985): Intrinsic projections within visual cortex: Evidence for orientation-specific local connections . Proc. Natl. Acad. Sci. USA **82**, 935–939

Mayer, M.L., Westbrook, G.L., Guthrie, P.B. (1984): Voltage-dependent block by Mg^{2+} of NMDA responses in spinal cord neurones. Nature **309**, 261–263

McGuire, B.A., Gilbert, C.D., Wiesel, T.N. (1985): Ultrastructural characterization of long-range clustered horizontal connections in monkey striate cortex. Soc. Neurosci. Abstr. **11**, 17

Price, D.J., Blakemore, C. (1985a): Regressive events in the postnatal development of association projections in the visual cortex. Nature **316**, 721–724

Price, D.J., Blakemore, C.(1985b): The postnatal development of the association projection from visual cortical area 17 to area 18 in the cat. J. Neurosci. **5**, 2443–2452

Ramachandran, V.S. (1988): Perception of shape from shading. Nature **331**, 163–166

Rauschecker, J.P., Singer, W. (1979): Changes in the circuitry of the kitten visual cortex are gated by postsynaptic activity. Nature **280**, 58–60

Rauschecker, J.P, Singer, W. (1981): The effects of early visual experience on the cat's visual cortex and their possible explanation by Hebb synapses. J. Physiol. (Lond.) **310**, 215–239

Rockland, K.S, Lund, J.S (1983): Intrinsic laminar lattice connections in primate visual cortex. J. Comp. Neurol. **216**, 303–318

Singer, W. (1979): Central-core control of visual cortex functions. In: *The Neurosciences*, Fourth Study Program, ed. by Schmitt, F.O. Worden, F.G. (MIT Press, Cambridge, MA) pp. 1093–1110

Singer, W. (1982): Central core control of developmental plasticity in the kitten visual cortex: I. Diencephalic lesions. Exp. Brain Res. **47**, 209–222

Singer, W. (1989): Ontogenetic self-organization and learning. In: *Brain organization and memory: Cells, systems and circuits* ed. by McGaugh, J.L., Weinberger, N.M; Lynch, G. (Oxford University Press, New York)

Singer, W. (1990): Search for coherence: a basic principle of cortical self-organization. Concepts in Neuroscience **1**, 1–26

Singer, W., Rauschecker, J.P. (1982): Central core control of developmental plasticity in the kitten visual cortex: II. Electrical activation of mesencephalic and diencephalic projections. Exp. Brain Res. **47**, 223–233

Singer, W., Tretter, F. (1976): Unusually large receptive fields in cats with restricted visual experience. Exp. Brain Res. **26**, 171–184

Singer, W., Rauschecker, J., Werth, R.(1977): The effect of monocular exposure to temporal contrasts on ocular dominance in kittens. Brain Res, **134**, 568–572

Stent, G.S. (1973): A physiological mechanism for Hebb's postulate of learning. Proc. Natl. Acad. Sci. USA **70**, 997–1001

Stryker, M.P., Harris, W.A. (1986): Binocular impulse blockade prevents the formation of ocular dominance columns in cat visual cortex. J. Neurosci. **6**, 2117–2133.

Szentagothai, J. (1973): Synaptology of the visual cortex. In *Handbook of Sensory Physiology*, ed. by Jung, R. (Springer, Berlin, Heidelberg)

Treisman, A. (1986): Properties, parts and objects. In: *Handbook of perception and human performances* ed. by Boff, K., Kaufman, L., Thomas, I. (Wiley, New York) pp. 1–70

Ts'o, D.Y., Gilbert, C.D., Wiesel, T.N. (1986): Relationship between horizontal interactions and functional architecture in cat striate cortex as revealed by cross-correlation analysis. J. Neurosci. **6**, 1160–1170

von der Malsburg, C., Bienenstock, E. (1986): A neural network for the retrieval of superimposed connection patterns. Europhys. Lett. **3**, 1243–1249

von der Malsburg, C., Schneider, W. (1986): A neural cocktail-party processor. Biol. Cybern. **54**, 29–40

Wiesel, T.N., Hubel, D.H. (1963): Single cell responses in striate cortex of kittens deprived of vision in one eye. J. Neurophysiol. **26**, 1003–1017

Wiesel, T.N., Hubel, D.H. (1965a): Comparison of the effects of unilateral and bilateral eye closure on cortical unit responses in kittens. J. Neurophysiol. **28**, 1029–1040

Wiesel, T.N., Hubel, D.H. (1965b): Extent of recovery from the effects of visual deprivation in kittens. J. Neurophysiol. **28**, 1060–1072

Wilson, J.R., Webb, S.V., Sherman, S.M.(1977): Conditions for dominance of one eye during competitive development of central connections in visually deprived cats. Brain Res. **136**, 277–287

9. Stimulus-Specific Synchronizations in the Visual Cortex: Linking of Local Features Into Global Figures?

Reinhard Eckhorn

Philipps University, Department of Biophysics, Renthof 7, D-3550 Marburg

Feature linking among multiple cortical maps in the visual system requires flexible mechanisms in order to cope with the immense variability of natural scenes. We recently discovered stimulus-specific dynamic interactions between cell assemblies in cat primary visual cortex that could function as a *global-linking mechanism* in sensory and motor systems: Stimulus-induced oscillatory activities of 30–80 Hz in remote cell assemblies in the same and different visual cortex areas mutually synchronize if the stimulus contains common features that activate the assemblies simultaneously. The results suggest that the stimulus-induced oscillations of those neural groups become synchronized, which represent the features to be linked in the current visual situation.

Our findings were facilitated by the fact that cells with similar receptive field (RF) properties are grouped in clusters (*local assemblies*) in the visual cortex. Easily recordable local mass signals, therefore, provide relevant information about activities of local assemblies. From each (individually moveable) fiber electrode of our multi-electrode array, we recorded in parallel single-cell spikes and two types of group activities; single unit activity (SUA) reveals specific RF properties, local slow-wave field potentials (LFPs) provide rough estimates of averaged local (excitatory) *mass inputs* on dendrites and somata near the recording positions and multiple unit activities (MUA) reveal average *mass output* spike activities of cell bodies close to the electrode tips. These versatile recording capabilities enable us to investigate neural signal couplings between all combinations of single cell and mass activities at various spatial recording constellations.

Synchronizations are present between all types of signals (SUA, MUA and LFPs) among assemblies in a single vertical cortex column (where similar local features are processed) and – important for our coding hypothesis – among spatially separated assemblies of the same cortical area and even between assemblies in two different areas, provided the assemblies share at least one visual feature that they process locally. Large moving patterns applied binocularly were particularly potent in evoking high degrees of synchronity. Coherence of stimulus-induced oscillations can vary in a single recording pair from well below 0.1 without stimulation to above 0.6 during stimulation. Synchronization between single-cell spike trains and oscillatory mass activity (LFPs or MUA) in

Springer Series in Synergetics
J. Krüger (Ed.)
Neuronal Cooperativity

spatially separated positions depends on the specific visual activation of the single cells.

Based on our neurophysiological findings, we simulated temporally coded feature-linking in various networks of model neurons. The networks contain either one or two neural layers ("cortex maps") with modulatory lateral and feedback *linking connections*. The performance of the model is demonstrated in examples of region linking, where the rhythmic activities of local assemblies, induced by spatio-temporally varying stimulus regions, become phase-locked.

In order to bring "neural feature linking" into correspondence with perceptual capabilities, we introduce the concept of the *linking field* of a local assembly of visual neurons. *Cortical linking features*, in this concept, are considered to be the neural representations of the corresponding perceptual linking features.

9.1 Search for Concepts of Global Cortical Processing

9.1.1 Processing of Visual Information in a Distributed System

From everyday visual experience we know that we have parallel and simultaneous access to the various visual features of an object: they are integrated into a unique and coherent percept. We know, on the other hand, that the different local features are processed separately in various cortical areas. The distributed cortical processing of local features has been extensively studied by analyzing single-cell receptive fields (RFs) and tuning curves in a variety of visual cortex areas. Candidate neural integration mechanisms for such distributed sensory representations have, however, not been found in neurophysiological experiments until recently (Eckhorn et al. 1988a–c; Singer et al. 1988).

For many years evidence for distributed processing has been collected by several groups, mainly from anatomical data. The appearance of intra- and inter-cortical connectivity patterns in combination with area-specific receptive field properties of single neurons, lead to the suggestion of distributed sensory processing (e.g. Mountcastle 1978; Zeki and Shipp 1988). From anatomical data and single-cell responses alone, however, it is only in simple cases that one can infer probable coding principles for sensory integration. Better support for the understanding of neural sensory processing might be given by models in which the linking of local visual features is carried out by *temporal codes*, as proposed by several groups including ours (Reichardt 1957, 1961; MacKay 1962; Grossberg 1976, 1980, 1983a,b; Reichardt and Poggio 1979; Reitboeck 1983c, 1988; Reichardt et al. 1983; Reitboeck et al. 1987, 1988: von der Malsburg 1983, 1988; Koenderink 1984; Legèndy and Salcman 1985; Toet et al. 1987; Eckhorn et al. 1986a, 1987, 1988a–c, 1989a,b,d–f, 1990a,b; Eckhorn 1987; Eckhorn and Reitboeck 1988; Singer 1988; Singer et al. 1988; von der Malsburg and Singer 1988; Damasio 1989a,b; Gray et al. 1989). In these models, objects can be defined as entities against the background by the spatiotemporal distribution of the correlation in evoked activities often across several levels of processing. MacKay was, to our knowledge, one of the first who argued that "temporal coding has the enormous advantage over spatial coding that a signal can be broadcast over a wide area, by the simplest means, without losing its identity," and . . "it can be picked out of a welter of other activity by a subsystem alert to it. The vulnerability of a temporally organized system to local damage could be relatively small, since the spatio-temporal patterns of activity are cooperative states of many elements" (MacKay 1962).

9.1.2 Pattern Recognition Requires "Early" Feature Linking Independent of Cognitive Aspects

Recognition of patterns and objects that are embedded in a complex visual scene requires preprocessing operations that have to function independently of cognitive aspects of the pattern to be recognized – it has to function "automatically". Since our investigations were made in the lightly anesthetized cat, we could not hope to discover general principles of sensory integration; instead, we would be lucky to find early (automatic) grouping mechanisms. Such pre-processing operations include the definition of regions within a visual scene (figure-ground separation), the definition of depth and the relative motion of objects in three dimensions.

Significant local visual features are arranged topographically in the primary visual cortex areas, so that we could search for mechanisms of *feature linking* among these elements. At the cortical stage of visual pre-processing, examples of linking features might be: proximity in visual space (similarity of position; compare, e.g. local correlators (Reichardt 1957, 1961) and the concept of local sign (Koenderink 1984; Toet et al. 1987)), local similarities of spatial frequency and texture (e.g. Reichardt and Poggio 1979; Reichardt et al. 1983; Julesz 1981; Green 1986; Reitboeck et al. 1987, 1988; Reitboeck 1988; Pabst et al. 1989), similarity of line orientation including coaxial alignment (e.g., Nelson 1985; Nelson and Frost 1985) and similarity of movement directions and velocities (for a review, see Horridge 1987). Near-simultaneity of events is also a linking feature (e.g. Wilson and Singer 1981; Altmann et al. 1986; Reitboeck 1983c, 1988; Eckhorn et al. 1986a, 1988c, 1989d; Singer 1988).

9.1.3 Figure Segmentation Defined by Correlated Activities

There have been widespread attempts to invoke temporal coding principles, especially for "automatic" feature grouping and segmentation in various sensory modalities (e.g. Freeman 1975, 1981, 1987; Grossberg 1976, 1980, 1983a,b; Basar 1980, 1983, 1988; Basar et al. 1988; Freeman and Skarda 1985; Johannesma et al. 1986; Freeman et al. 1988; von der Malsburg 1983; Singer 1988) including models from our own group (Reitboeck 1983c, 1988; Reitboeck et al. 1987,

1988; Eckhorn et al. 1986a, 1987, 1988a–c, 1989a,b,d, 1990a,b, Pabst et al. 1989). We started our considerations with the proposal that temporal signal correlations are a possible and efficient mechanism for the linking of regions and attributes that define a pattern (Reitboeck 1983c). Reitboeck further proposed that the partial coherence of signals generated by a neural assembly could be an operating principle for pattern definition and for correlating pattern attributes. On this basis we developed a model of texture definition based on temporal signal parameters (Reitboeck et al. 1987, 1988; Reitboeck 1988; Pabst et al. 1989). The basic concept of the model is that retinal image slips convert spatial texture intervals into temporal spike intervals. Temporal gating of those signals then generates texture region separation. A limiting property of this model in its initial form was its rigid stimulus-response coupling. It should be possible, however, that internal neural mechanisms can provide the temporal marking required to link regions according to more global criteria. We recently discovered a linking mechanism in cat primary visual cortex that would fulfill perfectly this requirement: *Stimulus-induced synchronisations* of local cortical oscillations in spatially separated but functionally close neural assemblies (Eckhorn et al. 1988 a,b,c, 1989 a,b,d,e,f; Singer et al. 1988; Gray and Singer 1989). In our current model of "automatic" feature grouping and segmentation of a visual scene, we assume that the local features belonging together are transiently linked by the self-organization process of synchronization of those local "oscillating neurons" that represent the related stimulus features by their RF properties.

9.1.4 Invariant Local and Flexible Global Pre-Processing

It is commonly accepted that pre-processing of local visual features by the primary visual cortex can be typified by stimulus-response tuning properties of single neurons. These properties are relatively stable and invariant (reactive properties), and they are similar in anesthetized and in wake animals. But grouping and integration of local coding features into linked neural representations requires a flexible (constructive) mechanism of self-organization (Johannesma et al. 1986). Otherwise, the enormous number of combinatorial possibilities of groupings of even few local features could not be handled by the visual system. It seems obvious to us, therefore, that region definition and object integration cannot be mediated simply by convergence of separately processed single-cell components onto a "cardinal neuron" or "cardinal group", as implied by hierarchic coding concepts. It seems more probable that widely processed features must be defined as belonging together by a special *cortical linking network*: it mediates (constructive) flexible linking of single-cell coding features by transiently marking the participating members by a temporal code; (for the definitions of "reactive" and "constructive" networks see Johannesma et al. (1986)). Below I will offer some arguments for why the types of connectivities and synapses in a linking network are probably different to those of a *feeding network* mediating the single-cell tuning properties and RFs. (Sect. 9.5.2).

9.2 Development of Experimental Techniques for the Analysis of Global Cortical Processing

9.2.1 Search for Automatic Grouping Mechanisms

At what level of neuronal organization should one look for mechanisms of transient grouping and integration, and what types of signals should be recorded? Obviously, new experimental strategies and methods had to be developed. The main difference with conventional single-cell analyses is that global processing is assumed to be based on the correlated activities of large numbers of neurons. Therefore one has to make multiple parallel recordings from as many of the presently interacting members of an activated assembly as possible. The recorded signals have to represent the activities from a considerable proportion of neurons being involved in the investigated visual task. This requirement is hard to meet with single microelectrode recording and data evaluation methods (Aertsen and Gerstein 1985; Aertsen et al. 1986, 1987, 1988; Johannesma et al. 1986; Krüger 1983; Aiple and Krüger 1988; Reitboeck 1983a,b,c; Schneider et al. 1983; Eckhorn et al. 1986a,b, 1987, 1988c, 1989c,d; Eckhorn and Reitboeck 1988; Epping and Eggermont 1987; Gerstein and Aertsen 1985; Melssen and Epping 1987; Palm et al. 1988).

The above considerations led us to the development of

1) the Marburg multiple-electrode technique (Reitboeck et al. 1981; Reitboeck 1983a,b,c; Mountcastle et al. 1987),
2) fast, multiple-channel RF measurements (Krause and Eckhorn 1983a,b; Krause et al. 1987, 1988; Lohmann et al. 1986, 1988; Eckhorn et al. 1986b, 1989c; Eckhorn 1987; Eckhorn and Reitboeck 1988),
3) several correlation analysis systems (Schneider 1980; Schneider et al. 1983; Eckhorn et al. 1986a; Eckhorn and Reitboeck 1988),
4) models for visual segmentation and integration via correlated activities in a distributed system of local feature analyzers (Reitboeck 1983c, 1988; Reitboeck et al. 1987, 1988; Eckhorn 1987; Eckhorn et al. 1987, 1988c; 1989a,b; 1990a,b; Pabst et al. 1989).

9.2.2 Recording of Different Signal Types with Multiple Electrodes

The author gained his first experience with multiple-electrode and signal-correlation methods in the early seventies. We manipulated two standard tungsten electrodes by independent microdrives in order to investigate the lateral geniculate synaptic circuitry (Eckhorn and Pöpel 1972a,b; Eckhorn and Pöpel 1975; Eckhorn et al. 1976). Our present recording and data evaluation techniques are much more sophisticated: we can record spike trains from up to 19 single cells with independently driveable fiber electrodes (Reitboeck et al. 1981; Reitboeck 1983a,b,c). In addition, we record several types of neural signals in parallel from each electrode: single-unit (SUA) and multiple-unit activities

(MUA), and local slow wave field potentials (LFP) (Eckhorn 1987; Eckhorn et al. 1987, 1988a,b,c, 1989c,d). Special fiber electrodes were developed to keep tissue damage as small as possible. They have thin shafts (70 μm diameter; Pt-Wo core, quartz insulation; 3-7 Megohms at 1 kHz; Reitboeck et al. 1981; Reitboeck 1983b), they are coated with an anti-adhesive film to minimize tissue displacements due to electrode movements and they can penetrate the intact dura of cat (Reitboeck et al. 1981; Eckhorn et al. 1987, 1988c, 1989cd; Bauer et al. 1989a,b) and monkey (Mountcastle et al. 1987). Our fiber electrodes pick up slow-wave field potentials down to less than 1 Hz and fast extracellular action potentials. This broadband recording capability opens up a wide range of possibilities for studies of the interactions of single cells with local group activities (MUA and LFPs). The electrode spacings in our manipulators were either 250 or 330 μm center-to-center, although of course the electrode tips were not always at the same depth (Eckhorn et al. 1988c, 1989d; Bauer et al. 1989a,b). This electrode separation combined with the recording of mass activities is well suited to sample the discrete cortical patch structures that generally extend over 100–500 μm (Bauer et al. 1989a,b; Eckhorn et al. 1989c,d).

9.2.3 Methods for Multiple-Channel Correlation Analyses Require Improvements

Twenty years ago the author started to analyze interactions in the lateral geniculate nucleus with cross-correlation analyses of pairs of spike trains under ongoing activity (Eckhorn and Pöpel 1972a,b). Vision is, however, a dynamic, nonstationary process that obviously cannot be analyzed adequately using "spontaneous" activity and conventional cross-correlation methods, because cross-correlation in its basic definition only comprehends stationary, time-invariant processes. A certain degree of time variance can, however, be evaluated with cross-orrelation by simply analyzing short epochs of responses and by displaying the correlograms in temporal succession (e.g., Schneider et al. 1983; Aertsen et al. 1988). We additionally developed real-time correlation methods that assess temporal variations of multiple-channel coupling (Eckhorn et al. 1986a). However, for multiple-channel neural data the correlation analysis tools have, in our opinion, not yet been developed to a satisfying state. The few genuine multiple-channel methods including dynamic aspects of neural interactions have, for various reasons, not proved to be practical, at least not for more than 4 channels (Eckhorn et al. 1986a; Gerstein and Aertsen 1985). Guiding principles offered for further improvements are: Analysis methods for neural data will extract more significant results, the better they mimic signal-processing strategies of the brain; neural network models adapted to the neural structures recorded from, therefore, might be powerful analysis instruments of neural multiple-channel data.

We made first steps in these directions by developing a "real-time covariance computer" for cell assemblies that is based on neuronal principles (Eckhorn et al. 1986a). This instrument includes low-pass filtering of spike trains (in order to model chemical synaptic transmission), adaptive thresholding and stochastic computing principles that serve to mimic temporal and spatial integration. Improved network models are now being developed in our group, which include dynamic thresholds, adaptation, different types of inputs on a single neuron (feeding and linking synapses), oscillation and stochastic synchronization capabilities and modifications of several of these properties by learning (Sect. 9.5.2 and Eckhorn et al. 1989a,b; 1990a,b; Pabst et al. 1989).

9.2.4 Difficulties with Single Cell Correlations in the Visual Cortex

The main technical problems associated with recording neural activity with our multiple-single electrodes have now been solved (Eckhorn et al. 1988c, 1989c,d; Bauer et al. 1989a,b). The probability of finding significantly coupled single-cell spike trains in the visual cortex manifesting narrow peaks in conventional correlograms is, however, low under our experimental conditions (minimal interelectrode distance was 250 μm). We assume that temporally precise and strong (monosynaptic) couplings between cortical neurons, separated horizontally by more than 250 μm, are generally not present in the visual cortex. If such couplings exist at all, presumably they are to be found only among direct neighbours or in the "feeding network" from afferent visual inputs (e.g. geniculate) and their projection targets (Toyama et al. 1981a; Tanaka 1983). If we consider the thousands of synaptic contacts on each cortical neuron, we should not be surprised that the effects of single excitatory connections are heavily masked by other inputs (Krüger 1983; Eckhorn and Reitboeck 1988; Aiple and Krüger 1988; Krüger and Aiple 1988, 1989). For these reasons, we decided to look at neural interactions on a broader scale of time and spatial neural organization, in order to assess the combined actions of large numbers of direct and indirect connections. Weak correlations in many individual signals might summate; they could then readily be detected with conventional cross-correlation methods (Aertsen and Gerstein 1985; see also Aiple and Krüger 1988; Shaw et al. 1982; van Kan et al. 1985; Eckhorn et al. 1988c, 1989c,d).

9.2.5 Local Mass Signals Provide Relevant Information About Cortical Grouping Mechanisms

The ease of finding cooperating neural assemblies was dramatically increased by recording neural mass signals such as MUA and LFPs in parallel with single-unit spike trains. This "triple recording" from each electrode offered us the opportunity to investigate neural couplings on different levels of organization and specificity, by analyzing all possible combinations of interactions between the different signal types. By this means we can study the interactions between single neurons possessing well-defined tuning properties, the average mass activity of the local dendritic PSPs (local slow-wave field potentials, LFPs; see Creutzfeldt et al. (1966)) and the spike activity of the local assembly (multiple-unit activity,

MUA). Correlations between mass activity of the same or different types were, of course, also evaluated. Using these different signal combinations, we generally found significant and often strong signal correlations. Our results show a variety of group interactions that have not been seen before and that can hardly be obtained from single-cell spike train recordings.

Why is such an approach successful? If neurons with similar tuning properties are locally clustered in the visual cortex, and if, in addition, the local neurons are mutually coupled either directly by electric fields (Schwartzkroin 1983) and/or via synaptic connections, then a reasonable signal-to-noise ratio should be obtained, because then the local neural group exhibits signal correlations among its members that are evoked by the common stimulus and enhanced by the internal connections (e.g. Freeman 1975; Basar 1980; Mitzdorf 1985, 1987; Shaw et al. 1982; Bullock 1988). Thus, the amplitudes of MUA and LFPs are expected to be higher the more coherently a stimulus activates the local group (e.g. Mitzdorf 1985, 1987; Bullock 1988; Bullock and Basar 1988; Petsche et al. 1988). Components not correlated with the present group activities will not summate and therefore will be of small amplitudes.

Evidence for clustering of single-cell tuning properties in sensory cortex was first provided by Mountcastle (1957, 1978) and Hubel and Wiesel (1962, 1963). Their findings indicate that even within relatively small volumes of cortical tissue, there exist many cells with similar responses to identical stimuli.

We consistently found the tuning properties of single cells to be grouped or clustered in A17 and A18. They remain almost unchanged over several 100 μm of electrode advancement, both in perpendicular and tangential recording tracks, and then change abruptly (e.g., orientation and direction specificities; Eckhorn et al. 1989d; Bauer et al. 1989a,b). "Tuning clusters" have been reported by other investigators using a variety of methods, e.g. Payne et al. (1981) by long recording tracks in A17; Weyand et al. (1987) by silencing certain A17 layers while recording from others; Ts'o et al. (1986) by double-electrode recordings and cross-correlation of SUA in A17 and Matsubara et al. (1985) by direct sampling with multiple penetrations in A18.

9.3. Local Properties of Stimulus-Induced Oscillations in the Visual Cortex

9.3.1 Stimulus-Specific Oscillations

Stimulus-induced oscillations were present in all three types of recorded neural signals; they were most obvious in local field potentials (LFPs) where they could be directly visualized with single sweeps on the oscilloscope (Fig. 9.1). The oscillations were strongest with those stimuli that were also preferred by the local neurons (Gray and Singer 1987a,b; Eckhorn et al. 1988a–c), i.e., the

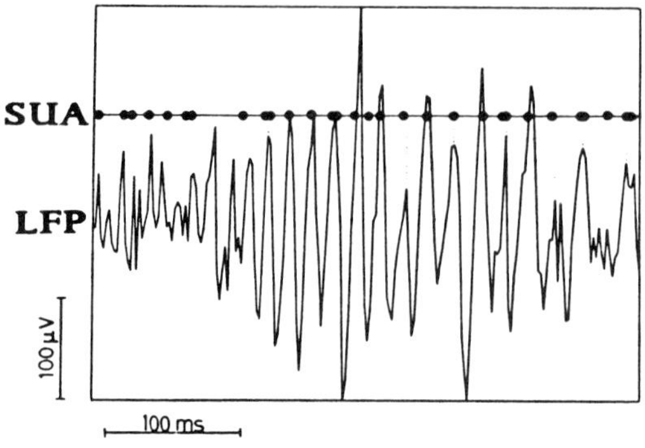

Fig. 9.1. Single-sweep stimulus-induced oscillations.
Single-unit spike train (SUA) and local (slow-wave) field potential (LFP) were simultaneously recorded with the same electrode in the upper layers of area 17. Stimulation with drifting grating of 0.7 cycles/deg moved at 8 deg/s in preferred direction. Note the transition from low amplitude stochastic to higher amplitude oscillatory activity about 100ms after start of stimulation (at t = 0) and the correlation of the spike occurrence times with the LFPs negativity during the oscillation spindle (negative LFP upwards). (From Eckhorn et al. 1988c)

oscillatory signals are linked with the specific tuning properties of the local cortical cells.

Stimulus induced rhythmic discharges of local neuron groups were also observed in multiple unit activities (MUA; Gray and Singer 1987a,b; Eckhorn et al. 1988a,b,c), but MUA oscillations were frequently weaker than those in LFPs. In single-cell spike trains of visual cortical neurons (A17 and A18), the rhythmic components are not usually directly visible. Their presence was, however, often evident when the auto-coincidence histograms (ACHs) or their frequency spectra were computed (Gray and Singer 1987a,b, 1989; Eckhorn et al. 1988c; for ACHs see e.g. Melssen and Epping 1987).

The spectral energy of stimulus-induced oscillations had its maximum in the frequency range between 40 to 80 Hz (often at 40–50 Hz; see also Figs. 9.4, 9.6, 9.10, 9.11, 9.13). Its periodic appearance contrasts other ongoing activities, which have spectral amplitude peaks at low frequencies (1–30 Hz; e.g. Fig. 9.11).

Stimulus-induced oscillatory (gamma range) group activations have obviously already been observed in the visual system by several investigators (e.g. cat LGNd: Arnett 1975; cat cortex: Krüger 1983; W. Singer 1985, personal communication; R. Eckhorn and S. Hansch 1985, unpublished observations in A17-A18 interaction experiments; monkey cortex: Freeman and van Dijk 1987). However, the oscillatory phenomena were not interpreted in a frame of global coding, as is the case in recent publications (e.g. Eckhorn et al. 1988c; Singer et al. 1988) including the present one.

9.3.2 Internal Temporal Structure of Stimulus-Induced Oscillations is not Stimulus-Correlated

The occurrence of oscillatory activities is waxing and waning during specific stimulation (Fig. 9.2, see also Figs. 9.6, 9.10, 9.11). Stimulus-induced oscillations of LFPs and MUA appear as *oscillation spindles* of about 80–250 ms duration, separated by intervals of stochastic activity, and their *response latencies*

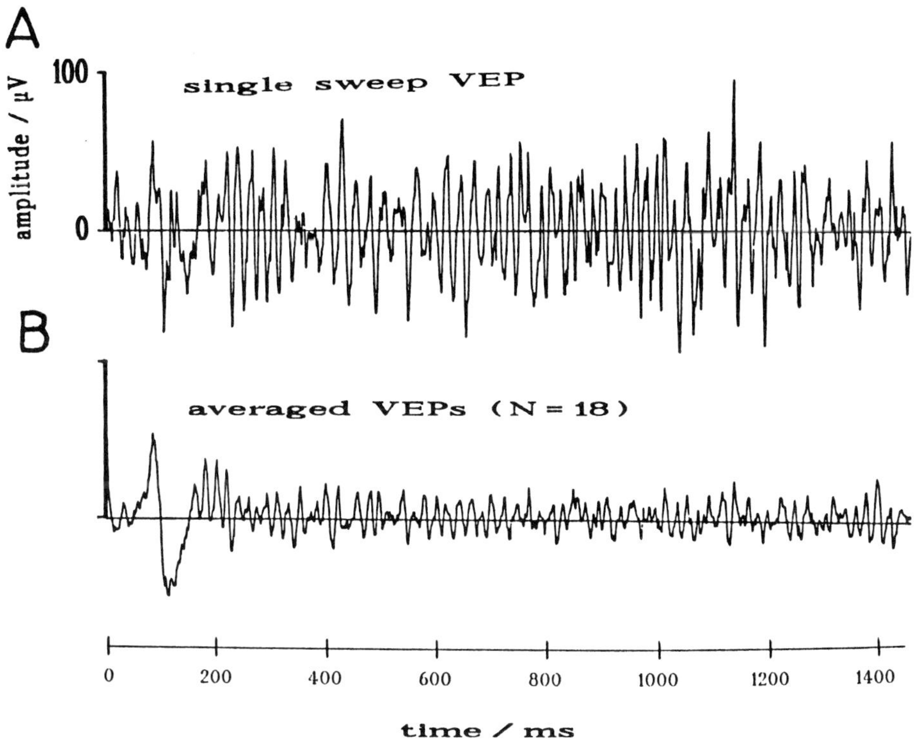

Fig. 9.2. Oscillation frequencies are not precisely coupled to the stimulus but intracortical early visually evoked potentials are.
A Single sweep local field potential (LFP, recorded intracortically with a microelectrode) in response to a moving grating stimulus. **B** Average of 18 LFP responses to identical stimulus repetitions. Recordings were made in the upper layers of A18. Stimulus movement in the preferred direction started at t = 0. Note that the early stimulus-locked components become clearer and the later oscillations are suppressed by stimulus-locked averaging. (From experiments of R. Eckhorn, M. Munk, W. Kruse and M. Brosch)

were found to be significantly longer than the fast components of the averaged visually evoked cortical potentials (VECPs). Averaged VECPs contain the stimulus-locked response components. This type of correlated activity is assumed to play a functional role in the definition of visual relations also (Sect. 9.5.3; Stimulus-Forced and Stimulus-Induced Synchronizations). Oscillation spindles occur with considerable temporal jitter in response to identical stimuli and their "momentary frequencies" are highly variant (Gray and Singer 1987a,b; Eckhorn et al 1988 a,b,c). Therefore it is not astonishing that the "micro-timing", as revealed by the phases of the high-frequency oscillations, was generally not coupled with the stimuli. The "momentary" frequency and the phases of the oscillatory components are obviously determined by interactions within the neuronal network and do not have rigid temporal coupling to the stimulus. This independence from the stimulus is demonstrated in Figs. 9.2 and 9.6, where single LFP sweeps are shown in the upper part and response averages to identical stimulus repetitions are plotted below. Averaging, obviously, uncovers the stimulus-locked VECP at short delays, but averages out the oscillating components. This seeming lack of stimulus-response coupling of oscillation frequencies might be one of the reasons why we and other groups have been unaware until recently of this type of reverberatory activity in the visual cortex. The pioneering work of Freeman in the rabbit olfactory bulb and the monkey visual cortex have obviously stimulated the investigations of the group of W. Singer and our group (Freeman 1975; Freeman and van Dijk 1987), leading to the discoveries of stimulus-specific synchronizations in the cat visual cortex.

The high-frequency stimulus-induced oscillations found in the visual cortex are different to the "*wavelet*" activity evoked by whole-field flashes in the retina, geniculate and visual cortex and by clicks in the auditory system: wavelets are strictly phase-locked to the stimulus and therefore are obtained with stimulus-locked averaging (Cracco and Cracco 1978; Whittaker and Siegfried 1983; Kraut et al. 1985; Mitzdorf 1987). We classify wavelets and other stimulus-locked signals as another type of synchronization that can define transitory sensory relations (Sect. 9.5.3).

Despite the "uncorrelation" of the oscillation frequencies to the stimulus phase, strong stimulus couplings can often be found with the *envelope* of oscillatory neural response. This means that the occurrence of spindles is itself stimulus-coupled, even though the frequencies of the oscillations may be statistically independent from stimulation. Stimulation-to-LFP-envelope couplings are quantified by "envelope PSTHs" (Fig. 9.3). This type of LFP-PSTHs generally shows clear separation between oscillatory responses and stimulus pauses.

9.3.3 Correlations Between LFPs and Single-Cell Spikes

Correlations between local oscillating slow-wave potentials (LFPs) in the gamma range and single-cell spike trains recorded in the same position, have been found in the visual cortex first by Gray and Singer (1987a,b) . Similar observations are already known from the rabbit olfactory bulb (e.g. Freeman 1975). These authors quantified the spike-to-LFP correlations by deriving the relation between the amplitude of the LFP and the spike-occurrence probability. Their and our results (Figs. 9.4, 9.13, 9.14) show in general that increasing LFP negativity is associated with increasing spike discharge probability.

In our recent analyses we used mainly the peak amplitude of the spike-triggered average (STA) as a measure for correlation strength between action potentials and LFPs. The correlation of more than 200 cells has been tested by recording spikes and LFPs in parallel from the same electrode. Correlation maxima with strengths between "not significant" and as high as 0.8 were found (peak of correlogram normalized to spike rate and LFP amplitudes). The correlation for a specific recording position was generally strongest when the cells were driven binocularly by their preferred whole-field stimuli (Sect. 9.5.5).

The observation that single-cell tuning properties (coding of local visual features?) and the activities of great numbers of neurons (coding of global features?) are specifically correlated, leads to the central conclusion that interac-

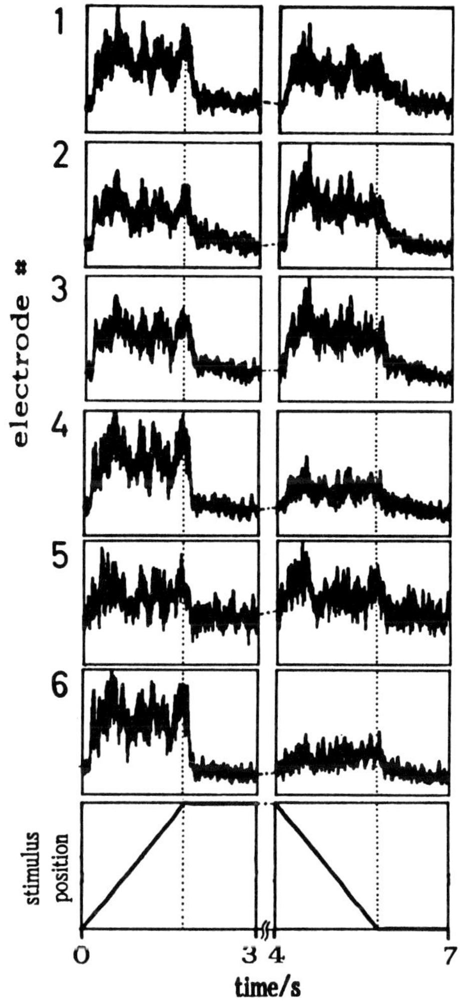

Fig. 9.3. Peri-stimulus time histograms of local field potential envelopes.
The LFPs were intracortically recorded with microelectrodes; they were bandpass filtered (−3dB at 40 and 60Hz) in order to suppress parts of the stochastic signal components and to leave mainly the oscillatory components. Envelopes were calculated before stimulus-locked averaging by subtracting the mean, rectification of the LFP signal and subsequent low-pass filtering with a −3dB frequency cutoff at 160 Hz. Simultaneous recordings with an array of 6 electrodes in A17 and A18 (inter-electrode distance: 250 μm). Stimulus movement (lowest panels) was in preferred and non-preferred directions for the neurons near electrodes 4 and 6. (From experiments of R. Eckhorn, M. Munk, W. Kruse and M. Brosch)

tions of local and global visual coding can be successfully analyzed by combined recordings of both signal types, and might lead to a deeper understanding of cortical sensory coding beyond the single cell RF concept.

9.3.4 Stimulus Dependencies of Oscillations: Local-Global Coding Relations

a) Orientation Tuning is Similar with Spikes and LFP Oscillations

The similarity of orientation tuning evaluated from action potential responses and LFPs (recorded at the same position) is an example for "local-global coding relations" on which our model is based. Gray and Singer (1987a,b) first showed this special relation in A17 of the cat visual cortex. In Fig. 9.5 we show examples of the local orientation/direction tuning properties derived from the same three

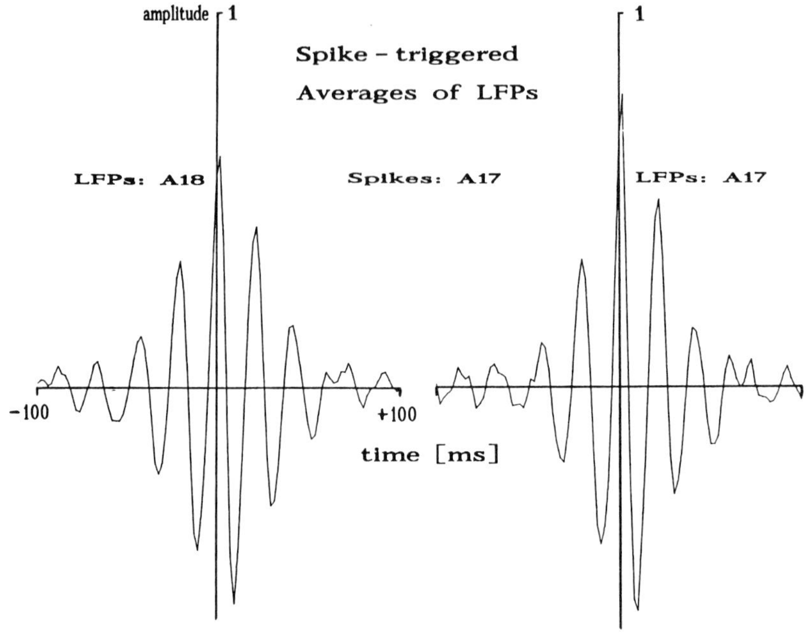

Fig. 9.4. Correlation between LFP oscillations and single-cell spikes (SUA), calculated as spike-triggered average (STA) of LFPs, normalized to the number of spikes and the energy of the respective LFPs. Negative LFP values plotted upwards; average oscillation frequency about 45 Hz. Spikes and LFPs were recorded in A17 with the same electrode (right STA), and LFPs with another electrode in A18 (left STA); i.e., the right STA shows the correlation between a single A17 cell spike train and the LFPs from the direct neighbourhood, while the left STA shows the correlation between the same A17 cell and LFPs from A18. (A17 cell and A18 RFs overlapped; stimulus: grating 0.7 cycles/deg drifting at 8 deg/s in and against preferred direction of cells in the A17/A18 recording positions). (From experiments of R. Eckhorn, M. Munk, W. Kruse and M. Brosch)

electrodes by separating SUA and LFP data from the compound signals. The orientation tuning of both SUA and oscillations of LFPs (measured in the entire gamma range) are often very similar. Directionality, however, is generally weaker in LFPs than in the respective single-cell responses (Bauer et al. 1989b). The LFP oscillations are thus closely related to the primary coding properties of single cortical neurons (here known as orientation tuning).

b) Binocular Stimulation Induces Oscillations of High Amplitudes

Another example for a "local-global coding relation" is demonstrated with binocular stimulation. Binocular stimulation was generally more effective in eliciting synchronized oscillations than monocular stimuli (Eckhorn et al. 1988a–c, 1989a,e). Fig. 9.13aA, dA show examples derived from simultaneous recordings in A17 and A18. Binocular stimuli were particularly potent in evoking intraareal LFP-to-spike synchronizations and interareal synchronization between A17 and A18 (see below). The binocularly induced coherence of the oscillations in both areas were up to 6 times greater than those induced with monocular stimuli,

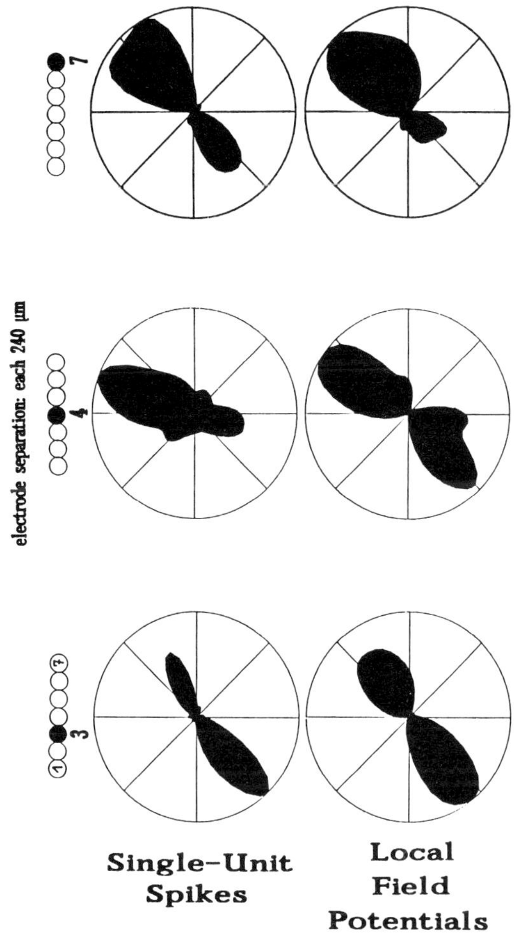

Fig. 9.5. Similar orientation tuning of local LFP oscillations and single cells (spike trains) at the same recording location.
Tuning curves calculated in response to a drifting grating (0.7 cycl/deg, 8 deg/s) moving in pseudo-random order in 16 different directions. The three SUA-LFP tuning pairs were recorded simultaneously, each with one microelectrode. Left column: SUA, Right column: LFPs. Stimulus: grating of 0.7 cycles/deg swept at 8 deg/s over the receptive field.
Tuning curves were derived from integrals over peri-stimulus time histograms (average of 10 stimulus repetitions). LFP histograms were derived in the following steps: (1) subtraction of mean amplitude calculated from the total data record, (2) rectification of the remaining signal, (3) low-pass filtering (Blackman-Harris window, −3dB at 100 Hz (Harris 1978), (4) calculation of average response histogram from single responses. In order to obtain the amplitude values of the tuning curves, each of the 16 PSTHs was integrated during the response period (after the mean value of the spontaneous activity was subtracted from the PSTHs); amplitudes for intermediate directions were obtained by interpolation as follows: (1) linear interpolation in Cartesian coordinates, (2) low-pass filtering (von Hann window (Harris 1978), width: 45° orientation), (3) transformation to polar coordinates. (Experiments of R. Eckhorn, R. Bauer and W. Jordan)

even though the monocular stimuli were applied to the dominant eye. An obvious suggestion derived from this result is that binocular correspondence might be evaluated by the primary visual cortex, with the help of the "linking-by-synchronization mechanism". (Might A18 evaluate coarse and A17 evaluate fine disparities with the help of this mechanism?)

9.4 Stimulus-Induced Synchronizations in Spatially Separated Assemblies

9.4.1 General Results

The appearence of stimulus-induced oscillatory synchronizations at functionally close but spatially distant locations of the visual cortex is considered by us as the most important result of our recent multiple electrode investigations. We showed experimentally the occurrence of transient groupings of spatially separated assemblies by signal synchronizations in the visual cortex (Eckhorn et al. 1988a,b). Stimulus-induced synchronizations were found throughout all layers of vertical columns of the iso- and cross-orientation type (Bauer et al. 1989a). Synchronizations were induced among vertical columns of the same cortical area (Eckhorn et al. 1988a–c; Singer et al. 1988) and even among assemblies of different cortical areas (A17 and A18; Eckhorn et al. 1988a–c) at those positions where common single-cell tuning properties were present. These observations are essential prerequisites for our model of global visual coding. Correlated activity can thus act as a label to link related features in a visual scene and to signalize a momentary association.

Our findings that stimulus-specific synchronized oscillations appear in spatially separated cortical regions differed at first from the initial observations of Gray and Singer (1987a,b; Gray et al. 1988). They found oscillatory activity to be limited to the dimensions of a single vertical orientation column, i.e., they reported finding no correlated components outside the column and considered the oscillations to be a local cortical phenomenon. Differences in experimental techniques are possible reasons for these different results. Later on, these authors confirmed that (within the same cortical area) correlated oscillatory activities occur also in more separated positions (Singer et al. 1988; Gray and Singer 1989).

9.4.2 Details About Stimulus-Induced Cortical Synchronizations

a) Observations from Single Response Sweeps

Stimulus-induced oscillations from neighbouring electrodes (250–330 μm separation) were nearly always highly synchronized when the cells in both cortical locations had similar stimulus preferences (at this electrode separation the RF properties were most often similar). An example is given in Fig. 9.6, which is a "superimposogram" of four simultaneously recorded LFPs from neighbouring electrodes. This figure also shows that the oscillations do not last continuously throughout visual stimulation. Instead, oscillation spindles occur with durations of about 100–250 ms, separated by pauses with stochastic activity of lower amplitudes and less inter-channel couplings. We found the response latencies of the spindles to be significantly longer than the fast components of the averaged visually evoked cortical potential (VECPs), calculated from intracortically recorded LFPs. In Fig. 9.6 this latency difference can be estimated to be about 90 ms.

b) Synchronization Within a Vertical Cortex Column

Correlation analysis between signals of supra- and infra-granular layers revealed coupled LFP waves of broad spectral content during ongoing activity and synchronized oscillations in the gamma rage during stimulus-activated states. Even between layers II and VI within the same column, a considerable fraction of the

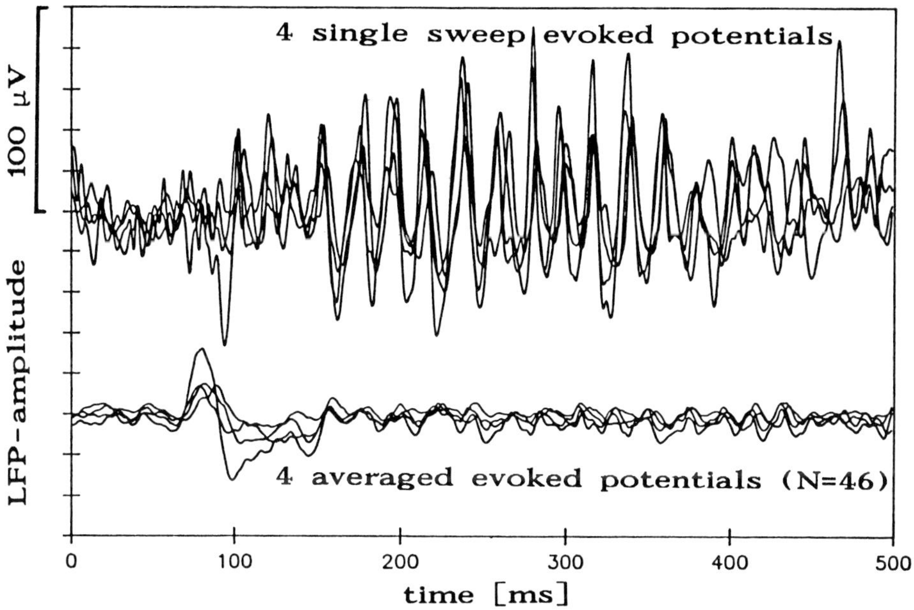

Fig. 9.6. Superimposograms of 4 simultaneously recorded LFPs in response to a drifting grating (0.7 cycles/deg swept at 8 deg/s; binocular stimulation).
Upper superimposograms: single-sweep LFP response epochs, recorded intra-cortically in parallel with neighbouring microelectrodes of our linear array; 3 recordings are from A 17 and one from A 18. Note the lack of correlation of the low amplitude stochastic activity directly after start of stimulus movement in contrast to the synchronized, high-amplitude oscillatory signals beginning about 150 ms after stimulus onset. Lower superimposograms are averages of 46 responses to identical stimuli. Note the suppression of the oscillations relative to the short latency evoked potentials at 70–120 ms. (Experiments: R. Eckhorn, M. Munk, W. Kruse and M. Brosch)

signal power of the stimulus-induced oscillations is synchronized. Moreover, in cross-orientation columns where single-cell orientation tuning is orthogonal in upper and lower layers (Bauer 1982; Bauer et al. 1983, 1987, 1989a; Dow and Bauer 1984; Bauer and Fischer 1987; Eckhorn et al. 1989d), a high degree of coherence of stimulus-induced oscillations was found.

c) Synchronization Within a Single Cortex Area

We have been using *correlation profiles* to illustrate spatial distributions of stimulus-induced synchronizations over larger cortical distances (Fig. 9.7). These are calculated by either of two methods: (1) spike-triggered averaging of LFPs during stimulus-evoked responses, which ensure that the triggering spikes are mainly stimulus-evoked (Figs. 9.4, 9.9, 9.13, 9.14) and (2) cross-correlation between LFPs (Figs. 9.7, 9.11, 9.12). The correlation profiles were obtained by plotting the amplitude of the main peak in the normalized cross correlograms versus cortical recording distance. A sequence of cross correlograms was calculated between a reference signal from one electrode and signals from each of the

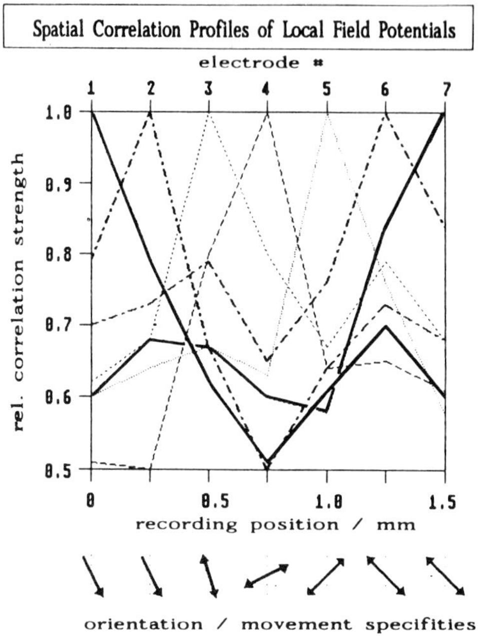

Fig. 9.7. Horizontal extent of stimulus-induced synchronizations within a single cortex area.
Simultaneous recordings of LFPs from layers II/III of area 17 with a linear array of microelectrodes. Stimulus: grating 0.7 cycles/deg drifting at 8 deg/s in and against preferred direction of cells on electrodes 1,2,3 and 6,7. The RFs of neighbouring recording positions overlapped. *Spatial correlation profiles* of LFPs; the peak amplitudes of the normalized cross correlograms were plotted over cortical distance. The different profiles were obtained by taking each electrode as reference for cross correlations with the signals on the remaining electrodes. The orientation /direction specificities of single units at each recording position is indicated below. (From Eckhorn et al. 1988c)

other electrodes of the linear array. In Fig. 9.7 such correlation profiles are shown for LFP recordings. With LFPs on electrode 1 as a reference signal, the peak correlation strength decreases to a minimum of about 0.5 at a distance of 0.75 mm (electrode 4). At a distance of 1.25 mm the correlation strength increases and drops again at 1.5 mm. Signals recorded by other electrodes were then taken in turn as the reference, in order to obtain the corresponding correlation profiles. Almost all profiles in Fig. 9.7 have a minimum at the center of the electrode array, i.e. they show stronger correlations between the more spatially separated outer recording positions. This common minimum might be related to the functional properties of the underlying cortical matrix of hypercolumns (Hubel and Wiesel 1962, 1963), which is defined by the local cortical RF properties such as orientation/direction tuning and ocularity. For this reason we determined some of the classical single-cell RF properties for all three signal types. The shape of the correlation profiles in Fig. 9.7 suggests that cell groups having similar orientation/direction specificities are more strongly linked (electrodes 1, 2, 3, 6, 7) than neighbouring cell groups having different orientation/direction properties (about orthogonal, electrodes 4 and 5).

To facilitate further discussion of stimulus-induced intraareal synchronizations, we integrated some of our findings into a simplified schema of cortical orientation hypercolumns (Fig. 9.8), taken from a model of Baxter and Dow (1988). In the three "iso-orientation patches" (middle row of "electrodes") that are simultaneously activated by the bright moving stimulus bar, large-amplitude synchronized activities are induced, while in nearby positions only small amplitude oscillations are present. Stimulus-specific activation, in this model, is medi-

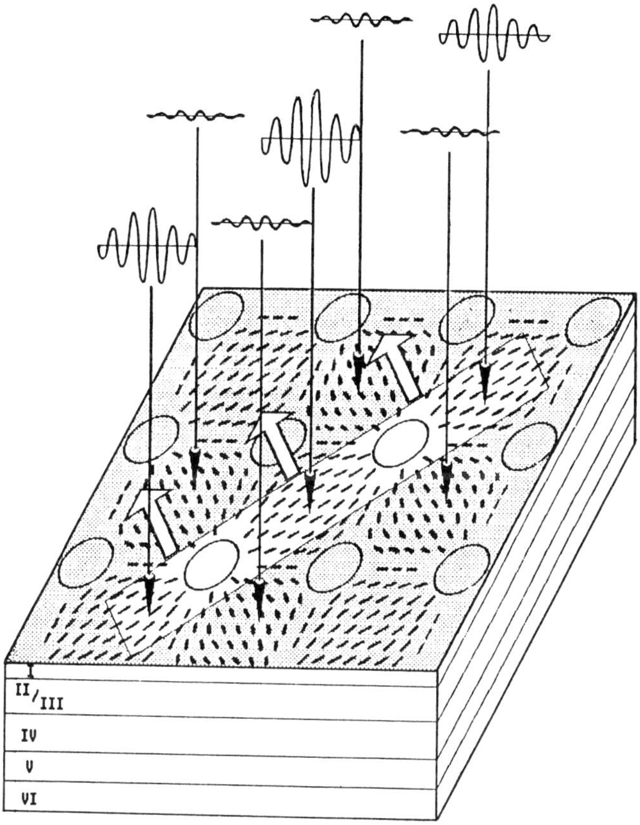

Fig. 9.8. Schema of intraareal stimulus-evoked synchronizations.
A piece of cortex showing 3 × 3 hypercolumns with centers of minimum orientation specificity surrounded by patches with cells of like orientation specificities (after Baxter and Dow 1988). A moving light bar indicates the cortical projection of a stimulus bar. Assemblies with like orientation preference exert stimulus evoked synchronizations of high amplitude, as is indicated at the ends of the arrows by the oscillation spindles. (Schema composed by W. Kruse)

ated via the afferent visual *feeding network*, while synchronization between neurons with similar local coding properties is mediated by a horizontal *linking network* (see also Sect. 9.5.2 and Eckhorn et al. 1989a,b; 1990a,b, Pabst et al. 1989).

9.4.3 Synchronizations Between Two Cortical Areas

Visual cortical areas are connected by a system of reciprocal projections arranged in a "patchy" way (e.g. Symonds and Rosenquist 1984; Bullier et al. 1984, 1988; Leferink et al. 1987; Ferrer et al. 1988; Salin et al. 1989). We were interested in mechanisms of interareal cooperativity, using the A17/18 connections as an example. Our question was: are the local visual features that are differently processed in the two areas also possibly integrated by a synchronization mechanism, similar to that observed inside a single cortex area?

a) "Correlation Patches" Characterize Cooperating Assemblies

Ongoing activities were used in some experiments in order to find "correlation patches" in the two primary visual areas A17 and A18 of the cat (Leferink et al. 1987; Reitboeck et al. 1988; Eckhorn et al. 1989d; Nelson et al. 1989). Patches of enhanced correlations were mainly found within that range of A17 where the RFs of A17 overlapped with the RF of a single A18 recording. We have found so far mainly synergistic correlation components of the medium and broad type (correlated signal components of medium (20–60 Hz) and low (2–10 Hz) frequency content) which were in phase in both areas (Eckhorn 1987; Leferink et al. 1987; Reitboeck et al. 1988b; Eckhorn et al. 1989d; Nelson et al. 1989).

Stimulus-induced oscillations were more often used by us to find interareal "correlation patches" in A17 and A18 (Eckhorn et al. 1988a,b,c, 1989d,e,f). Synchronizations of oscillatory activities, generally with near-zero phase differences, were observed. Figure 9.9 shows an example for the spatial distribution of the coherence of stimulus-induced oscillations in the two visual areas. We scanned A17 along recording tracks down the medial bank of the central sulcus over about 2.5 mm while the activity in A18 was recorded with two stationary electrodes. A "whole-field" stimulus grating induced strongly coherent oscillations of the LFPs in A17 and A18. Along the tracks, single-unit spikes were recorded at multiple positions in A17 and spike-to-LFP correlations were calculated using spike-triggered averaging. Highly correlated oscillations were found in this example at two "columnar recording positions" in A17 (about 1.2 mm apart). At electrode positions 1 and 2 (Fig. 9.9) the single-cell A17 RFs had, however, no overlap with the A18 fields, but their orientation preference was similar to that in A18 (these A17 and A18 RFs had low directionality values). In the second "synchronization patch" (position 7) the RFs in both areas did overlap, but a movement direction perpendicular to that in A18 was preferred (we could not decide with certainty whether at this position the vertical column had iso- or cross-orientation properties).

Our data on A17–A18 interactions show that stimulus induced synchronizations occur in positions where the neurons in both areas had at least one of their primary coding properties in common. It remains to be shown whether common coding properties are a sufficient condition for the synchronization of stimulus-induced oscillations between different areas and how certain local coding properties dominate synchronization.

b) Frequency Variation During Interareal Synchronizations

The degree of variability of interareal synchronizations can be guessed from Fig. 9.10 (spectral variability can be seen in the left panels). It shows amplitude spectra of single 250 ms responses (two lower rows), calculated from responses to the

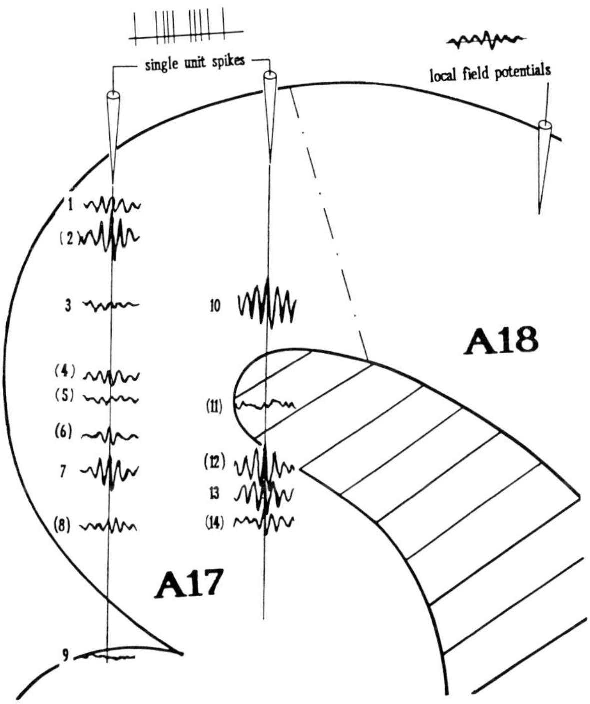

a

Receptive
Fields (RFs)

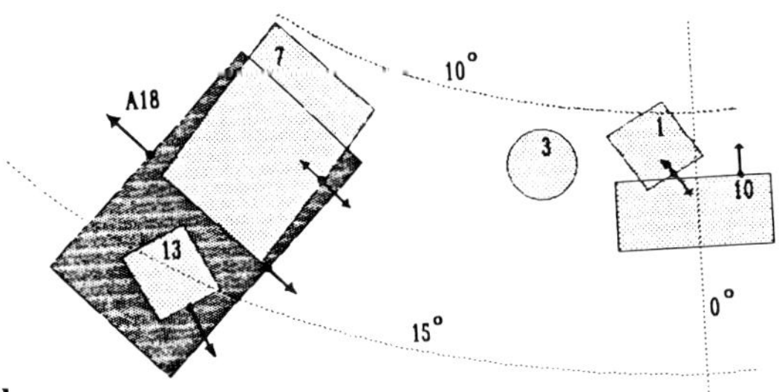

b

Fig. 9.9. Coherent oscillations in visual areas A17 and A18.
Intracortical recordings of local field potentials with a microelectrode in A18 at a fixed position and of single-unit spikes at successive positions in two electrode tracks in A17. Signal correlations between the 2 cortical areas were derived with spike-triggered averaging (STA) of LFPs. **a** In each position in A17, where spikes were recorded, the respective STA correlogram is plotted. Recording positions are drawn to scale in the frontal section at Horseley-Clarke coordinates of posterior 2 of the left hemisphere. (Amplitude scaling of STAs: max. amplitude of STA in position 10 is about 100 μV). **b** Relative RF positions and areas; numbers denote the recording positions, arrows the directionality of the cells. Binocular stimulation with a grating of 0.7 cycles/deg drifting in preferred direction at a drift rate of 8 deg/s. (From Eckhorn et al. 1988c)

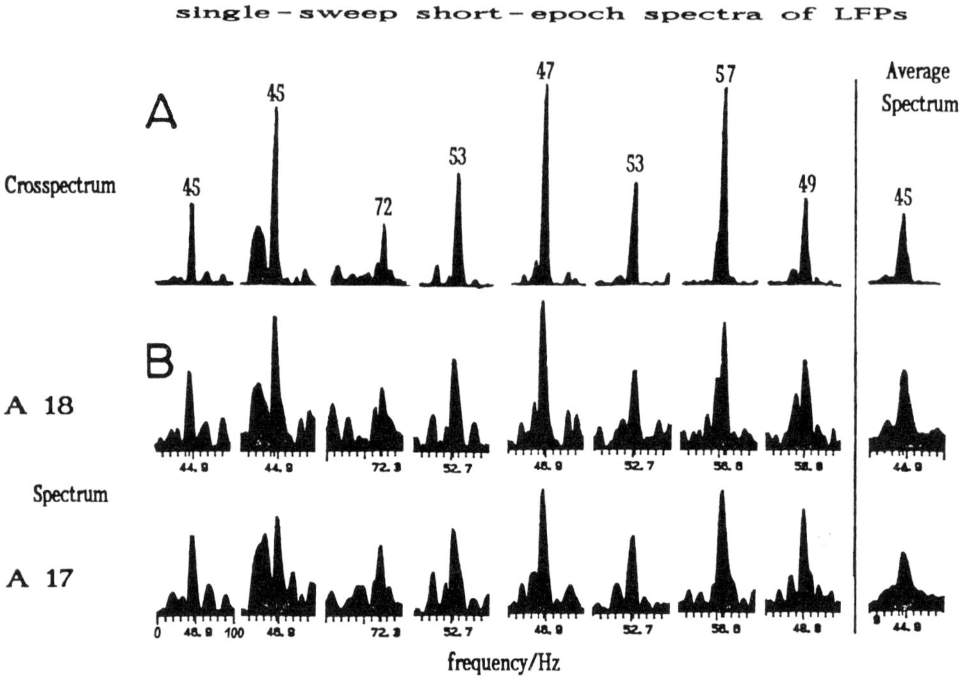

Fig. 9.10. Stimulus-evoked synchronizations in two visual cortical areas (A17 and A18). **A** Cross-amplitude spectra of short-epoch (250 ms) local field potentials recorded in parallel intracortically with microelectrodes at A17 and A18 positions with overlapping receptive fields. Spectra of responses to the same stimulus interval; (stimulus: grating of 0.7 cycles/deg drifting at 8 deg/s); **B** short-epoch (250 ms) amplitude frequency spectra of local field potentials. Left panel single-epoch spectra. *Right panel*: average of 19 single-epoch spectra. Numbers below and above the spectra denote the frequencies at the peak amplitudes (ordinate scalings arbitrary but equal for spectra of same type). Note: (1) the wide frequency range of the peak amplitudes in the single-epoch spectra and (2) the high degree of interareal coherence at the oscillation frequencies compared to the remaining frequency range. (B from Eckhorn et al. 1988c)

same interval of the stimulus cycle and the related single-epoch cross spectra. The narrow spectral peaks are distributed over a relatively wide range (44–73 Hz) but the reasonably large frequency variations occurred in parallel in the two visual areas (resolution of frequency analysis about 1.9 Hz). Averages of the single-epoch spectra show lower and broader peaks due to inter-trial variations, but the peaks remain a prominent feature after averaging.

c) Time Course of Stimulus-Induced Synchronization Between Two Cortical Areas

How do the cooperative states of synchronization between A17 and A18 develop in time? An example of such temporal behavior is shown in Fig 9.11 for two positions in A17 and A18 having overlapping receptive fields. Binocular stimulation with a whole-field grating induced large-amplitude oscillations of about 45

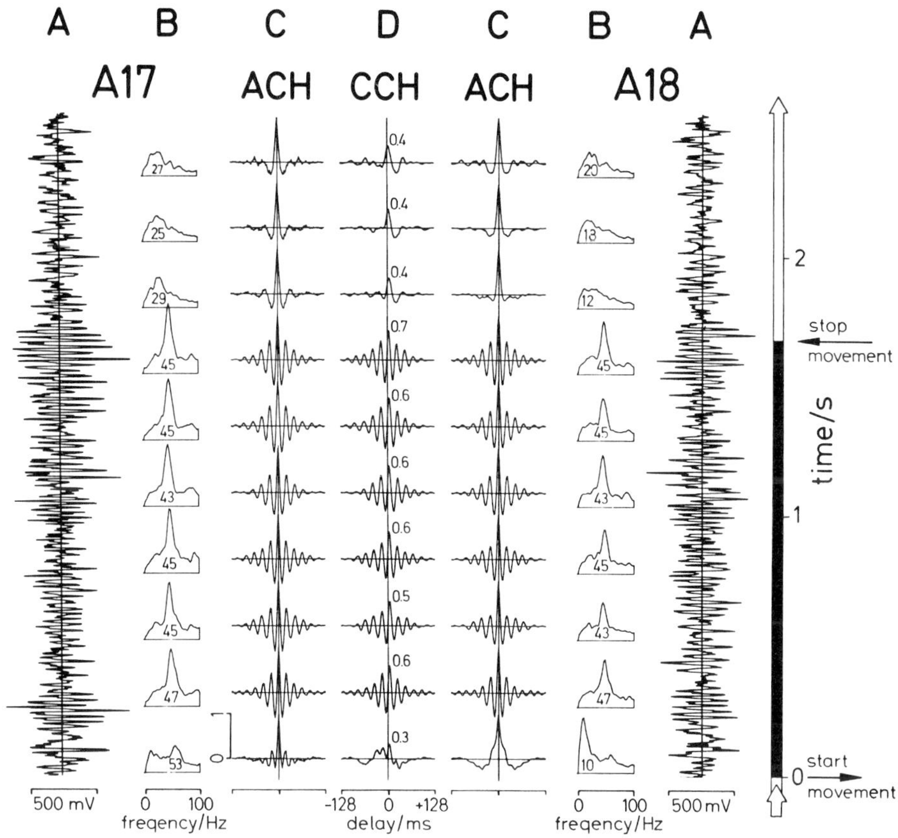

Fig. 9.11. Temporal sequence of stimulus-evoked synchronizations between A17 and A18 at positions with RF overlap.
Intracortical local field potential recordings with microelectrodes from A17 and A18 of the cat (position 7 of Fig. 9.9). Recording bandwidth (−3 dB) for A17: 13–250 Hz, for A18: 1–100 Hz. Binocular stimulation with grating (0.7 cycles/deg.) swept at optimal direction with 8 deg/s ("optimal" for local single-unit spike responses). Black bar and arrows denote the period of stimulus movement. Left data from A17; right data from A18. **A** Single-sweep LFP recordings. Note the oscillatory activation during stimulus movement and the transition to a more stochastic state after stimulus cessation; **B** amplitudes of the LFP frequency spectra, calculated for successive 256 ms epochs; averaging in the frequency domain of 19 responses. Note the peak at about 44 Hz appearing at start of stimulus movement and disappearing at stop. Numbers in the spectrograms denote the frequency of the maximum; **C** auto-correlation functions, calculated by FFT transformation from the data of **B**. Note the stimulus-evoked oscillations; **D** normalized cross-correlation functions calculated from the respective 256 ms LFP epochs. Numbers at the correlation peaks denote their amplitudes. (From Eckhorn et al. 1988c)

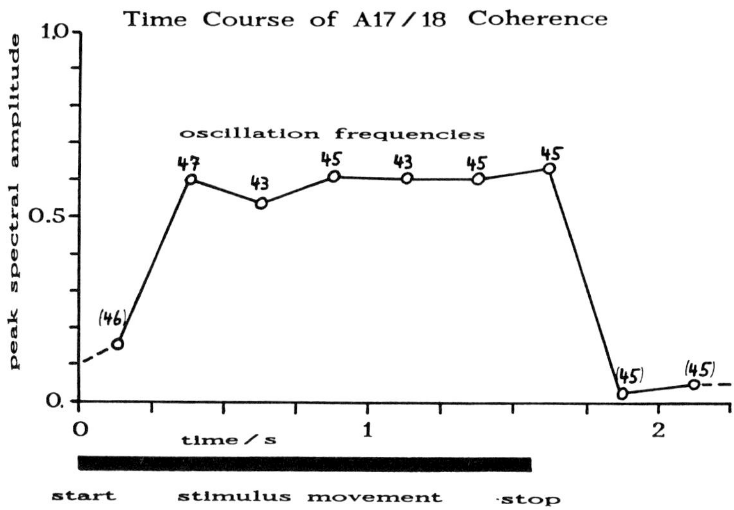

Fig. 9.12. Time course of stimulus-evoked A17/18 synchrony.
Degree of synchronization of the intracortically recorded LFPs is evaluated by the coherence of the interareal synchronization frequency (maximum of coherence function; mean frequencies of maxima are indicated above the curve). Coherence values were determined from averages of 19 short-epoch (250 ms) responses recorded in parallel with microelectrodes from A17 and A18 positions with overlapping RFs; (amplitudes of LFPs during stimulus-induced synchronization were in the 100 μV range; see also legends of Figs. 9 and 11). (Experiment by R. Eckhorn, M. Munk, W. Kruse and M. Brosch)

Hz (Fig. 9.11A). After the stimulus stopped moving, the oscillatory component vanished and broadband frequency components remained (Fig. 9.11B,C). The maxima of the short-period (250 ms) normalized cross-correlograms can be taken as a measure for the overall correlation strength between the LFPs of both areas (Fig. 9.11D). Correlation was relatively high during the entire stimulus period, changing from about 0.4 during stationary presentation to about 0.6 during stimulus movement. The shape of the cross-correlograms reveals, however, that totally different types of signal components determine the correlograms derived under moving and stationary stimulation: high-frequency oscillatory synchronization during stimulation changes to broadband signal coupling after cessation of movement.

Because we wanted to clarify the effect of stimulus-induced interareal synchronization of oscillatory activity over time, we calculated short-epoch (250ms) cross-spectral densities, which give an estimate of *coherence in the frequency domain*. For the data shown (Fig. 9.11), the succession of amplitude peaks of the short-epoch coherence (at the oscillation frequencies) is plotted over the stimulation period as a synchronization index in Fig. 9.12. Coherence at the oscillation frequencies varied in both areas from well below 0.1 without stimulation to above 0.6 during stimulus movement.

This example clearly shows the enormous stimulus influence on the synchronicity among two local assemblies in different cortical areas, which can be explained neither by far field volume conduction nor by a direct entrainment by frequency components contained in the common stimulus. The latter statement can be tested by subtraction of the appropriate statistical expectation value for the stimulus influence on the response correlations ("shift predictor": Gerstein and Perkel 1972; Aertsen and Gerstein 1985; Gerstein et al. 1986; Aertsen et al. 1979, 1986, 1987, 1988; Schneider et al. 1983; Melssen and Epping 1987; Palm et al. 1988). Application of this statistical test did not cause significant changes in our cross-spectra, because (1) the stimulus did not contain the high-frequency components of the oscillations and (2) because their frequencies and phases were highly variable and were not locked to the stimulus (Eckhorn et al. 1988a–c, 1989d). We therefore conclude that stimulus-specific synchronization is a true neural self-organization process having degrees of freedom from the stimulus in the frequency and phase of the oscillations.

9.4.4 Other Stimulus-Specific Variations of Interareal Synchronizations

We showed above that stimulation versus no stimulation can cause large differences in the synchronization of the oscillatory activities in A17 and A18. A general observation, not only for interareal correlations, was that spatially extensive stimuli that cause changes in single-cell responses also influence coherence. Variations in A17-A18 synchronization due to stimulus changes are demonstrated by two types of stimulus variations: (1) monocular versus binocular stimulation and (2) stimulus movement in directions perpendicular to each other.

a) Monocular Versus Binocular Stimulation Alter A17-A18 Synchronizations

Synchronization was quantified by spike-triggered averaging of local field potentials (Fig. 9.13). With monocular stimulation of the dominant eye, the single-cell spikes of one area were only loosely coupled with the field potentials in the other area and oscillatory components are barely visible, although the RFs of the recording positions in both areas overlapped. Binocular stimulation, on the other hand, induced oscillatory activities in both areas that were strongly synchronized. The corresponding correlation spectra reveal these differences, in particular by their peaks at the synchronization frequency (Fig. 9.13B, outer panels). Such coordination between the activity of both areas, which is particularly strong during binocular stimulation, must be important for many aspects of binocular vision; it could, for example, improve the specificity and speed of inter-eye alignment.

b) Stimulus Movement Direction Alters A17–A18 Synchronization

We used in this example (Fig. 9.14) spike-triggered averaging of LFPs to assess the stimulus specificity of interareal synchronization between single cells in one cortical area and LFPs in another. With the preferred stimuli, the local (in-

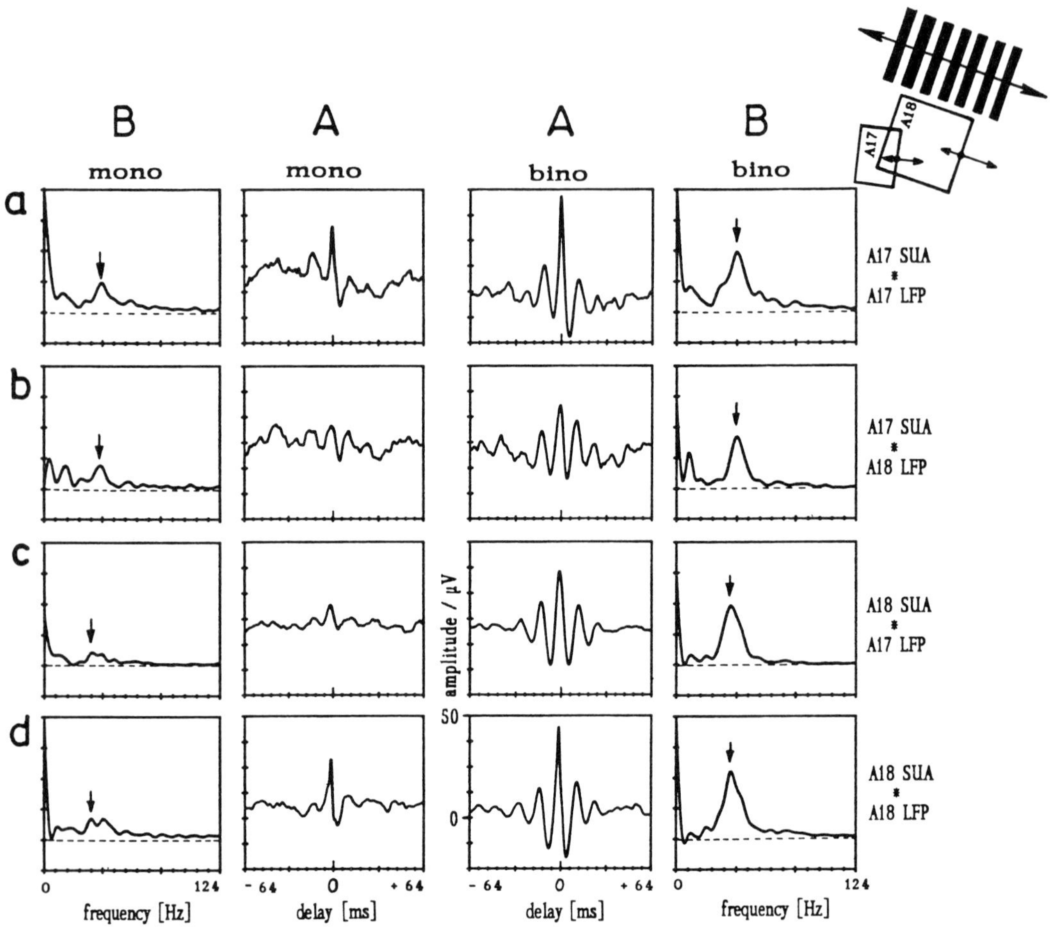

Fig. 9.13. Binocularly evoked synchronizations between A17 and A18 are much stronger than monocular ones.
A Spike-triggered averages of LFPs, calculated for all combinations of A17 and A18 spikes and LFPs (symbols shown at right side; ordinate scale shown in the middle is equal for all STAs of A); **B** correlation spectra calculated by Fourier transformation from the respective spike-triggered averages of A. Receptive fields at the A17 and A18 recording positions slightly overlapped (RF symbols at upper right). Stimulus: Grating, 0.7 cycles/deg moving in and against preferred direction at 8 deg/s. Note the synchronization due to binocular (*bino*) compared with monocular (*mono*) stimulation. The latter was applied to the dominant eye. (Experiment by R. Eckhorn, M. Munk, W. Kruse and M. Brosch)

Fig. 9.14. Stimulus-movement direction can induce synchronization between A17 and A18 single-cell and LFP-oscillatory activities
Synchrony is revealed by spike-triggered averaging of LFPs. Single-cell spike trains and LFPs were both recorded in parallel from the same three microelectrodes of a 19-channel linear matrix; shaft distance of neighboring electrodes was 330 μm. RFs on electrode 6 (in A17) and electrode 13 (in A18) overlapped and had orthogonal direction preferences while the RF on electrode 3 (in A17) did not overlap but had the same directional preference as the A18 RF (RFs shown schematically at the

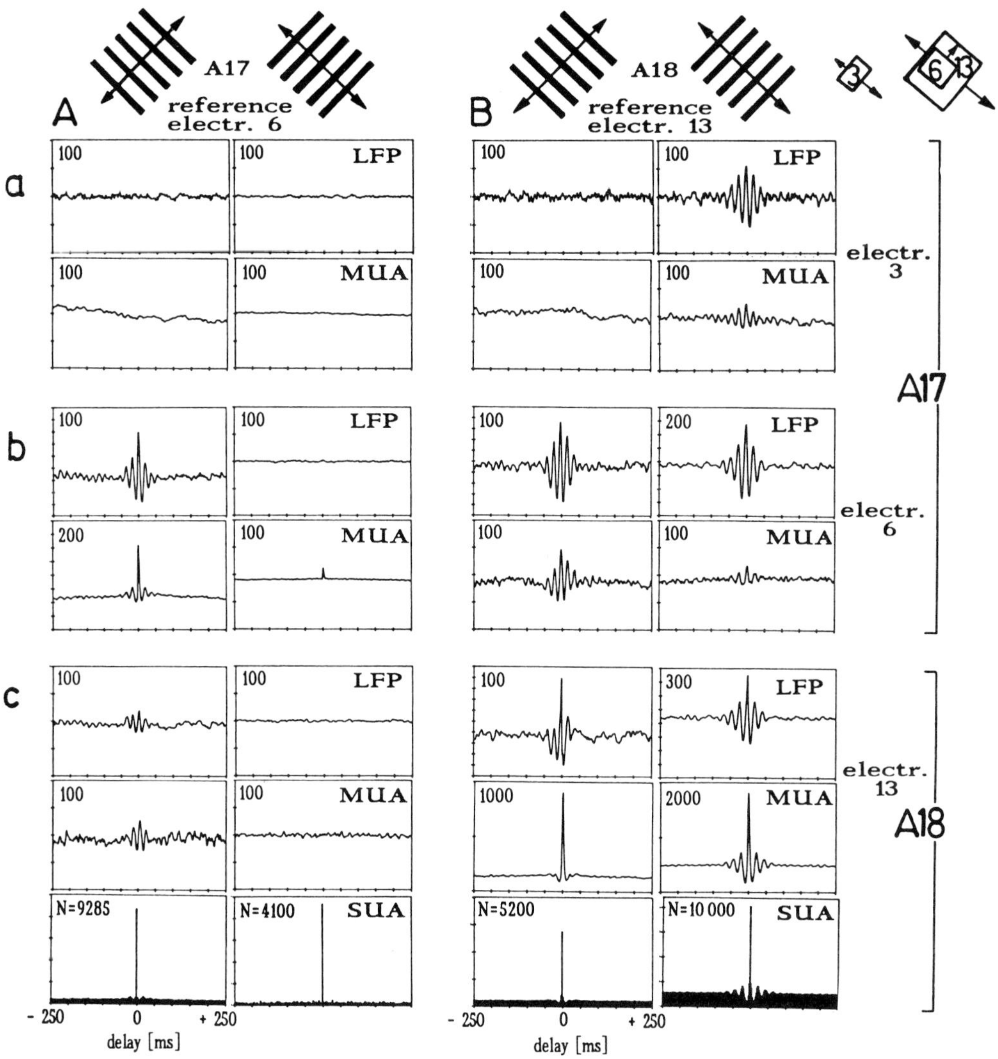

right top). **A** All STAs in left double column: spikes on electrode 6 (A17) as reference for spike-triggered averaging of LFPs or MUA; **B** All STAs in right double column: spikes on electrode 13 (A18) as reference. **a** Averaged LFPs (upper) and MUA (lower) from electrode 3 in A17; **b** signals from el. 6 in A17; **c** signals from el. 13 in A18. Stimuli: Grating 0.7 cycles/deg drifting at 8 deg/s in the directions sketched above. SUA: single-unit spike activity, LFP: local field potentials, MUA: multiple-unit spike activity. The lowermost correlograms in row c are auto-coincidence functions of the spike trains on the reference electrodes 6 (Ac) and 13 (Bc). Numbers in the upper left corner of the correlograms denote the amplitude scales in μV: note that they are different.

Note the different degrees of synchronization between SUA and LFPs with the two different stimulus directions; e.g., strong *intraareal spike-to-LFP synchronizations* were induced with the preferred single-cell movement directions; *interareal synchronization* is strong between A18 spikes on el. 13 and A17 LFPs and MUAs (panels Ba and Bb), while the A17 spikes from el. 6 were only loosely coupled with A18 LFPs and MUAs (panels Ac), and only when the stimulus direction preferred by the cell on el. 6 was applied (more general considerations about the shown effects are mentioned in the text and they will be discussed in more detail in a forthcoming paper). (Experiment by R. Eckhorn, M. Munk, W. Kruse and M. Brosch)

traareal) SUA-LFP correlations are strong; there is a high degree of synchronity. SUA-LFP correlations are considerably smaller and oscillatory synchronization is weak or even absent when not-preferred stimuli were applied. These correlograms characterize single-cell spike coupling with the local LFPs, probably recorded from the direct neighbourhood of its soma. A17/18 and A18/17 correlograms showed different degrees of synchronization when spikes were recorded in one area and LFPs in the other. Interesting in this example are the observations that the A17 single cell activities (electrodes 3 and 6) show only some synchronization with A18 LFPs when their preferred stimuli are applied, while the A18 cell (electrode 13) is synchronized to some degree with A17 LFPs with both stimulus directions. Changes in effective signal couplings with changing stimuli proved to be a general property of neural assemblies and has been reported by a variety of groups (e.g. Gerstein et al. 1983, 1986; Aertsen et al. 1986, 1987, 1988; Epping and Eggermont 1987; Eckhorn et al. 1987, 1988a–c, 1989a,b,d–f, 1990a,b; Reitboeck et al. 1988b; Gray and Singer 1989).

9.5 Mechanisms for Stimulus-Induced Synchronizations Between Remote Cortical Assemblies

9.5.1 Neurophysiological Arguments for Synchronized Local Oscillators

What could be the underlying mechanisms for synchronization of stimulus-induced oscillations in spatially distant structures? Recurrent excitatory connections can induce phase-locking between non-linear local oscillators (e.g. Freeman 1975, 1979, 1987; Othmer 1984). Freeman and his coworkers have performed extensive investigations of oscillatory phenomena in the rabbit olfactory system with EEG recordings (e.g. Freeman 1975, 1981, 1987; Freeman and Skarda 1985; Gray and Skinner 1988). They developed a computer model of the olfactory system that was able to explain most of their experimental observations (Freeman and Skarda 1985; Freeman 1987; Freeman et al. 1988). In particular, their simulations indicated that the crucial parameter for state changes in the network (from stochastic to oscillating behaviour) is *gain of mutually excitatory feedback* among elements representing local subsets of excitatory neurons (Freeman and Skarda 1985). It seems plausible that a network with similar properties might be responsible for coherent oscillations in cat visual cortex. We want to note, however, that phase-locking among remote assemblies in the visual cortex is probably not realized via "classic additive" synaptic connections as in the models of the olfactory system by Freeman, because summation of horizontal excitatory inputs with feeding inputs (mediating the receptive field properties) would cause superpositions of RF properties; e.g., the spatial resolution of the system would deteriorate with such linking connections. Non-linear "modulatory" interactions such as those chosen in our model networks would be more appropriate for mediating phase-locking without the "negative" influence on RFs (Sect. 9.5.2).

In order to build up coherent oscillations within a local assembly, the assembly's neurons have to be activated to the oscillatory state by specific visual feeding inputs. The smallest *oscillator unit* is the key module; it might well be a single neuron, but is more probably a local circuit with feedbacks via other nearby neurons (Freeman, e.g., 1975, 1985, 1987; Freeman et al. 1988).

Coupled cortico-geniculate oscillations have been proposed in models of preattentive vision by Grossberg (1980, 1983a,b), who postulated that synchronization between cortical and thalamic circuits might be a powerful mechanism for supporting stereoscopic vision, for the improvement of the signal-to-noise ratio and for "filling-in" phenomena. In several recording tracks through the lateral geniculate we have failed as yet to observe oscillatory activity, and Singer and coworkers have reported similar negative findings (Singer et al. 1988).

Anatomical evidence that connections with probably an excitatory influence play a primary role in the visual cortex is well documented in the literature. Clustered intrinsic excitatory connections extending horizontally over up to several millimeters were reported to be present in the primary visual cortex (e.g. Gilbert 1985; Gilbert and Wiesel 1983, 1987). Interareal reciprocal connections were found to be distributed in isolated patches that terminate at correspondent positions of the visual representation, and they are also considered to be mutually excitatory (e.g. Symonds and Rosenquist 1984; Leferink et al. 1987; Salin et al. 1989; for a review see Zeki and Shipp 1988). The direct influence of these connections for visual function has, however, not yet been demonstrated.

Physiological evidence for excitatory intracortical connections has been reported for the cat primary visual cortex. Using cross-correlation techniques Ts'o and coworkers found that distribution and range of the "excitatory" interactions do correspond to the clustering and extent of the horizontal connections observed anatomically (Ts'o et al. 1986). They stated that "the extent of the horizontal connections and their facilitatory nature suggests that they may contribute to properties beyond those determined by classical receptive field analyses". We argue that the long intra- and inter-areal connections belong, at least in part, to a linking network. The axons of intracortical connections have small diameters and low conduction velocities (about 1 m/s). Assemblies that are intracortically connected and that simultaneously exhibit stimulus-induced oscillations could theoretically, therefore, be phase-locked within a quarter of an oscillation cycle up to distances of 6.25 mm (assuming an oscillation frequency of 40 Hz). We have shown, on the other hand, that areas 17 and 18 can synchronize their stimulus-induced oscillations with about zero phase shift (Eckhorn et al. 1988a–c); This finding suggests that synchronizations among remote A17 assemblies might significantly be caused by recurrent modulatory connections from other visual cortical areas (Eckhorn et al. 1990a). Such recurrent inter-areal connections are faster conducting than intra-areal fibers and they are connected with high degrees of convergence and divergence, much wider than would be necessary for the "construction" of classical RF properties (e.g. Salin et al. 1989). Nelson and Frost (1985) have already argued that the intraareal connections are quantitatively too broadly ramifying to account for any classical RF regions (except for periodic complex cells, reported by Pollen and Ronner 1975).

Role of the A17/18 Connections. We have seen that assemblies representing common linking features of a visual scene can transiently label the activities of the participating neurons with a temporal code, which denotes their actual associations. For the primary visual areas of the cat with their parallel visual inputs, no other plausible functional role has been claimed so far for the massive reciprocal interconnections (Eckhorn et al. 1988a–c; 1989d–f; 1990a,b). Functional elimination of either A17 or A18 in the cat by cooling or ablation of one of the areas did not basically alter the primary coding properties of single units in the intact area (Donaldson and Nash 1975; Dreher and Cottee 1975; Sherk 1978). We assume that the momentary stimulus configuration and thus the distribution of synchronized oscillations across each single area determines the mode of cooperation between A17 and A18 (cooperative parallel, cooperative serial, independent parallel or local mixtures of combinations). If, for example, the activated assemblies in both areas are involved in the coding of features of the same moving rigid object, then their oscillating activities might be synchronized, i.e., these patches in the different areas cooperate at that moment. Simultaneously, other parts of the two areas might be involved in the coding of different, more independent, aspects of a visual scene such as figure and ground. The signals of the assemblies representing these different aspects are not correlated.

9.5.2 Arguments from Network Models for Synchronized Local Oscillators

In the simulations of feature linking via synchronisation we use a *model neuron* containing leaky integrators at its synapses and a dynamic threshold mechanism with negative feedback (Fig. 9.15A; Eckhorn et al. 1989a,b). Our model neuron has an important feature that assures linking between neurons without major degradation of the "receptive field" (RF) properties. The "RF-preserving" property is due to the modulatory effect that *linking inputs* exert on *feeding-inputs*: The integrated signals from the linking inputs, together with a constant offset term (+1), interact multiplicatively with the integrated signals from the feeding inputs. Afferent "visual" signals drive the neurons mainly via the feeding inputs, which largely determines the RF properties. Without linking signals, the output of the "neural multiplyer" is identical to the (integrated) feeding signal. With activity at the linking inputs, the amplitude of the feeding inputs is modulated by the multiplier and the threshold discriminator will switch at different times, thereby shifting the phase of the output pulses (Eckhorn et al. 1989a,b, 1990a,b). Such receptive field-preserving linking property has not been taken into account in other models of oscillatory feature linking (e.g., Kammen et al. 1989; Sporns et al. 1989).

The *neural networks* used in the simulations consist either of one or two one-dimensional layers of "visual neurons" (Fig. 9.15C; Eckhorn et al. 1990a,b). The "receptive field" properties within each neural array are assumed to be identical (all cells respond, for example, optimally to contrast edges of the same orientation; RF centers are equidistantly aligned). Each neuron of layer 1 has a single feeding input (continuous amplitude signal) that defines the neuron-

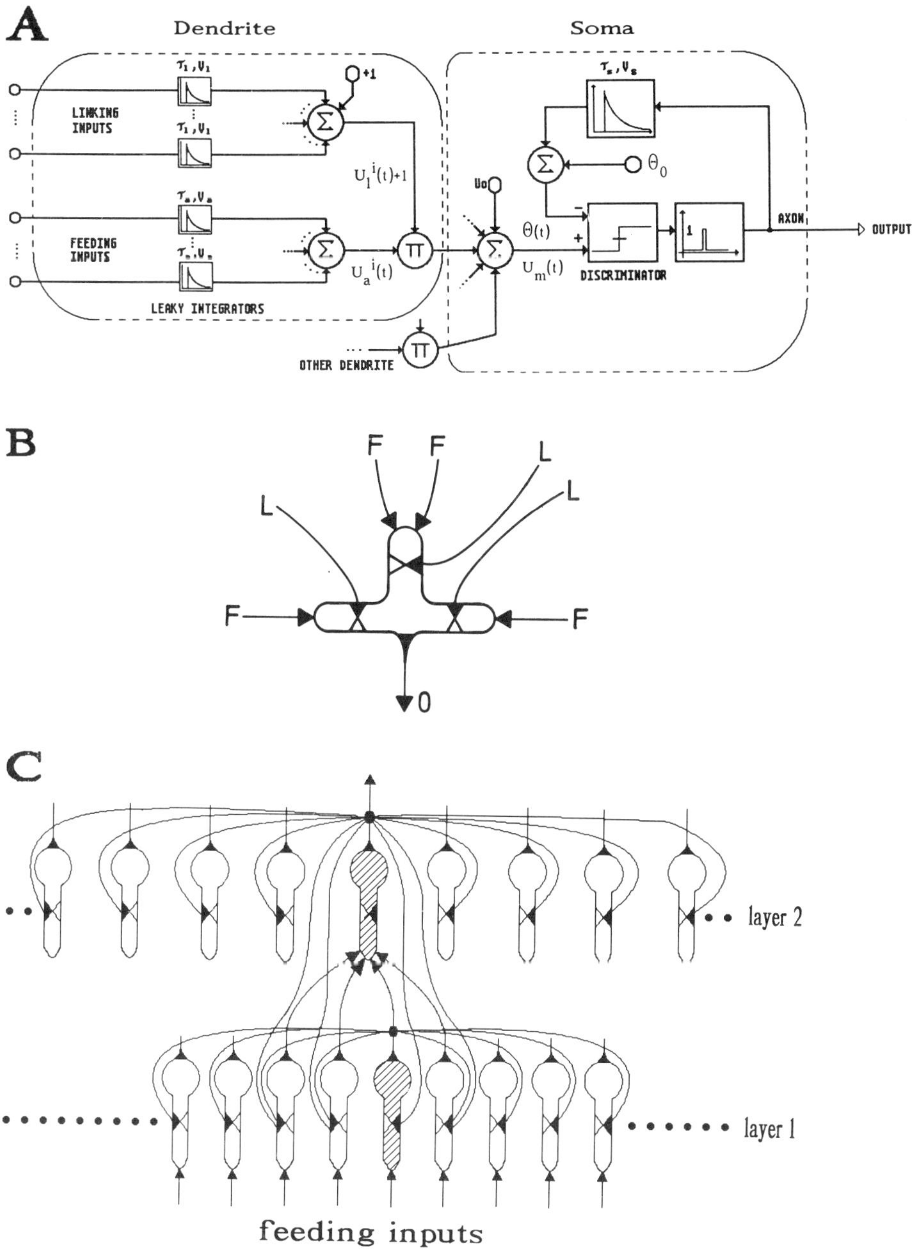

Fig. 9.15. The model neuron.
A *Circuit diagram*. Linking and feeding inputs with leaky integrators on a single "dendrite" interact multiplicatively. Dendritic output signals are summated and fed into the spike generation mechanism with dynamic threshold in the "soma". **B** *simplified symbol for neuron in A*. *F* feeding connection; *L* linking connection; *O* output (axon); **C** *neural network* used for simulation Fig. 9.16B. Full output connectivity is only shown for one layer-1 and one layer-2 neuron (hatched symbols). From (Eckhorn et al. 1990b)

al "RF properties", while the RFs of layer 2 neurons are defined via convergent feeding inputs from four neighboring layer 1 neurons. In terms of RFs this means that layer 2 RFs completely overlap four layer 1 RFs. In both layers each neuron is connected to 4 neighbors at each side. The (positive) strength of the linking synapses laterally declines with distance. Forward-feeding and backward-linking connections are each of equal strength. (The model parameters, such as time constants, coupling strengths and number of convergent and divergent connections, do not crucially influence the simulation results).

The *dynamic response of the network* is shown in the following examples of moving stimuli. In the first example a "patch of higher intensity" moves with constant velocity across the inputs of layer 1 while the feedback from layer 2 is disconnected (Fig. 9.16A). During the stimulation period the synchronized oscillations are maintained without major phase jumps. The relatively constant extent of the synchronized assembly following the moving stimulus is clearly visible, although a high level of stochastic background activity was applied to the inputs.

In the simulation for Fig. 9.16B the second layer is included. Each layer 2 neuron projects back to linking inputs of four layer 1 neurons. Neurons in the two "stimulated regions" in layer 1 respond with phase-locked rhythmic burst discharges, although they are separated by a gap of spontaneously active neurons. Thus, the two layer 1 regions are transiently linked from the "higher visual map" with its broader RFs. This synchronized activity in separated regions can be interpreted to serve the definition of spatial unity and temporal continuity of an object's "neural representation" during the synchronization period (Eckhorn et al. 1988a–c, 1989a,b,d–f; Eckhorn et al. 1990a,b).

9.5.3 Breaking a Link

Inclusion in and exclusion from a synchronized assembly would be necessary at region boundaries within the cortical representation of a visual scene. Thus, any type of cell that is able to code spatial or temporal differences (local contrasts, end-stopping, movement etc.) should be able to contribute to the inclusion or release of linked local cells. There must be local wiring adapted to this task: Activation of such "contrast detectors" by a stimulus discontinuity could inter-

Fig. 9.16. Simulation of stimulus-induced synchronizations.
Each "action potential" is marked by a dash at the time of its occurrence. Stochastic input activity was applied at all layer-1 feeding inputs during the entire sweep. **A** *Single layer simulation* with 50 neurons; "stimulation" with a constantly moving patch of increased intensity at the feeding inputs of 10 neurons, beginning at the lower end of the array (indicated at the left side); STIM: stimulus duration indicated by black bar; MUA: "multiple unit activity", calculated as spatial average of spike outputs with laterally declining weight distribution at the center of the array; LFP: "local field potential", spatial average of "soma potentials", same weighting as for MUA; **B** *double layer simulation* with 50 neurons in layer 1 and linking feedback from the 23 neurons of layer 2 (Fig. 16C). Note the synchronization of bursts in layer-1 subregions with those in layer 2, due to lateral and feedback linking connections of layer 2. From (Eckhorn et al. 1990b)

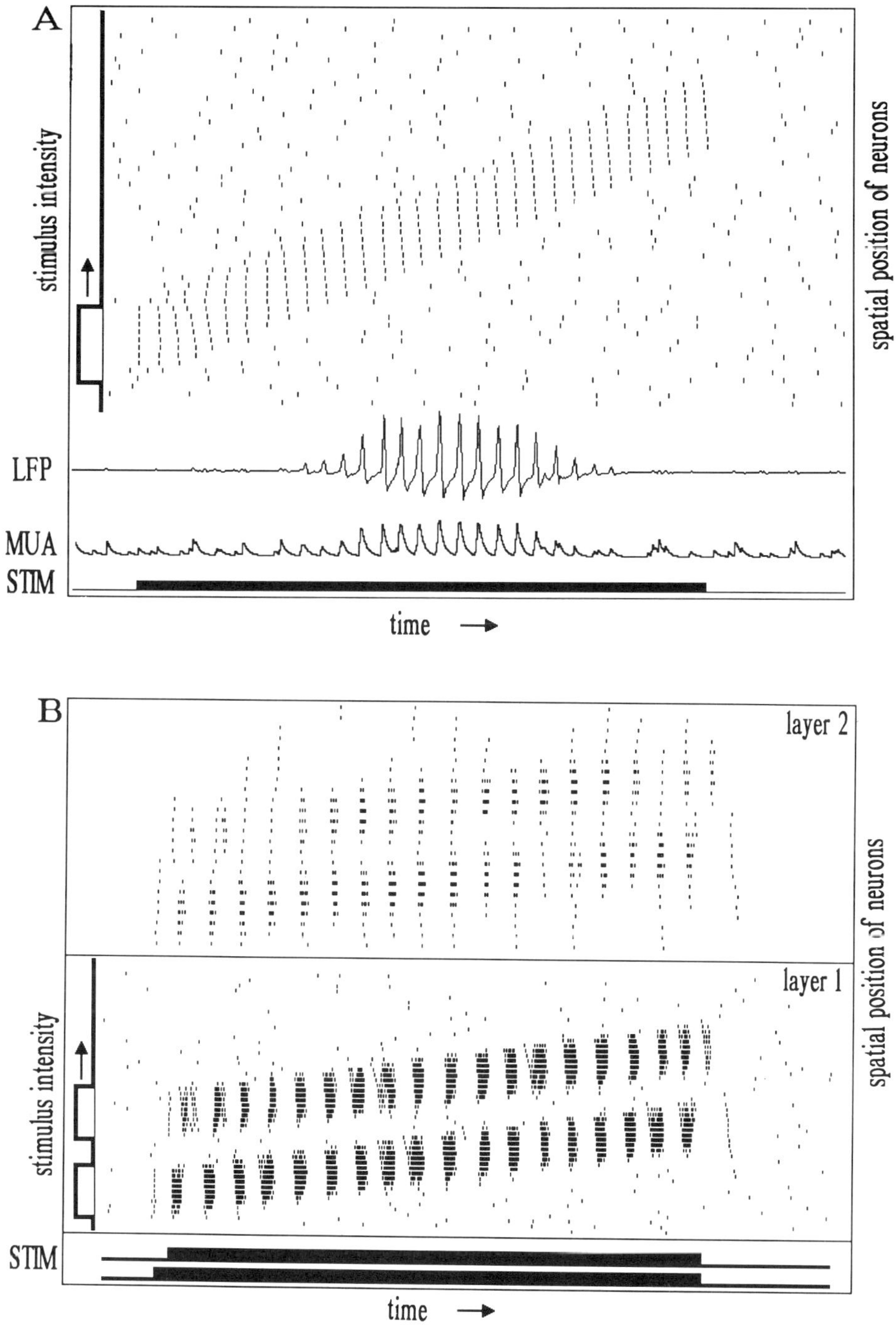
A
stimulus intensity
spatial position of neurons
LFP
MUA
STIM
time
B
layer 2
layer 1
stimulus intensity
spatial position of neurons
STIM
time

rupt the linking, e.g., by shunting inhibition on the local connections of the linking network. In all positions of a visual scene, where no detectable discontinuities are present, the "contrast detectors" are not activated so the linking network is functionally intact and might support "perceptual filling-in" by mediating the synchronization with the actual "main assembly" (see also the models of Grossberg (1976, 1983a,b)). *Inclusion* of a particular "unit oscillator" into an already internally synchronized assembly is thus mediated by synchronizing that unit with the other ones. Speed and strength of such a phase-coupling depends on the type and efficacy of the neural connections between structures that are synchronizing each other. *Exclusion* of an oscillator unit, however, could be mediated by functionally cutting off its linking connections to the reverberating assembly. Inclusion in and release from a synchronized group is assumed to function within a single domain of feature representation or between different feature domains; both might happen within the same or across different cortical areas.

9.5.4 Stimulus-Forced and Stimulus-Induced Synchronization

Two different types of synchronizations were observed in cat visual cortex and in our neural networks: stimulus-forced and stimulus-induced synchronizations (Eckhorn et al. 1989a,b; Pabst et al. 1989; Eckhorn et al. 1990a,b).

Stimulus-forced synchronizations are driven directly by stimulus transients; the spatially distributed signal components are temporally locked to the time course of a common stimulus and they are only oscillatory with extremely strong transient stimuli (such as light flashes). The degree of synchrony, although mainly generated by the common stimulus transients, can be enhanced via lateral and/or feedback linking connections of the same type as those mediating the stimulus-induced oscillatory synchronizations. Stimulus-forced synchronizations are likely to play a role already at primary processing stages (retina and lateral geniculate body) and in all areas of the visual cortex.

Stimulus-forced synchronizations have been observed in the visual cortex since the early beginnings of cortical electrophysiology: their overall manifestation is visually evoked potentials. Detailed results about such stimulus-locked synchrony of intracortical signals were obtained with current source density analyses (CSD; e.g., Mitzdorf 1987). Most interesting with respect to the "linking-concept" is the fact that synchronized CSD responses are not only evoked via direct RF activation at the recording positions, but also by large-area stimuli that are far away from the RFs of neurons at the recording position. Another important finding is that a characteristic activation sequence of synchronized CSD signals traverses the cortical layers. Both effects are largely independent of the type of stimulation (Mitzdorf 1987).

Stimulus-induced synchronizations (recorded as oscillatory mass activities) are probably generated internally via a self-organizing process among stimulus-driven local "oscillators" that are mutually connected. (The interaction is supposed to be via the same linking network that supports stimulus-forced synchrony). During fixation periods of the eye and with slowly drifting retinal stimuli (and perhaps also during mental visualization and "attentive drive"), retinal signals do not provide strong synchronous transients. In such "temporally ill-defined" visual situations, our model network (and obviously the cortex too) can generate the synchronous signal transients necessary for the generation of temporal linking (association) by generating sequences of synchronized bursts of 35 to 80 Hz. Sensory systems that are never supplied by well-timed signals from their receptors, such as the olfactory system, would always have to use the latter mechanism of self-generation of repetitive synchronized activities in order to form temporal association codes (for oscillations in olfactory system see, e.g., Freeman 1975).

We assume that stimulus-forced (event-locked) synchronization and stimulus-induced oscillatory synchronization are *complementary mechanisms* in temporal feature linking. In natural vision, both types of synchronization occur in a given assembly, generally in temporal succession. This is supported by our observation that oscillatory activities never occurred during or immediately after strong-evoked potentials (VECPs), i.e. VECPs obviously suppress oscillations. Stimulus-forced synchronizations are assumed to establish fast but "crude sketches" of association codes in the visual cortex, whereas repetitive internally generated synchronization allows the formation of complex and sophisticated linking codes in iteration steps with 35 to 80 Hz (see also the models of Grossberg 1983a,b; Damasio 1989a,b; Sporns et al. 1989).

9.5.5 Linking Fields

In order to bring neuronal mechanisms of feature linking, as discussed above, into correspondence with perceptual capabilities, we introduce the concept of the *linking field* (association field) of a local neural assembly. The linking field extends the concept of single cell receptive fields (local coding) to neural ensemble coding:

As the *linking field of a local reference assembly of visual neurons*, we define that area in visual space where appropriate global stimulus features can induce synchronization between the activities of the reference assembly and other assemblies (that have their aggregate RFs within the range of the linking field). Linking fields are transiently "constructed" by the properties of their RFs and constituent neurons, according to the linking interconnections (feed-forward, lateral, feedback).

Linking fields of local assemblies in visual cortex can roughly be estimated from recordings of stimulus-induced LFP oscillations, because LFPs, recorded in a single location, are influenced by synchronized activities from remote assemblies in their amplitudes, if these signals are transmitted via axonal connections to the recording location. Such "gamma-spindle fields" or "oscillations fields" were recently measured by us. LFP oscillation fields were found to be generally much broader and less stimulus-specific than the classical receptive fields of the assemblies' neurons, (and even much broader than the aggregate RFs of local assemblies). Oscillation fields of cat A18 assemblies were found to extend to 30 by 30 degree visual angle, which indicates influences from synchronized locations over many millimeters of cortical distance. We assume that such long-range stimulus specific synchronizations are a main task of the "association fiber systems" that connect non-correspondent locations of the visual cortical representation within the same and between different cortical areas. Such "association" connections with "receptive field mismatch" are known since many years, but their functional roles ahd remained unclear (e.g., Bullier et al. 1984, 1988; Ferrer et al. 1988; Gilbert and Wiesel 1983, 1985; Zeki and Shipp 1988).

9.6 Data from Awake Cats

Some of our findings in the lightly anesthetized cat are now supported by first results from the awake cat: Gray and Singer (1989) and our group have found stimulus-induced synchronizations of local oscillatory activities with implanted electrodes in awake cats that are specific with respect to the movement direction of a grating (unpublished experiments of R. Bauer and R. Eckhorn 1987, 1988). The similarity of the stimulus-induced oscillatory states between anesthetized and awake cats (recorded in the same cat with chronically implanted electrodes at recording positions with local tuning properties specified in the anesthetized state) suggests that stimulus-induced oscillatory synchronizations are obviously not due to anesthesia, but they indeed occur in more natural states of the cortex. A logic progression of our experimental work and proof of the hypotheses for global coding requires, however, the trained, behaving, conscious monkey (Freeman and van Dijk 1987). We can then correlate the occurrence of stimulus-induced spatio-temporal synchronizations between single-cell and mass activities, with the perceptually guided reactions of the animal, to global aspects of a figure.

Acknowledgements. I am especially grateful to Prof. H. J. Reitboeck, Head of our Working Group, for his continuous support in developing both concepts and the multiple-electrode instrumentation. This article could not have been written without help in experiments, data acquisition and processing by my colleagues, as credited in the figure captions, and including Drs. R. Bauer, M. Brosch, W. Jordan, W. Kruse, J. Leferink, M. Munk and T. Schanze. Neural modelling was carried out by M. Arndt and P. Dicke to whom I am grateful. Many thanks to my colleague Dr. J.I. Nelson for improving the English. I am also grateful to Prof. T.H. Bullock for helpful comments on this manuscript. Expert help in the care and preparation of our animals and in experimental techniques was supplied by U. Thomas, W. Gerber, W. Lenz, D. Mischke, P. Muth, and J.H. Wagner. This project received financial support from the Deutsche Forschungsgemeinschaft Ba 636/4-1,2 (to R. Bauer), Re 547/2-1 (to H. J. Reitboeck and R. Eckhorn), Ec 53/4-1 (to R. Eckhorn and J. I. Nelson) and from Stiftung Volkswagenwerk VW I/35695 (to H. J. Reitboeck), and I/64605 (to H. J. Reitboeck and R. Eckhorn).

References

Aertsen, A.M.H.J., Gerstein, G.L. (1985): Evaluation of neuronal connectivity: sensitivity of cross correlation. Brain Res. **340**, 341–354

Aertsen, A.M.H.J., Smolders, J.W.T., Johannesma, P.I.M. (1979): Neural representation of the acoustic biotope: on the existence of stimulus-event relations for sensory neurons. Biol. Cybern. **32**, 175–185

Aertsen, A.M.H.J., Gerstein, G.L., Johannesma, P.I.M. (1986): From neuron to assembly: neuronal organization and stimulus representation. In: *Brain Theory* Palm, G; Aertsen, A.M.H.J. (eds.), (Springer, Berlin, Heidelberg)

Aertsen, A.M.H.J., Bonhoeffer, T., Krüger, J. (1987): Coherent activity in neuronal populations: analysis and interpretation. In: Physics of Cognitive Processes Caianiello E (ed) (World. Scientific, Singapore) pp. 1–34

Aertsen, A.M.H.J., Gerstein, G.L., Habib, M.K., Palm, G. (1989): Dynamics of neuronal firing correlation: Modulation of "effective connectivity" J. Neurophysiol. **61**, 900–917

Aiple, F., Krüger, J. (1988): Neuronal synchrony in monkey striate cortex: interocular signal flow and dependency on spike rates. Exp. Brain Res. **72**, 141–149

Altmann, L., Eckhorn, R., Singer, W. (1986): Temporal integration in the visual system: influence of temporal dispersion on figure-ground discrimination. Vis. Res. **26**, 1949–1957

Arnett, D.W. (1975): Correlation analysis of units recorded in the cat dorsal lateral geniculate nucleus. Exp. Brain Res. **24**, 111–130

Basar, E. (1980): *EEG-Brain Dynamics.* (Elsevier, Amsterdam)

Basar, E. (1983): Synergetics of neuronal populations. A survey on experiments. In: Synergetics of the Brain, ed., E. Basar, H. Flohr, H. Haken, A. Mandell (Springer, Berlin, Heidelberg) 183–200

Basar, E. (1988): EEG-dynamics and evoked potentials in sensory and cognitive processing by the brain. In: Dynamics of sensory and cognitive processing by the brain, E. Basar (ed) (Springer, Berlin, Heidelberg) pp. 30–55

Basar, E., Basar-Eroglu, C., Röschke, J., Schütt, A. (1989): The EEG is a quasi-deterministic signal anticipating sensory-cognitive tasks. In *Brain Dynamics*, ed. by E. Basar, T.H. Bullock, Springer Ser. Brain Dynamics, vol. 2: (Springer, Berlin, Heidelberg) pp. 43–71

Bauer, R. (1982): A high probability of an orientation shift between layers 4 and 5 in central parts of the cat striate cortex. Exp. Brain Res. **48**, 245–255

Bauer, R., Fischer, W.H. (1987): Continuity or incontinuity of orientation columns in visual cortex: A critical evaluation of published and unpublished data. Neurosci. **22**, 841–847

Bauer, R., Dow, B.M., Snyder, A.Z., Vautin, R.G. (1983): Orientation shift between upper and lower layers in monkey visual cortex. Exp. Brain Res. **50**, 133–145

Bauer, R., Eckhorn, R., Jordan, W., Arndt, M. (1987): Iso- and cross-orientation columns in cat striate cortex: A re-examination with simultaneous single- and multi-unit recordings. Proc. EBBS Conf. Novi Sad, Jugoslavia, T10

Bauer, R., Eckhorn, R., Jordan, W. (1989a): Iso- and Cross-orientation columns in cat striate cortex: A re-examination with simultaneous single- and multi-unit recordings. Neurosci **30**, 733–740

Bauer, R., Eckhorn, R., Jordan, W. (1989b): Different directional anisotropies for texture and grating stimuli in supragranular cells of cat striate cortex: A basis for figure-ground separation? In: Elsner N, Singer W (eds), Dynamics and Plasticity in Neuronal Systems, Thieme-Verlag, Stuttgart, New York, p. 350

Baxter, W., Dow, B.M, (1988): Centric models of the orientation map in primate visual cortex. In Neural Information Processing Systems – Neural and Synthetic. Anderson DZ (ed), (Amer. Inst. Phys., New York) pp 62–72

Bullier, J., Kennedy, H., Salinger, W. (1984): Branching and laminar origin of projections between visual cortical areas in the cat. J. Comp. Neurol. **228**, 329–341

Bullier, J. McCourt, M.E., Henry, G.H. (1988): Physiological studies on the feedback connection to the striate cortex from cortical areas 18 and 19 of the cat. Exp. Brain Res. **70**, 90–98

Bullock, T.H. (1988): Compound potentials of the brain, ongoing and evoked: perspectives from comparative neurology. In: Basar E (ed), Dynamics of sensory and cognitive processing by the brain, Springer-Verlag, pp. 3–18

Bullock, T.H., Basar, E. (1988): Comparison of ongoing compound field potentials in the brains of invertebrates and vertebrates. Brain Res. Rev. **13**, 57–75

Cracco, R.Q., Cracco, J.B. (1978): Visual evoked potentials in man: early oscillatory potentials. Electroencephal. Clin. Neurophysiol. **45**, 731–739

Creutzfeldt, O.D. Watanabe, S., Lux, H.D. (1966): Relation between EEG-phenomena and potentials of single cells. Part I and II. Electroenceph. Clin. Neurophysiol. **20**, 1–37
Damasio, A.R. (1989a): The Brain Binds Entities and Events by Multiregional Activation from Convergence Zones. Neural Computation **1**, 121–129
Damasio, A.R. (1989b): Time-locked multiregional retroactivation: A systems level proposal for the neural substrates of recall and recognition. Cognition **33**, 25–62
Donaldson, I.M.L., Nash, J.R.G. (1975): The effect of a chronic lesion in cortical area 17 on the visual responses of units in area 18 of the cat. J. Physiol. **245**, 325–332
Dow, B.M., Bauer, R. (1984): Retinotopy and orientation columns in the monkey: a new model. Biol. Cybern. **49**, 189–200
Dreher, B., Cottee, L.J. (1975): Visual receptive-field properties of cells in area 18 of cat's cerebral cortex before and after acute lesions in area 17. J. Neurophysiol. **38**, 735–750
Eckhorn, R. (1987): Relating cortical structure and function with multi-electrode/correlation techniques: Investigation of column and patch-systems in cat's visual cortex. Abstr Vol Confer on Dynamics of Sensory and Cognitive Processing by the Brain, West-Berlin
Eckhorn, R., Pöpel, B. (1972a): A contribution to neuronal network analysis in cat lateral geniculate body: simultaneous recordings of maintained activity. Intern. J. Neurosci. **4**, 137–142
Eckhorn, R., Pöpel, B. (1972b): Statistical network analysis of cat lateral geniculate body: cross-correlation and rate of transinformation. Eur. J. Physiol. (Suppl) **335**, R85
Eckhorn, R., Pöpel, B. (1975): Rigorous and extended application of information theory to the afferent visual system of the cat. II. Experimental results. Biol. Cybernetics **17**, 7–17
Eckhorn, R., Grüsser, O.-J., Kröller, J., Pellnitz, K., Pöpel, B. (1976): Efficiency of different neuronal codes: Information transfer calculations for three different neuronal systems. Biol. Cybernetics **22**, 49–60
Eckhorn, R., Schneider, J. Keidel, R. (1986a): Real-time covariance computer for cell assemblies is based on neuronal principles. J. Neurosci. Meth. **18**, 371–383
Eckhorn, R., Krause, F., Nelson, J.I. (1986b): The receptive field cinematogram: a multiple receptive field mapping technique for multiple microelectrode studies. Soc. Neurosci. Abstr. 36.3
Eckhorn, R., Bauer, R., Jordan, W., Epping, W., Arndt, M. (1987): Correlated activity and coding properties of neuron groups within vertical columns of the visual cortex: A multi-electrode study. 2nd IBRO-Congrs, Neurosci. 22 Suppl. 1306P
Eckhorn, R., Reitboeck, H.J. (1988): Assessment of cooperative firing in groups of neurons. In: Springer Series in Brain Dynamics E Basar, edit. Vol 1, Springer Verlag, Berlin, Heidelberg, pp. 219–227
Eckhorn, R., Bauer, R., Jordan, W., Brosch, M., Kruse, W., Munk, M., Reitboeck, H.J. (1988a): Are form- and motion-aspects linked in visual cortex by stimulus-evoked resonances? Multiple electrode and cross-correlation analysis in cat visual cortex. EBBS-Workshop on Visual Processing of Form and Motion, Tübingen, Confer. Vol, p. 7
Eckhorn, R., Bauer, R., Brosch, M., Jordan, W., Kruse, W., Munk, M. (1988b): Functionally related modules of cat visual cortex show stimulus-evoked coherent oscillations: A multiple electrode study. Invest. Ophthalmol. Vis. Sci. **29**, 331 (12)
Eckhorn, R., Bauer, R., Jordan, W., Brosch, M., Kruse, W., Munk, M., Reitboeck, H.J. (1988c): Coherent oscillations: A mechanism of feature linking in the visual cortex? Multiple electrode and correlation analysis in the cat. Biol. Cybernetics **60**, 121–130
Eckhorn, R., Reitboeck, H.J., Arndt, M., Dicke, P. (1989a): Feature linking via stimulus-evoked oscillations: Experimental results from cat visual cortex and functional implications from a network model. Intern. Conf. Neural Networks, Washington. Conf. Vol. I: 723–730
Eckhorn, R., Reitboeck, H.J., Arndt, M., Dicke, P.(1989b): A neural network for feature linking via synchronous neural activity. In: Models of Brain Function. R.M.J. Cotterill (ed.). Cambridge Univ. Press, pp. 255–272
Eckhorn, R., Bauer, R., Reitboeck, H.J. (1989c): Global feature processing in the visual cortex can be analyzed by combining multiple single-cell and mass-activity recordings with signal correlation methods. Perception **18**, A67
Eckhorn, R., Bauer, R., Reitboeck, H.J. (1989d): Discontinuities in visual cortex and possible functional implications: Relating cortical structure and function with multi-electrode/correlation techniques. In Basar E, Bullock T.H. (eds), Springer Series in Brain Dynamics Vol 2, pp. 267–278
Eckhorn, R., Munk, M., Kruse, W., Brosch, M., Jordan, W., Bauer, R. (1989e): Stimulus-evoked synchronizations between two visual cortex areas of the cat: A linking mechanism for different visual features? In: Elsner N, Singer W (eds), Dynamics and Plasticity in Neuronal Systems, Thieme-Verlag, Stuttgart New York, p. 349

Eckhorn, R., Kruse, W., Brosch, M., Munk, M., Jordan, W., Schanze, T., Bauer, R. (1989f): Pre-attentive feature linking between different cortex areas by stimulus-specific synchronizations? A multiple single-cell and mass-activity analysis in cat visual cortex. Perception **18**, A29

Eckhorn, R., Reitboeck, H.J., Dicke, P., Arndt, M., Kruse, W. (1990a): Feature linking across cortical maps via synchronization. In: Parallel Processing in Neural Systems and Computers, Eckmiller R (ed), Elsevier, Amsterdam, pp. 100–104

Eckhorn, R., Reitboeck, H.J., Arndt, M., Dicke, P. (1990b): Feature linking via synchronization among distributed assemblies: Simulations of results from cat visual cortex. Neural Computation **2**, 293–307

Epping, W.J.M., Eggermont, J.J. (1987): Coherent neural activity in the auditory midbrain of the grassfrog. J. Neurophysiol. **57**, 1464–1483

Ferrer, J.M.R., Price, D.J., Blakemore, C. (1988): The organization of cortico-cortical projections from area 17 to area 18 of the cat's visual cortex. Proc. Roy. Soc. (London) B **233**, 77–98

Freeman, W.J. (1975): Mass action in the nervous system. Academic Press, New York

Freeman, W.J. (1979): Nonlinear gain mediating cortical stimulus-response relations. Biol. Cybernetics **33**, 237–247

Freeman, W.J. (1981): A Physiological Hypothesis of Perception. Prospectives in Biology and Medicine **24**, 561–592

Freeman, W.J., (1987): Simulation of chaotic EEG patterns with a dynamic model of the olfactory system. Biol. Cybern. **56**, 139–150

Freeman, W.J., Skarda, C.A. (1985): Spatial EEG patterns, non-linear dynamics and perception: the Neo-Sherringtonian view. Brain Res. Rev. **10**, 147–175

Freeman, W.J., Van Dijk, B.W. (1987): Spatial patterns of visual cortical fast EEG during conditional reflex in a rhesus monkey. Brain Res. **422**, 267–276

Freeman, W.J., Yao, Y., Burke, B. (1988): Central pattern generating and recognizing in olfactory bulb: A correlation learning rule. Neural Networks **1**, 277–288

Gerstein, G.L., Aertsen, A.M.H.J., (1985): Representation of cooperative firing activity among simultaneously recorded neurons. J. Neurophysiol. **54**, 1513–1528

Gerstein, G.L., Perkel, D.H. (1972): Mutual temporal relationships among neuronal spike trains. Statistical techniques for display and analysis. Biophys. J. **12**, 453–473

Gerstein, G.L., Bloom, M., Espinoza, I., Evanczuk, S., Turner, M. (1983): Design of a laboratory for multi-neuron studies. IEEE SMC **13**, 668–676

Gerstein, G.L., Aertsen, A., Bloom, M., Espinosa, I., Evanczuk, S., Turner, M. (1986): Multineuron experiments: observation of state in neural nets. Proceed Symposium Synergetics, Bavaria, W. Germany, Springer-Verlag, Berlin, Heidelberg, New York

Gilbert, CD. (1985): Horizontal integrations in the neocortex. Trends Neurosci. **8**, 160–165

Gilbert, C.D., Wiesel, T.N. (1983): Clustered intrinsic connections in cat visual cortex. J . Neurosci. **3**, 1116–1133

Gilbert, C.D., Wiesel, T.N. (1987): Relationships between cortico-cortical projections, intrinsic cortical connections and orientation columns in cat primary visual cortex. Soc. Neurosci. Abstr. Vol. **13**, 5.9

Gray, C.M., Singer, W. (1987a): Stimulus-dependent neuronal oscillations in the cat visual cortex area 17. 2nd IBRO-Congrs, Neurosci Suppl, 1301P

Gray, C.M., Singer, W. (1987b): Stimulus specific neuronal oscillations in the cat visual cortex: a cortical functional unit. Soc. Neurosci. Abstr. 404.3

Gray, C.M., Skinner, J.E. (1988): Centrifugal regulation of neuronal activity in the olfactory bulb of the waking rabbit as revealed by reversible cryogenic blockade. Exp. Brain Res. **69**, 378–386

Gray, C.M., Koenig, P., Engel, A., Singer, W. (1988): Nonlinear cooperativity mediating oscillatory responses in cat visual cortex: local and global interactions. In: Sense Organs, Interface Between Environment and Behaviour, Elsner N, Barth FG (eds). Proc 16th Göttingen Neurobiol. Conf. Thieme Stuttgart New York, 375

Gray, C.M., König, P., Engel, A.K., Singer, W. (1989): Oscillatory responses in cat visual cortex exhibit inter-columnar synchronization which reflects global stimulus properties. Nature **338**, 334–337

Gray, C.M., Singer, W. (1989): Stimulus-specific neuronal oscillations in orientation columns of cat visual cortex. Proc. Nat. Acad. Sci. USA. **86**, 1698

Green, M. (1986): What determines correspondence strength in apparent motion? Vision Res. **26**, 599–607

Grossberg, S. (1976): Adaptive pattern classification and universal recording: II. Feedback, Expectation, Olfaction, Illusions. Biol. Cybern. **23**, 187–202

Grossberg, S. (1980): How does a brain build a cognitive code? Psychol. Rev. **87**, 1–51

Grossberg, S. (1983a): The quantized geometry of visual space: The coherent computation of depth, form, and lightness, Behavioural Brain Sci. **6**, 625–692
Grossberg, S. (1983b): Neural substrates of binocular form perception: filtering, matching, diffusion and resonance. In Basar E, Flohr H, Haken H, Mandell A.J. (eds), Synergetics of the Brain, Springer-Verlag, Berlin, Heidelberg, New York, Tokyo, pp. 274–298
Harris, F.J. (1978): On the use of windows for harmonic analysis with the discrete fourier transform. Proc IEEE. **66**, 51–83
Horridge, G.A. (1987): The evolution of visual processing and the construction of seeing systems. Proc. R. Soc. (London) B, **230**, 279–292
Hubel, D.H., Wiesel, T.N. (1962): Receptive fields, binocular interaction, and functional architecture in the cat's visual cortex. J. Physiol. **160**, 106–154
Hubel, D.H., Wiesel, T.N. (1963): Shape and arrangement of columns in cat's striate cortex. J. Physiol. **165**, 559–568
Johannesma, P., Aertsen, A., van den Boogaard, H., Eggermont, J., Epping, W. (1986): From synchrony to harmony: ideas on the function of neural assemblies and on the interpretation of neural synchrony. In: Brain Theory, Palm G, Aertsen A (ed), Springer-Verlag, Berlin Heidelberg, pp. 25–47
Julesz, B. (1981): A theory of preattentive texture discrimination based on first order statistics of textons. Biol. Cybern. **41**. 131–138
Kammen, D.M., Holmes, P.J., Koch, C. (1989): Cortical architecture and oscillations in neuronal networks: feedback versus local coupling. In: Models of Brain Function. R.M.J. Cotterill (ed.). Univ. Press. Cambridge, New York, Melbourne, pp. 273–284
Koenderink, J. (1984): The concept of local sign. In van Doorn AJ, van de Grind WA, Koenderink J (eds), Limits in Perception. Utrecht, VNU Sci Press, pp. 495–549
Krause, F., Eckhorn, R. (1983a): Receptive fields for motion stimuli (RFM) determined for different types of cat visual neurons. Neurosci. Lett. (Suppl) **14**, 209
Krause, F., Eckhorn, R. (1983b): A new method for the measurement of visual receptive fields. Perception 12, Suppl.
Krause, F., Gonzalez, F., Nelson, J.I., Eckhorn, R. (1987): A fast method for predicting coding properties of visual cortical simple cells . Perception 16, Suppl., 262
Krause, F., Nelson, J.I., Gonzalez, F., Eckhorn, R. (1988): Simultaneous signalling of form and motion in cat visual cortex? Conf. Abstr. Vol. "Visual Processing of Form and Motion": P8
Kraut, M.A., Arezzo, J.C., Vaughan, H.G., (1985): Intracortical generators of the flash VEP in monkeys. Electroenceph clin Neurophysiol **62**, 300–312
Krüger, J. (1983): Simultaneous individual recordings from many cerebral neurons: techniques and results. Rev. Physiol. Biochem. Pharmacol. **98**, 177–233
Krüger, J., Aiple, F. (1988): Multimicroelectrode investigation of monkey striate cortex: spike train correlations in the infragranular layers. J Neurophysiol. **60**, 798–828
Krüger, J., Aiple, F. (1989): The connectivity underlying the orientation selectivity in the infragranular layers of monkey striate cortex. Brain Res **477**, 57–65
Leferink, J., Eckhorn, R., Nelson, J.I., Wagner, H-J. (1987): Characterization of patchy projection systems between cortical areas: A combined multi-electrode and HRP study in cat's visual cortex. 2nd IBRO-Congrs, Neurosci. 22 Suppl. 1305P
Legéndy, C.R., Salcman, M. (1985): Bursts and recurrences of bursts in the spike trains of spontaneously active striate cortex neurons. J. Neurophysiol **53**, 926–939
Lohmann, H., Eckhorn, R., Reitboeck, H.J. (1986): Receptive fields evaluated for visual evoked intracortical potentials. Proc. ENA Sat. Vision Conf. Lyon, France, p. 40
Lohmann, H., Eckhorn, R., Reitboeck, H.J. (1988): Visual receptive fields of local intracortical potentials. J. Neurosci. Meth. **25**, 29–44
MacKay, D.M., (1962): Self-organization in the time domain. In: Self-Organizing Systems, Yovits MC, Jacobi GT, Goldstein GD (eds), Spartan, Washington, pp. 37–48
Matsubara, J., Cynader, M., Swindale, N.V., Stryker, M.P. (1985): Intrinsic projections within visual cortex: evidence for orientation-specific local connections. Proc. Natl. Acad. Sci. USA. **82**, 935–939
Melssen, W.J., Epping, W.J.M. (1987): Detection and estimation of neural connectivity based on crosscorrelation analysis. Biol Cybern **57**, 403–414
Mitzdorf, U. (1985): Current source density method and application in cat cerebral cortex: investigation of evoked potentials and EEG phenomena. Physiol. Rev. **65**, 37–100
Mitzdorf, U. (1987): Properties of the evoked potential generators: current source-density analysis of visually evoked potentials in the cat cortex. Intern. J. Neurosci. **33**, 33–59

Mountcastle, V.B. (1957): Modality and topographic properties of single neurons of cat's somatic sensory cortex. J. Neurophysiol. **20**, 408–434

Mountcastle, V.B. (1978): An organizing principle for cerebral function: The unit module and the distributed system. In Edelman GM, Mountcastle VB (eds), The Mindful Brain, MIT Press, Cambridge, Mass

Mountcastle, V.B., Poggio, G.F., Reitboeck, H.J., Georgopoulos, A.P., Johnson, K.O., Longerbeam, M.B., Steinmetz, M.A., Phillips, J.R., Brandt, D.K., Habbel, C.G., (1987): A system for multiple microelectrode, recording from the neocortex of waking monkeys. Proc. Soc. Neurosci. p. 629

Nelson, J.I. (1985): The cellular basis of perception. In Rose D, Dobson VG (eds), Models of the Visual Cortex. Wiley, New York, pp. 108–112

Nelson, J.I., Frost, B.J., (1985): Intracortical facilitation among cooriented, co-axially aligned simple cells in cat striate cortex. Exp. Brain Res. **61**, 54–61

Nelson, J.I., Munk, M., Bullier, J., Eckhorn, R. (1989): Functional connectivity revealed in and outside of receptive field overlap by 3 cross-correlation techniques. Soc. Neurosci. Abstr. Vol. 15, Part 2, p. 157 (119.16)

Othmer, H.G. (1984): Synchronization, phase-locking and other phenomena in coupled cells. In: Temporal Order, Rensing L, Jaeger NI (eds), Springer-Verlag, Berlin, Heidelberg, New York, pp. 130–143

Pabst, M., Reitboeck, H.J., Eckhorn, R. (1989): A model of pre-attentive region definition in visual patterns. In: Models of Brain Function, R.M.J. Cotterill (ed.), Cambridge Univ. Press.

Palm, G., Aertsen, A.M.H.J., Gerstein, G.L. (1988): On the significance of correlations among neuronal spike trains. Biol Cybern **59**, 1–11

Payne, B.R., Berman, N., Murphy, E.H. (1981): Organization of direction preferences in cat visual cortex. Brain Res. **211**, 445–450

Petsche, H, Pockberger, H., Rappelsberger, P. (1988): Cortical structure and electrogenesis. In: Basar E (ed), Dynamics of sensory and cognitive processing by the brain, Springer-Verlag, pp. 123–139

Pollen, D.A., Ronner, S.F. (1975): Periodic excitability changes across the receptive fields of complex cells in the striate and parastriate cortex of the cat. J. Physiol. (London) **245**, 667–697

Reichardt, W. (1957): Autokorrelations-Auswertung als Funktionsprinzip des Zentrainervensystems (bei der optischen Wahrnehmung eines Insektes). Z. Naturforsch. **12b**, 448–457

Reichardt, W. (1961): Autocorrelation, a principle for the evaluation of sensory information by the central nervous system. In: Rosenblith WA (ed) Sensory Communication. MIT Press, Cambridge, MA, pp. 303–317

Reichardt, W. Poggio, T. (1979): Figure-ground discrimination by relative movement in the visual system of the fly. Biol. Cybern. **35**, 81–100

Reichardt, W., Poggio, T., Hausen, K. (1983): Figure-ground discrimination by relative movement in the visual system of the fly. Part II: Towards the neural circuitry. Biol Cybern. **46** (Suppl): 1–30

Reitboeck, H.J. (1983a): A 19-channel matrix drive with individually control-lable Fiber microelectrodes for neurophysiological applications. IEEE SMC **13**, 676–682

Reitboeck, H.J. (1983b): Fiber microelectrodes for electrophysiological recordings. J. Neurosci. Meth. **8**, 249–262

Reitboeck, H.J., (1983c): A multi-electrode matrix for studies of temporal signal correlations within neural assemblies. In Basar E, Flohr H, Haken H, Mandell A (eds), Synergetics of the Brain. Springer-Verlag, Berlin Heidelberg, New York, pp. 174–182

Reitboeck, H.J. (1988): Neural mechanisms of pattern recognition. In Sensory Processing in the Mammalian Brain: Neural Substrates and Experimental Strategies. Lund J (ed), Oxford Univ. Press, pp. 420–465

Reitboeck, H.J., Adamczak, W., Eckhorn, R., Muth, P., Thielmann, R., Thomas, U. (1981): Multiple single-unit recording: Design and test of a 19-channel micro-manipulator and appropriate fiber electrodes. Neurosci. Lett. **7**, S148

Reitboeck, H.J., Eckhorn, R., Pabst, M. (1987): A model of figure/ground separation based on correlated neural activity in the visual system. In: Synergetics of the Brain, H Haken, ed, Springer-Verlag, Berlin, Heidelberg, New York pp. 44–54

Reitboeck, H.J., Pabst, M., Eckhorn, R. (1988): Texture description in the time domain. In Computer Simulations in Brain Science. Cotterill MJ (ed), Cambridge University Press, pp. 479–494

Salin, P.A., Bullier, J., Kennedy, H. (1989): Convergence and divergence in the afferent projections to cat area 17. J. Comp. Neurol. **283**, 486–512

Schneider, J., (1980): Correlator: Pollux. Philipps-Univ, Deptm Biophysics, Internal Report
Schneider, J., Eckhorn, R., Reitboeck, H.J. (1983): Evaluation of neuronal coupling dynamics. Biol. Cybernetics **46**, 129–134
Schwartzkroin, P.A. (1983): Mechanisms of cwell synchronization in epileptiform activity. Trends in Neurosci. **6**, 157–160
Shaw, G.L., Harth, E., Scheibel, A.B. (1982): Cooperativity in brain function: Assemblies of approximately 30 neurons. Exper. Neurol. **77**, 324–358
Sherk, H. (1978): Area 18 cell responses in cat during reversible inactivation of area 17. J. Neurophysiol. **41**, 204–215
Singer, W. (1988): Pattern recognition and self-organization in biological biological systems. In: Processing Structures for Perception and Action, Marko H, Hauske G, Struppler A (eds). VCH Verlagsgesellschaft, Weinheim, pp. 1–18
Singer, W., Gray, C.M., Engel, A., König, P. (1988) Spatio-temporal distribution of stimulus-specific oscillations in the cat visual cortex II: Global interactions. Soc. Neurosci. Abstr. Vol. 14, 899
Sporns, O., Gally, J.A., Reeke, G.N., Edelman, G.M. (1989): Reentrant signaling among simulated neuronal groups leads to coherency in their oscillatory activity. Proc. Natl. Acad. Sci. (USA) **86**, 7265–7269
Symonds, L.L., Rosenquist, A.C. (1984): Laminar origins of visual cortical connections in the cat. J. Comp. Neurol. **229**, 39–47
Tanaka, K. (1983): Cross-correlation analysis of geniculostriate neuronal relationship in cat. J. Neurophysiol. **49**, 1303–1318
Toet, A., Blom, J., Koenderink, J.J., (1987): The construction of a simultaneous functional order in nervous systems. I. Relevance of signal covariances and signal coincidences in the construction of a functional order. Biol. Cybern. **57**, 115–125
Toyama, K., Kimura, M., Tanaka, K. (1981): Organization of cat visual cortex as investigated by cross-correlation technique. J. Neurophysiol. **46**, 202–214
Ts'o, D.Y., Gilbert, C.D., Wiesel, T.N. (1986): Relationships between horizontal interactions and functional architecture in cat striate cortex as revealed by cross-correlation analysis. J. Neurosci. **6**, 1160–1170
van Kan, P.L.E., Scobey, R.F., Gabor, A.J. (1985): Response covariance in cat visual cortex. Exp. Brain Res. **60**, 559–563
von der Malsburg, C. (1983): How are nervous structures organized? In: Basar E, Flohr H, Haken H, Mandell AJ (eds) Synergetics of the Brain, Springer-Verlag, Berlin, Heidelberg, New York, Tokyo, 238–249
von der Malsburg, C. (1988): Pattern recognition by labeled graph matching. Neural Networks **1**, 141–148
von der Malsburg, C., Singer, W. (1988): Principles of cortical network organization. In: Rakic P, Singer W (eds) Neurobiology of Neocortex. Wiley & Sons, Chichester, New York, Brisbane, Toronto, Singapore, pp. 69–99
Weyand, T.G., Malpeli, J.G., Lee, C., Schwark, H.D. (1987): Cat area 17: IV. Two types of corticotectal cells defined by controlling geniculate inputs. J. Neurophysiol. **56**, 1102–1108
Whittaker, S.G., Siegfried, J.B., (1983): Origin of wavelets in the visual evoked potential. Electroencephal. Clin. Neurophysiol. **55**, 91–101
Wilson, J.T.L., Singer, W. (1981): Simultaneous visual events show a long-range spatial interaction. Perception & Psychophys. **30**, 107–113
Zeki, S., Shipp, S. (1988): The functional logic of cortical connections, Nature **335**, 311–317

10. Cross-Correlation Analyses of Motoneuron Inputs in a Coordinated Motor Act

P.A. Kirkwood and T.A. Sears

Sobell Department of Neurophysiology, Institute of Neurology, Queen Square, London WC1N 3BG, England

The neuronal co-operativity of active movements occurs at several levels. At the first level most movements involve the coordinated action of several muscles. For instance in mammalian respiratory movements, the subject of this chapter, diaphragm contractions are assisted by activation of inspiratory intercostal muscles. This assistance is needed not only to produce expansion of the thorax but to prevent its collapse, which the negative intra-thoracic pressure could otherwise produce. At the second level, within a given muscle several motoneurons are usually coactivated and are recruited in a stereotyped sequence corresponding to the appropriate level of control at different force levels. Finally, for a given motoneuron, many inputs must cooperate to produce a discharge.

Motoneurons are probably extreme in the degree of convergence of active inputs required to bring them up to threshold. Typical single-fibre excitatory post-synaptic potentials (unitary EPSPs) are of the order of $100\mu V$ in amplitude, meaning that something like 12,000–20,000 EPSPs per second of non-synchronized inputs are required for the cell to be brought into its firing range (Sears 1977; Kirkwood et al. 1984). Thus a given movement is represented by a complex spatio-temporal pattern of activity in many motoneurons, which is itself the result of patterns of activity in a network of input neurons whose connections are both convergent (to reach threshold) and divergent (to ensure coactivation). Because the motoneurons are the output of the nervous system (Sherrington's final common path), a very extensive part of the nervous system must cooperate in producing a given movement.

The aim of the experiments described here has not been to try to identify patterns of activity *de novo* (defined by their spatio-temporal relationships) and then to give them meaning; instead a well-defined stereotyped pattern is the starting point of the analysis. That pattern of activity consists of the respiratory discharges of the external intercostal motoneurons of an anaesthetized paralyzed cat in the prone position, and the task has been to analyse the pattern of inputs which produce that output, knowing, of course, that the pattern of inputs is only one of an infinity of combinations that might pertain in different circumstances. The analysis of the pattern of inputs consists of identifying spatial aspects of connections of neurons with known temporal patterns. Recognition of features in these spatial aspects, particularly changes in them in different circum-

Springer Series in Synergetics
J. Krüger (Ed.)
Neuronal Cooperativity

stances, should give insight into the nature of the central control of the respiratory movements and their relationship to other movements. In trying to identify features in the patterns of activity in multi-neuronal populations, our aim is similar to other studies in this volume but we are not asking "what is the significance of such and such a pattern in its own right?", but rather "how does this pattern generate the known pattern of motoneuron activity?". In some ways our aim is thus the reverse of that of some of the other authors of this volume, who are mostly concerned with sensory functions. Further, we do not ascribe much functional significance to the moment-by-moment (i.e. ms time-scale) aspects of the pattern, suggesting instead that the motoneuron and its associated circuits, acting as an integrator, are specifically designed to filter out such information. However, the methods we employ, with multiple neuron recordings and cross-correlations between spike discharges, together with the problems of interpreting such correlations in terms of connections (interpretations which certainly involve the ms time-scale), are aspects which we have very much in common with the other authors.

10.1 Patterns of Activity and of Motoneuron Sychronization

The motoneurons which we have been concerned with are situated in twelve segments of the cat thoracic spinal cord, those in each segment innervating the external intercostal muscle layer of one intercostal space (there are 13 ribs in the cat). The muscle fibres of each space run from rib to rib and the angle of the fibres is such that, acting in concert, their action at its most simple is to move the ribs outward and forward when activated during inspiration and hence to aid in increasing thoracic volume (Da Silva et al. 1977). Their actions are, of course, more complicated than that and are still the subject of considerable debate. One needs to take into account the posture of the animal and interactions with other inspiratory muscles, such as the diaphragm which inserts into the more caudal ribs, the levator costae muscles which run from vertebra to rib and the internal intercostal layer, whose fibres run at a different angle and which is generally regarded as having an expiratory action, though it is sometimes assigned a postural role (as is part of the external layer by some authors). These details do not concern us here, but it should be remembered that the mechanics of this shell-like structure are not neccessarily simple and are certainly not uniform across the surface of the thorax.

During inspiration, broadly simultaneous activation of the external intercostal motoneurons of the different segments occurs with a common ramp-like pattern, also shared with the diaphragm. This is generally regarded as a feature of the medullary pattern generator and has a basic mechanical correlate in the force required to oppose the static elastic recoil of the lungs and, dependent on lung volume, the rib cage. The ramp may appear, of course, in a more fractionated form in the activity of different motoneurons according to their recruitment order (as for all motoneurons), but also according to the position of the

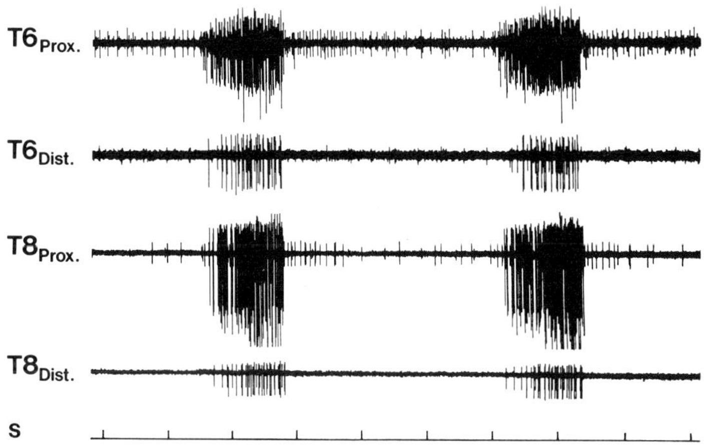

Fig. 10.1. Discharges recorded from naturally occurring filaments of the external intercostal nerves of segment T6 (upper two traces) and T8 (lower two traces) in an anaesthetized, paralyzed cat (artificially ventilated at a pump rate of 52/min). The bursts of large spikes occur during the inspiratory phase of the central respiratory cycle and come from alpha motor axons. The small spikes, which also appear in expiration, come from gamma motor axons (Sears 1964). The first and third traces were recorded from the most proximal (dorsal) filament in each segment, the second and fourth from a distal filament. As is always the case in these preparations, the distal filaments show less activity even with a raised level of CO_2, as here (end-tidal value 7.4%). Previously unpublished data of Kirkwood et al. (1982a)

muscle fibres of the motor units, which only occupy restricted regions of a given intercostal space. Over the whole thorax there thus appears a spatio-temporal pattern of activity, which is remarkably stereotyped when factors such as posture and levels of anaesthesia are controlled. This is partly illustrated in Fig. 10.1, which shows multi-unit recordings from small, naturally occurring branches (filaments) of the external intercostal nerves of two different intercostal spaces. There is coactivation of alpha motoneurons during inspiration at all four sites, but less activity at some sites than at others. There are generally two gradients, with more activity in rostral spaces than in caudal ones (see Kirkwood et al. 1982a) and more activity in the proximal (dorsal) areas than more distally (Fig. 10.1). In addition, another focus of activity is found ventrally in the parasternal, interchondral region (not illustrated).

In the early 1970s Sears and his colleagues embarked on a programme of cross-correlation studies of motoneuron discharges, in the belief that useful information about the organization of the inputs to these motoneurons could thereby be obtained (Sears and Stagg 1976). A particular attraction of the method was that it could also be readily used on e.m.g recordings and in the human (for any synergistic muscles), thus giving, it was hoped, information about disorders of such connections in neurological conditions. It has turned out that these respiratory discharges have proved to be remarkably fertile ground both for developing ideas on the interpretation of cross-correlations and for revealing several aspects of the organization of the motoneuron inputs, even though one of the original aims, that of identifying the majority of the cells that form the inputs generating the pattern of activity, has still not been achieved. The methods are now being employed in more than one laboratory in the study of neurological disease (Datta et al. 1985a,b; Farmer et al. 1990; Davey et al. 1986, 1990).

Drawing on the timing of the monosynaptic reflex discharge of motoneurons, Sear's hypothesis was that an increased probability of motoneuron firing would be seen during the rising phase of any unitary EPSP. Because most individual motoneuron inputs (i.e. pre-synaptic axons) *from any source* were expected to diverge to many motoneurons, these branched axons, even in the absence of any synchrony between inputs, would give near-synchronous unitary EPSPs in many motoneurons and thus a raised probability of synchronous firing above that expected by chance. However, at that time very little information was available on the correlative behaviour of mammalian neurons; it was not known how strong such an effect might be, nor how it might compare with other mechanisms which could synchronize motoneurons. Not least was the likelihood, via essentially the same hypothesis, that the last order interneurons, which provide the above-mentioned branched axons, could themselves have synchronized discharges because of their own shared inputs, and so on back along a chain of interneurons. For the first of these issues, the synchronization turned out to be relatively weak, at least as compared to that resulting from the large EPSPs of invertebrate preparations reported in the literature then. Possible problems in the measurement of such weak effects were readily overcome by the use of technology and experimental design appropriate for dealing with large numbers of spikes (Stagg 1973; Davies et al. 1983). As to the second issue, the existence of "presynaptic synchronization" (Kirkwood, et al. 1982b) cannot now be denied, although in their initial study Sears and Stagg (1976) were probably lucky in choosing to study this particular system in moderately deeply anaesthetized animals, where the effects of presynaptic synchronization seem to be minimal. Our current view is that, in general, motoneuron synchronization (and by extension the synchronization of most neuronal pairs) is highly likely to contain components of this sort and particular care must always be taken in drawing conclusions that do not allow for these effects. Our measurements since 1976 have included a continuous re-assessment of these effects, as will become apparent.

Some of the mechanisms producing motoneuron synchronization can be differentiated by the time-courses of the synchronization produced. Figure 10.2 shows cross-correlograms constructed from the discharges of paired groups of intercostal motoneuron discharges. All except d and e come from multi-unit recordings of external intercostal nerve filaments or whole nerve inspiratory discharges in our standard preparation of the pentobarbitone-anaesthetized, paralyzed cat. In a–c the correlation was performed on one set of recordings from a pair of proximal nerve filaments in adjacent intercostal spaces, using three different time scales and bin widths. The strongest correlation of all, a, represents the periodic respiratory drive and needs little further discussion, except to point out the clear difference between it and the stochastic elements in the correlations seen on the shorter time scales of panels b, c, f–i. This respiratory periodicity has an obvious functional significance, but its appearance in the correlograms gives us no more (indeed less) information about the connections involved than does visual inspection of the original recordings. However, the presence of this component must be acknowledged since it often appears as a non-stationarity, giving a slight curvature or slope to the base-line as is particularly apparent in b.

Of the stochastic components, one is explicable by the branched axon hypothesis of Sears and Stagg (1976) and there are at least three types of pre synaptic synchronization that may be distinguished. One of these is known as high-frequency oscillation (HFO) (Fig. 10.2f). This form of highly periodic synchronization in the range 60–120Hz is known to be transmitted to the motoneurons from medullary respiratory neurons (Cohen 1979). HFO is potentiated by the chemical stimuli for respiration (hypercapnia or hypoxia) and may be abolished by some lesions of the medulla (Davies et al. 1986) or by barbiturates. In some cases it may obscure signs of connections for individual neurons, but because of its undoubted origin in respiratory neurons its relative timing for different groups of motoneurons can be used to give at least a general idea of the number of synapses and/or the conduction velocities in the descending respiratory pathways (Cohen 1973; Kirkwood et al. 1982b; Kirkwood and Sears 1989).

A second type of presynaptic synchronization was denoted as "broad-peak" by Kirkwood et al. (1982b) and corresponds to the triangular-shaped peak extending to about $\pm$20ms in Fig. 10.2c. The assigning of this component to presynaptic synchronization was made on the basis of the duration of the peak, which is too long to be accounted for by any known unitary PSPs (post-synaptic potentials) in motoneurons. In normoxic or hyperoxic animals the effect apparently arises from non-respiratory inputs, and was ascribed by Kirkwood et al. (1982b) to the action of interneurons, synchronized to each other or perhaps firing in bursts. Broad-peak synchronization is a particular feature of the behaviour of these motoneurons after chronic spinal lesions, (Kirkwood et al. 1984) and since synchrony of this sort can lead to tremor (Dietz et al. 1976) it may represent pathological function. Kirkwood et al. (1984) hypothesised that the organization of the normal spinal cord is specifically adapted to reduce this synchrony to a minimum (cf. Davey and Ellaway 1988). In man, both stroke (Datta et al. 1985a,b) and Parkinsonism (Davey et al. 1986) can lead to increased synchronization of motoneurons, and for the stroke patients the tendency was towards a time course comparable to these broad peaks in the cat. It should be stressed that the appearance of broad peaks cannot itself be regarded as pathognomic; such synchronization is present in some normal situations or individuals in both cat and man. The issue is a quantitative one needing considerably more study. The degree to which these components are a feature of normal motoneuron sychronization is likely to be dependent on the muscles concerned and the task involved (Sears and Stagg 1976; Berger et al. 1983; Adams et al. 1989).

For intercostal activation during quiet respiration, the distinction between the functionally relevant and the (assumed) disadvantageous stochastic forms of

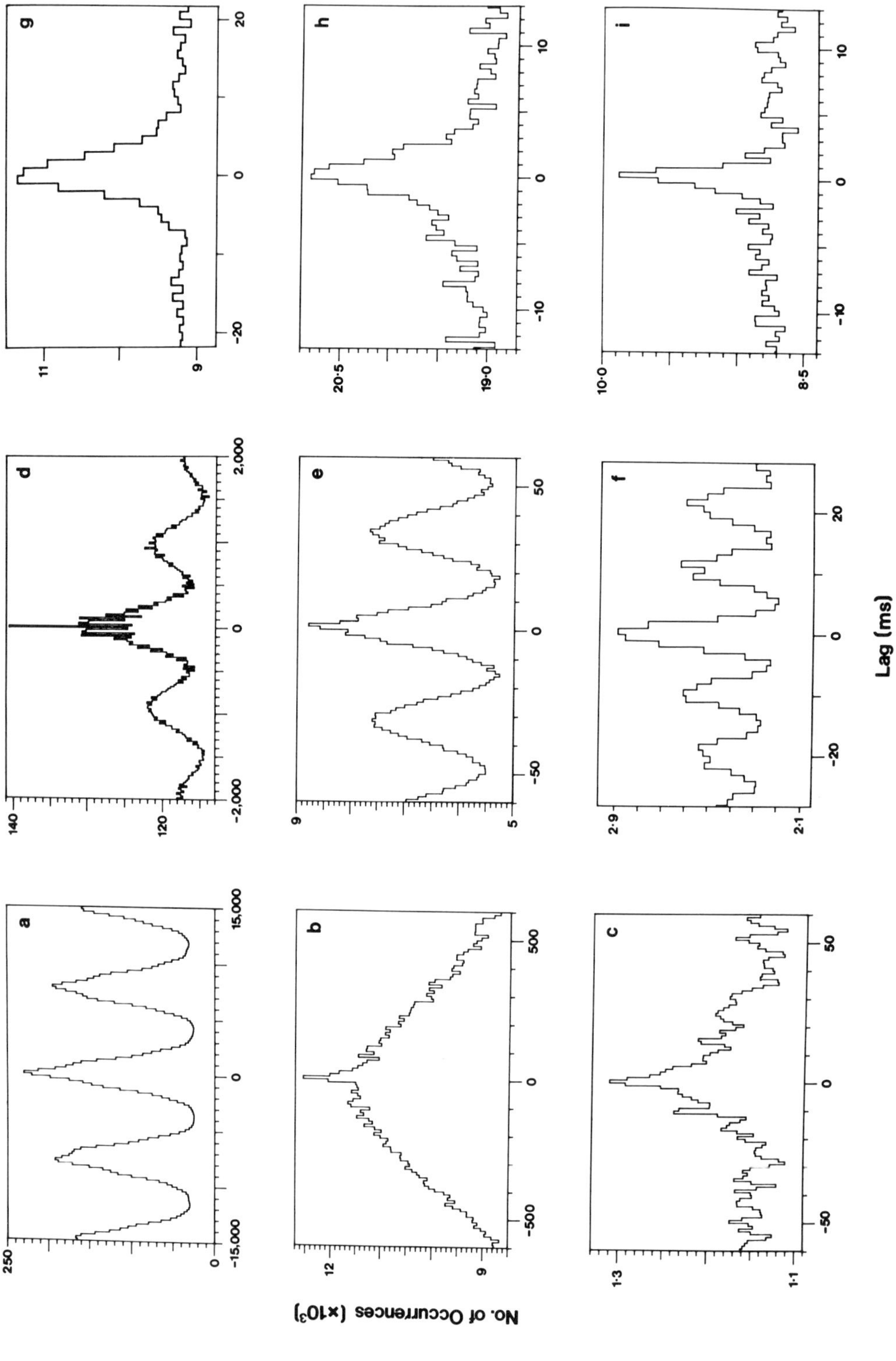
a
b
c
d
e
f
g
h
i
250
0
-15,000
0
15,000
12
9
-500
0
500
1·3
1·1
-50
0
50
140
120
-2,000
0
2,000
9
5
-50
0
50
2·9
2·1
-20
0
20
11
9
-20
0
20
20·5
19·0
-10
0
10
10·0
8·5
-10
0
10
No. of Occurrences (×10³)
Lag (ms)

synchronization (Fig. 10.2,c,f) can easily be made because the time course of the task involved is much slower than the time course of the synchronization. When the intercostals are used for a different task, such as human speech where the prosodic qualities of the utterances appear to demand short pulses of subglottal pressure (Draper et al. 1960), time courses of synchronization which are similar or faster than broad-peak synchronization are seen (Sears and Stagg 1976) where the difference between stochastic and functionally meaningful synchronization is not so clear.

In the *awake* cat the intercostals provide yet another example of synchronization with a similar time course, but now periodic at about 30Hz, during purring (Fig. 10.2d,e). This frequency is the purring frequency, so this form of synchronization is presumably functionally useful, but it also may be viewed as originating in unavoidable presynaptic synchronization, as for the stochastic forms, because peripheral afferent fibres, presumably synchronized by the purring vibration, contribute a major component (Kirkwood et al. 1987).

The remaining three panels in Fig. 10.2, forming the right-hand column (g–i), represent some of the variations seen in relatively narrow central peaks in the correlograms. Because of the clear distinction between these forms and broad-peak synchronization, they were, perhaps rather uncritically, lumped together by Kirkwood et al. (1982b) under the heading of "short-term synchronization", which had been defined by Sears and Stagg (1976) and analysed by Kirkwood and Sears (1978) as originating in non-synchronized impulses in branched presynaptic axons. However, the use more recently of longer segments of data (Davies et al. 1985a) and narrower bin widths has made it quite clear that the variation in the durations of these peaks may be systematic (Kirkwood and Sears 1989). Our present view, supported by various items of indirect evidence, is that only the narrowest of these peaks, such as in Fig. 10.2i, can confidently be ascribed to the branched axon hypothesis alone and the others should be regarded as containing a greater or lesser component resulting from presynaptic

◁

Fig. 10.2. Cross-correlograms showing a wide variety of time-courses for the synchronization of intercostal motoneuron discharges. **a–c** Analyses of the synchronization of external intercostal alpha motoneurones of T3 (reference spikes) and T4 under light anaesthesia and paralysis. One segment of data analysed with three different bin widths, 200ms, 10ms and 1ms respectively. Histogram in **c** smoothed with a three-point triangular filter. The periodic wave in **a** represents the respiratory period (7.5s). **d,e** Synchronization of intercostal EMG discharges (mostly internal intercostal) recorded via implanted wires in segments T8 (reference) and T9 in an awake cat, lying quietly and purring. The same segment of data analysed with bin widths of 20ms (**d**) and 1ms (**e**). The periodicity in **d** (about 1Hz) corresponds to the respiratory period, that in **e** (imperfectly sampled in **d**) corresponds to the purring (about 30Hz). **f** Synchronization between external intercostal alpha motoneuron discharges of T7 (reference) and T6. Light anaesthesia and paralysis as in **a–c**, but this time the synchronization is of the "high-frequency oscillation" (HFO) form. Bin width 1ms, HFO frequency 98HZ. **g–i** Synchronization in moderately deeply anaesthetized, paralysed animals. **g** external intercostal alpha motoneurons in T7 (reference) and T8. **h,i** Similar measurements from segments T3 (reference) and T4 of a different animal: **h** external intercostal nerve discharges; **i** inspiratory motoneuron discharges of the internal intercostal nerves of the same segments. Bin widths: **g** 1ms; **h,i** 0.384ms. Half-widths of peaks: 5.5ms, 3.4ms and 1.4ms respectively. Previously unpublished data of (**a–c**) Davies et al. (1985b), (**d,e**) Kirkwood et al. (1987), (**f,g**) Kirkwood et al. (1982b) and (**h,i**) Kirkwood and Sears (1989)

synchronization. The evidence for this will be presented later but the known presence of extensive convergence/divergence among motoneuron inputs would strongly support an *a priori* case that anything wider than the narrowest regularly observed peaks would include such a component. The only exception to this argument would be if the narrowest peaks arose from a different mechanism, for which there are two possible weak candidates: (i) the existence of direct excitatory effects between motoneurons, for which there is some experimental support (Nelson 1966; Khatib et al. 1989) but little evidence for a strong effect on synchronization, or (ii) the (hypothetical) existence of motoneurons discharged only by inputs employing only somatic EPSPs and/or having particularly low values of temporal dispersion at the different terminals of their axons.

10.2 The Primary Correlation Operator

During the 1960s and early 1970s pioneering analyses of cross-correlograms in terms of synaptic connections were made in invertebrate preparations by interpreting time courses of peaks, most importantly via features of their periodicities (Bryant, et al. 1973). In the correlograms derived from motoneurons, periodicities are weak or absent (excluding HFO or purring). This may be so for a variety of reasons: firstly, at a single synapse the basic excitatory effect is weak and, in our experiments which involve non-stationary spike trains, the firing rates of the neurons vary; secondly, for common input effects large numbers of such inputs are involved, each with its own different firing rate. For these correlograms, deductions about connections must therefore be made from features of the central or primary peak (or trough). For a correlogram between the input and output spike trains at a single synapse with neurones firing randomly or at low rates, the form of this peak was denoted as the primary correlation kernel by Knox (1974). The form of the peak from more complicated connections (e.g. "branched-axon" common inputs) can easily be derived by convolution from this kernel, but the reverse is of course not always the case.

At the start of our analysis the appropriate form of the kernel for the small EPSPs of motoneurons was not known, although Knox (1974) and Knox and Poppele (1977) had provided evidence favouring a time course given by the derivative of the rising phase of the EPSP, at least for the large EPSPs arising from electrical stimulation. Thus, in analysing the motoneuron synchronization measured by Sears and Stagg (1976), some assumption had to be made. The underlying causal factor in all correlated discharges are the unitary PSPs. A reasonable amount of data were available on the shapes and sizes of these for motoneurons and so Kirkwood and Sears (1978) made them their starting point, and the assumption they made was that the primary correlation kernel could be described as a linear differential function of the PSP, at least for small PSPs. The relationship between the PSP and the correlation kernel was denoted as the primary correlation operator. The usefulness of the assumption was shown by

Kirkwood and Sears (1978) by their measurements of an intracellular analogue of motoneuron synchronization, namely, the average depolarization seen in one motoneuron at the time of firing of another. This waveform was denoted the average common excitation (ACE) potential. A relatively simple form of the primary correlation operator was assumed, given by

$$f'(t) = Af(t) + B\frac{df(t)}{dt} \qquad (10.1)$$

where t is time, f(t) is the time course of the EPSP, f′(t) that of the primary correlation kernel and A,B are constants. When this acted on EPSPs of appropriate shapes for motoneurons the time courses of both the measured ACE potentials and the motoneuron synchronization of Sears and Stagg (1976) (Fig. 10.3) could be accurately predicted. Moreover, because the operator was a linear one it also expressed the relationship between the ACE potential and time course of the motoneuron synchronization. Thus these two measurements could be used to estimate the values of the coefficients in the operator and allow, at least to a first approximation, the interpretation of any correlogram peak between input and the output of a motoneuron in terms of the shape and size of the underlying EPSP.

This model was verified by measuring a population of unitary EPSPs generated by muscle spindle afferents in external intercostal motoneurons, and comparing these with cross-correlation peaks constructed from (sometimes the same) afferent discharges and the efferent discharges of other motoneurons of the same pool (Kirkwood and Sears 1982a,b). The measurements showed that the effect of the EPSP was a raised probability of firing mostly during the rising phase of the EPSP, but extending as a weak effect into the tail of the EPSP (Fig. 10.3), as predicted from (10.1). The strength of the effect was estimated as the peak value of the increase in firing probability as a proportion of the baseline (pre-trigger) value. For an EPSP of nominal 100μV amplitude and a population of motoneurons, where the usual range and variation of EPSP rate-of-rise may be assumed (as was seen in the spindle-evoked EPSPs), this peak value may be taken as 0.5 (Davies et al. 1985b) i.e. the mean of the values derived from the ACE potential and from the spindle effects. A sufficiently narrow bin width is assumed in order to estimate the peak amplitude.

This result has been confirmed by more exact measurements on lumbar motoneurons, where the added security of the intracellular recording allows both the EPSP and its effect on the firing probability to be measured in the same motoneuron (Fetz and Gustafsson 1983; Gustafsson and McCrea 1984; Cope et al. 1987). The rather larger EPSPs and low synaptic noise used by Fetz and Gustafsson (1983) gave correlogram peaks related to the EPSP derivative, including a trough due to its falling phase, whereas the smaller ones with larger synaptic noise in the two later studies gave peaks which included a component outlasting the rising phase, as is implied in (10.1)

The effect of the synaptic noise in allowing threshold to be crossed during the falling phase of the EPSP was specifically noted by Kirkwood (1979), the two-component expression of (10.1) itself being justified by a theoretically de-

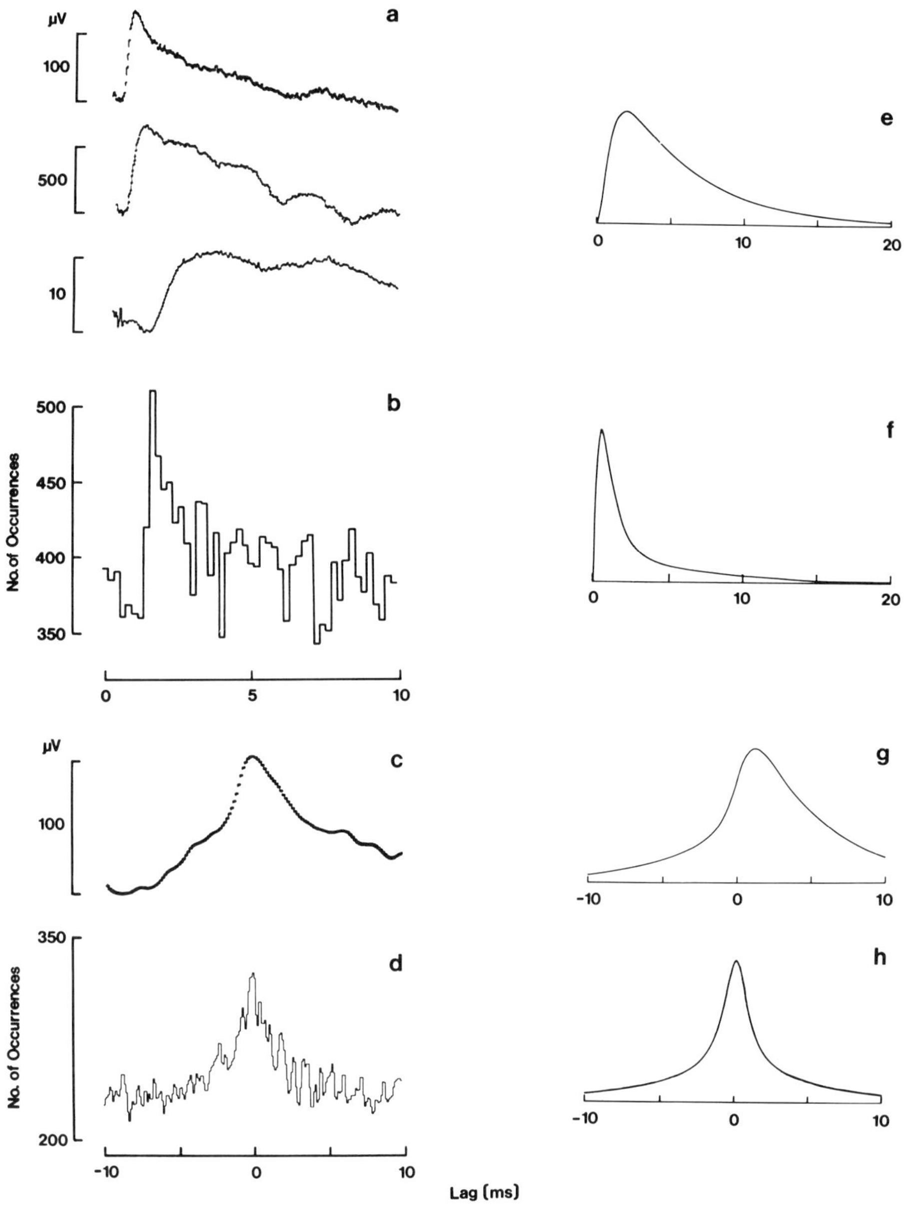

Fig. 10.3. Measured (a–d) and theoretical (e,h) representations of the synaptic events in neuronal synchronization. **a** Single-fibre EPSPs recorded in three different external intercostal motoneurons from a single primary muscle spindle afferent. **b** Post-spike histogram (equivalent to the cross-correlogram at positive lags) for the same spindle afferent (reference spikes) and external intercostal nerve alpha discharges of the same segment. **c** Average common excitation (ACE) potential in an external intercostal nerve motoneurone (the average depolarization at times relative to reference spikes, in this case external intercostal nerve alpha discharges of the same segment). **d** Synchronization between nerve discharges of two branches of the external intercostal nerve of one segment. **e** Assumed shape for an EPSP. **f** Transformation of this into a probability of firing via (10.1). **g,h** Predicted time courses of the ACE potential and cross-correlogram for the same conditions as **f**. Figures reproduced as in the Journal of Physiology: **a,b**, Fig. 1 of Kirkwood & Sears (1982a); **c–h**, Figs. 11,12 of Kirkwood & Sears (1978) redrawn for the conditions $b = a/2$, $\alpha = 10\beta$ and $1/\beta = 5$ms (for definitions, see original reference)

rived expression from H.B. Bostock. The derivation of this is still unpublished, so it is worth giving here a few more details than are in Kirkwood (1979), in order to aid discussion. The expression was derived analytically as the probability of crossing a threshold from below, during in EPSP with a time-course y(t) and derivative y′(t), in the presence of synaptic noise with Gaussian distributions of membrane potential (standard deviation σ) and of rate of change of membrane potential (standard deviation σ'). The mean membrane potential was stationary at $\theta\sigma$ mV below threshold. The expression for the probability of threshold crossing was then given by R, where:

$$R = \frac{1}{2\pi\sigma}\exp\left(\frac{-(\theta\sigma - y)^2}{2\sigma^2}\right)\left\{\sigma' \exp\left(\frac{-y'^2}{2\sigma'^2}\right) + y'\sqrt{\frac{\pi}{2}}\left[1 + \operatorname{erf}\left(\frac{y'}{\sigma\sqrt{2}}\right)\right]\right\}. \tag{10.2}$$

This may be expanded as a power series which, when y is small with respect to $\theta\sigma$ and when ignoring terms of the order or higher than $(y/\sigma)^3$ or $(y'/\sigma')^3$, reduces to

$$R = \frac{1}{2\pi}\frac{\sigma'}{\sigma}\exp\left(\frac{-\theta^2}{2}\right)\left[1 + \frac{\theta y}{\sigma} + \frac{(\theta - 1)}{2}\left(\frac{y}{\sigma}\right)^2\right] \left[1 + \frac{y'}{\theta'}\sqrt{\frac{\pi}{2}} + \left(\sqrt{\pi} - \frac{1}{2}\right)\left(\frac{y'}{\sigma'}\right)^2\right]. \tag{10.3}$$

If terms of the order of $(y/\sigma)^2$ etc. are ignored and only those related to the EPSP are considered, the expression reduces to

$$R_{PSP} = \frac{1}{2\pi}\frac{\sigma'}{\sigma}\exp\left(\frac{-\theta^2}{2}\right)\left[\frac{\theta}{\sigma}y + \frac{\pi/2}{\sigma'}y'\right], \tag{10.4}$$

which may be taken as the two-component expression already quoted in (10.1).

These analytically derived expressions for threshold crossing have been tested in a computer simulation, in which the synaptic noise was made up of high rates of randomly occurring small EPSPs and found to be accurate (Fig. 10.4a,b). When a spike followed by realistic after-hyperpolarization was included in the simulation, the amplitudes of the kernels measured for these spikes were remarkably close to those derived from the threshold crossings, though the peaks were rather wider (Fig. 10.4c,e,f,g,i). This was so despite the considerable changes in membrane potential distribution that occur near threshold when crossing and recrossing of threshold at short intervals is prevented by spike occurrence. Other factors which also come into play when the spike mechanism is added may compensate for this, such as the lack of independence of membrane potential and its rate of change near to threshold. In any case, the simulations show the important result that the non-linearities predicted by the equation for threshold crossing (10.2) still appear to apply to the firing motoneuron (Fig. 10.4c,i).

Several points can be made from these equations. Firstly, the higher order terms become more important the larger the ratios (y/σ) or (y'/σ') become.

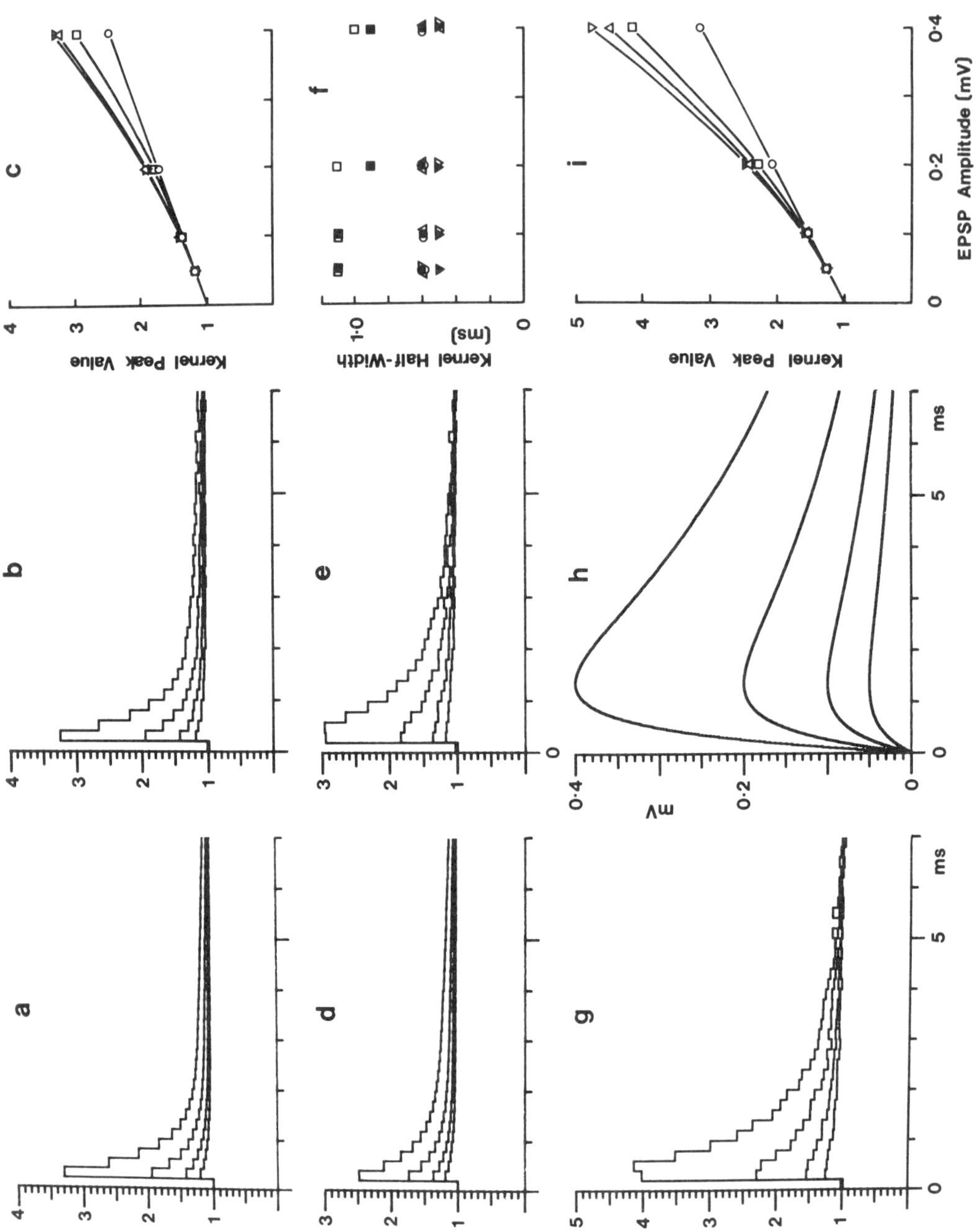
a
b
c
d
e
f
g
h
i
Kernel Peak Value
Kernel Half-Width (ms)
Kernel Peak Value
EPSP Amplitude (mV)
mV
ms

Hence some of the deviations from the two-component expression seen by Gustafsson and McCrea (1984) and Cope et al. (1987) for EPSPs larger than 100μV may be readily explained (cf. Fig. 10.4). Secondly, for the threshold-crossing estimations, which give the firing rate independent of the after-hyperpolarization, the background firing rate is given by the expression outside the square brackets in (10.3,4). Firing rate could thus change as a result of changes either in the quality of the synaptic noise (σ'/σ) or by change in the threshold, but these would have different effects on the time course of R_{PSP}. Thirdly, if the firing rate is kept constant, most simply by holding θ constant and σ/σ' constant but allowing the total amount of synaptic input to increase (i.e. σ and σ' increasing together), the overall magnitude of R_{PSP} decreases.

In certain experimental situations variability of synaptic noise may well be an important (if uncontrolled) parameter and its effect should not be ignored in comparing correlograms under different circumstances. For this reason, although the measurements of Gustafsson and McCrea (1984) and Cope et al. (1987) are internally more exact than the indirect measurements on the intercostals, they may still not fully represent the situation of motoneurons fired by their natural synaptic drives, because a considerable part (or all) of the depolarization in those studies was provided by current passed through the microelectrode and

Fig. 10.4. Theoretical and simulated primary correlation kernels. **a** Kernels predicted by (10.2) for the EPSPs shown in **h**, with $\sigma = 1.07$mV, $\sigma' = 0.618$mV/ms and $\theta = 1.132$. **b** Kernels describing the threshold crossings measured from computer simulation of a motoneurone where synaptic noise was constructed by summing a randomly occurring mixture of EPSPs (time courses as in **h** with the following amplitudes and mean rates (1000s of EPSPs/s): 50μV, 7.6; 100μV, 3.8; 200μV, 1.9; 400μV; 0.95. The mean depolarization was 10.789mV and the threshold was set at 12mV. The membrane potential distribution approximated to Gaussian with the values used in **a**. The kernels in **b** were calculated for the EPSPs of each amplitude and are virtually identical to those in **a**. **c** Amplitudes of kernels for these same conditions plotted against EPSP amplitude: ▽, kernels from **a**; △, kernels from **b**; ○, kernels from **d**; □, kernels from **e**.

d Kernels calculated from (10.4), the linearized form of (10.2), using the same parameter values as in **a** (in 10.1), B/A = 4.91 ms). Agreement with the kernels of **a** and **b** is only seen for the smaller EPSPs. **e** Correlation kernels for computer simulation using the same conditions as **b**, but now allowing a motoneurone spike, followed by a hyperpolarization to occur when threshold is crossed. The kernels represent the cross-correlations between the start times of the EPSPs of each amplitude and the motoneuron spikes. **f** (Open symbols) half-widths of the correlation kernels of **a**, **b**, **d**, **e**, symbols as in **c**; (closed symbols), similar measurements, but with the parameters used in **g**.

g Kernels derived from the simulations of a firing motoneurone as in **e**, but this time the EPSP rates were halved. The mean membrane potential in the absence of spikes was 5.28mV, σ was 0.764mV and σ' was 0.436mV/ms. The threshold was set at 6.22mV ($\theta = 1.226$) to give the firing motoneurone a similar spike rate to that in **e** (9.8 imp/s for **e**, 9.6 imp/s here). Thus for **g**, about half of the depolarization (5.18 out of 12mV) was derived non-synaptically. 10,000 motoneuron spikes (1,000 seconds of simulated data) were used for each of **e** and **g**; **h** EPSPs used for all the above, which were calculated as the difference between two negative exponentials of time constant 6ms and 0.5ms. Although of relatively long half-width compared to many real motoneuron unitary EPSPs, these EPSPs have their most important feature, the rate of rise, in a realistic range (about 0.5mV/ms measured over 10–90% for the 400μV example, proportionately less for the others; cf. 0.11–0.63mV/ms for EPSPs in the range of amplitude 86–310μV from Cope et al., 1987. **i** Amplitudes of the kernels derived as in **c** (same symbols) but with the synaptic noise parameters as in **g** (for (10.1), B/A = 5.39ms). Note overall larger amplitudes than in **c** and a greater degree of non-linearity.

Previously unpublished data from H.B. Bostock. Simulations represent refinements of the model of Tuck (1977), which may be consulted for details of the after-hyperpolarization

only a fraction by stretch-evoked synaptic activity. Figure 2 of Gustafsson and McCrea (1984) should be compared with Fig. 10 of Kirkwood and Sears (1982b). For this reason, when interpreting cross-correlograms in terms of EPSPs, Davies et al. (1985b) chose to use the indirectly derived values from the intercostals (a peak increase of 0.5 times the baseline probability for a 100μV EPSP) rather than the more directly observed values of Gustafsson and McCrea (1984) (cf. Cope et al. 1987), which would have been about twice as large. The value for the intercostals is also close to that derived from either Bostock's analytical expression or his simulations for small EPSPs, whereas for rather larger EPSPs or for less noise the simulations corresponded well with the larger values from the other authors. The slope of the curves above 0.2mV in Fig. 10.4i are more than twice the slope of those below 0.1mV in Fig. 10.4c.

Finally in this section some comment on the general use of (10.1) in order to fit observed correlogram peaks should be made. One might be tempted to use this equation to interpret rather broad peaks in motoneuron correlograms as branched-axon effects, needing the coefficients A to be large and B to be small. However, no observations of well-verified monosynaptic connections giving kernels of this form have been seen so far. In terms of (10.4) it would need the maximum value of y/σ to be much larger than normal and at the same time the maximum value of y'/σ' to be much smaller than usual. This would mean that the common inputs should employ large, very slow-rising PSPs, but that the synaptic noise as a whole should be made up of relatively low amplitude, fast-rising components. In our view the most likely way this should arise is if the slow-rising common PSPs actually represent groups of fast-rising components, i.e. if pre-synaptic synchronization is present.

10.3 Patterns of Input Connections to Intercostal Motoneurones

The measurements of Sears and Stagg (1976) and Kirkwood et al. (1982a) revealed several features of the distribution of last-order inputs to the motoneurons. This information was obtained on the assumption that synchronization peaks which were mostly as wide as Fig. 10.2h (some slightly wider) represented only the branched-axon mechanism. Although our current view is that this assumption is not entirely correct, the conclusions may not be greatly in error (see below) and will therefore be described as originally deduced.

Firstly, the synchronization could be seen for any two groups of motoneurons within any segment, as long as sufficiently large numbers of spikes were used in the measurement. No obvious differences were seen between rostral and caudal segments. The synchronization was strongest for motoneurons of the same segment, and showed a gradual decline in strength with the segmental separation. The strength of synchronization declined to a low level, typically 30% or less of the strength measured within a segment (around the threshold of statistical significance), by a segmental separation of three (Fig. 10.5). Thus

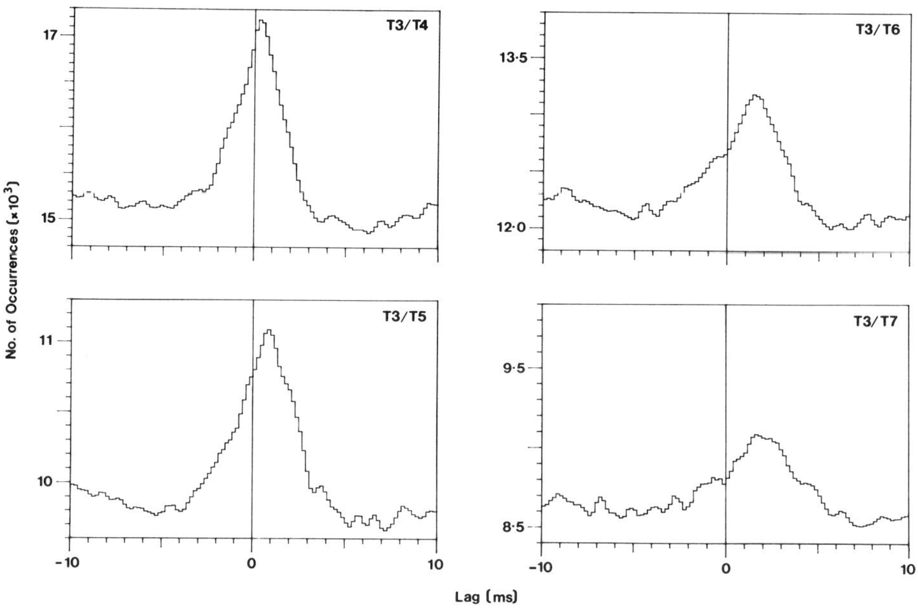

Fig. 10.5. Variation in the synchronization of external intercostal alpha motoneuron discharges with segmental separation. Four cross-correlograms from a single data segment from a moderately deeply anaesthetized cat, using discharges in the segments indicated, the first-named providing the reference spikes. Bin width 0.192ms. Histograms smoothed with a five-point triangular filter and scaled in proportion to the baseline counts. Previously unpublished data of Davies et al. (1985b)

most of the input axons generating this synchronization only make connections to motoneurons over a few adjacent segments. Indirect evidence suggested an association between short-term synchronization (as opposed to the broad-peak form) and the respiratory drive, implying that it was the axons conveying the respiratory drive that provided most of the input (in at least moderately deeply anaesthetized animals), and that it was these axons, whichever cells they arose from, that had this distribution of connections. Thus, despite a wide synergy of activation by the respiratory input to at least 10 and perhaps all 12 intercostal spaces, most of the *last order* connections are organised in a more restricted way. If pre-synaptic synchronization was involved in some of the measurements then it seems unlikely that the component of synchronization due to the branched-axon mechanism could actually have a *wider* distribution than measured here; if anything the reverse is likely.

Although this distribution of inputs was assigned mainly to the respiratory inputs, it should be noted that two other inputs (both fairly weak) were already known to have similarly restricted segmental spreads, namely individual muscle spindle afferents (Kirkwood and Sears 1982b) and the Renshaw axons (as a group) activated from the motoneurons of one segment (Kirkwood et al. 1981). Two to four adjacent segments thus appear to form some sort of "unit" of activation within the wider synergy of the whole external intercostal layer.

The second main result, also clear in Fig. 10.5, was that a time shift away from zero was seen in the position of the peak for short-term synchronization. This time shift was approximately proportional to the segmental separation, and was in the direction corresponding to the spikes in the more caudal segment occurring after those more rostrally. It is thus consistent with the origin of the synchronization being a set of descending fibres. Moreover, the conduction velocity calculated from the time shifts was about 28m/s, which corresponded well with the then current estimates of the mean conduction velocity of inspiratory bulbospinal axons. (More recent estimates have given a larger value for this conduction velocity, around 50m/s, but this figure applies to the cervical regions. For the mid-thorax, Davies et al. (1985a) arrived at a figure of 33m/s, close to the original estimate.) Kirkwood et al. (1982a) therefore suggested that it was the bulbospinal axons that were responsible for generating the synchronization and thus providing the respiratory drive.

10.4 Direct Measurements of the Respiratory Input to Intercostal Motoneurons

The suggestion above was not supported by subsequent measurements of the connections between bulbospinal neurones and intercostal motoneurons. Spike-triggered averaging of synaptic potentials in intercostal motoneurones from bulbospinal neurones yielded no examples of definite monosynaptic excitation (Lipski and Merrill 1983; Merrill and Lipski 1987; but cf. Duffin and Lipski 1987). Davies et al. (1985a,b) used cross-correlation between the discharges of single bulbospinal neurons and efferent external intercostal nerve discharges in up to six different segments simultaneously and, presumably because of the greater sample of motoneurons available with this technique, were able to show that there actually was a monosynaptic connection, in which 50% of the bulbospinal neurons participated. Nevertheless, there was overall agreement with Lipski and Merrill in that the observed connection was relatively weak. The properties deduced by Kirkwood et al. (1982a) to belong to bulbospinal axons were therefore then reassigned to (unknown) interneurons. Subsequently, a large population of respiratory interneurons, many with descending axons of appropriate conduction velocities, have been described (Kirkwood et al. 1988).

It is worth dealing with the results of Davies et al. (1985a,b) in some detail because there are features in that study which are not represented in other cross-correlation studies. To our knowledge this was the first use of cross-correlation to assess the overall strength of a pathway. Many authors might consider there to be too many unknowns to make such an assessment reliable (e.g. Merrill and Lipski 1987) and indeed we ourselves have issued warnings against the over-interpretation of cross-correlation results (Kirkwood and Sears 1978; Kirkwood 1979). However, we feel that our experience since then with these same motoneurons in the same conditions (very important, given the state dependency of synchronization within the system), in particular the similarity between the correlation kernels derived from the ACE potentials, from the spindle afferent connections (results both from our laboratory and from other authors) and from the theoretical results (Fig. 10.4), together with the use of multi-unit motoneuron populations to give an adequate sample, justify our translation of cross-correlation peaks into an average size of single-fibre EPSP. The firing rates of the bulbospinal neurons were known for the assessment but other assumptions had to be made about the number of these neurons and the integration of the trains of EPSPs in the motoneurons. Overall, the accuracy of the assessment was limited to something like an order of magnitude, but since the assessment was that the average depolarization was at least an order of magnitude less than was needed to generate the motoneuron discharge, we feel that the conclusion that the monosynaptic pathway provided only a small fraction of the respiratory input (cf. Lloyd 1944) is reasonably secure. The strength of the monosynaptic connections to phrenic motoneurons assessed in the same way was about three times as strong. This comparison depends very little on the assumptions, the techniques being identical. However, the conclusion that the monosynaptic pathway therefore provides only a third (or thereabouts) of the respiratory input to the phrenic motoneurons, i.e. still a minority of the input, is probably within the margin of error of those assumptions. This margin would allow the monosynaptic input to be the main one in this case.

An important factor in making the above assessments was the use of restrictive criteria for the acceptance of peaks as indicative of a monosynaptic connection. These criteria consist of statistical tests for the significance of a peak, together with criteria based on latencies and durations of the peaks. The statistical criteria will not be dealt with here. With regard to the others, it is our belief that in general these criteria cannot be stated as precise numerical limits. They must be independently argued, often from an empirical base, for the particular circumstances of a particular measurement. For instance, the duration of a peak, representing (via the analysis previously described) mostly the rising phase of an EPSP, will be longer either for oligosynaptic connections or for presynaptic synchronization than for a monosynaptic connection. This is because of temporal dispersion at the extra synapse(s) involved. The maximum duration that would be *totally* safe as a criterion for a monosynaptic connection is given by the duration of a peak consisting of the convolution of the correlation kernels for the *fastest-rising* EPSPs at the (minimum) two synapses involved in the other connections. This is very short, probably around 0.5ms. Although actual correlation peaks for monosynaptic connections *may* be narrower than this (Kirkwood and Sears 1982b; Davies et al. 1985b), the majority are wider, as would be expected from the known range of rise times of single-fibre monosynaptic EPSPs in motoneurons (e.g. Jack et al. 1971). Thus, setting such a low limit can reject a large number of correlation peaks which could result from a monosynaptic connection. Realistically, any disynaptic connection is likely to involve divergence to and convergence from multiple interneurons, inevitably involving temporal dispersion due to variable PSP rise times and pre-terminal slowing. Reported rise times of disynaptic spike-triggered average PSPs have been relatively long (Watt et al. 1976; Takakusaki et al. 1989).

If discrimination against a disynaptic or oligosynaptic connection is the issue, then of course latency can be used as a criterion and all peaks with latencies less than a given limit can be accepted as monosynaptic, which may rescue many peaks rejected on the grounds of duration. If discrimination against presynaptic synchronization is an issue, then things are more difficult. No additional acceptance criterion applicable to individual peaks is available, though some peaks may be *rejected* if their latency is too early.

Special pleading and empirical judgements thus come into play. Davies et al. (1985a) reasoned that the main cause for concern in their experiments was the synchronization of individual bulbospinal neurons with other bulbospinal neurons. Because significant effects would only be expected if many such neurons were synchronized (previously published measurements of medullary respiratory neuron synchronization showed relatively weak or uncommon effects), they argued that neurons over the whole conduction velocity range known to represent the bulbospinal axons would be involved. Temporal dispersion in the long descending axons from medulla to thorax would then give predictably long durations to the peaks resulting from the synchronization. The observed peaks fell into two groups, one with half-widths all less than 1.1ms, the other all greater than 1.7ms, as is illustrated in Fig. 10.6. The shapes and latencies of the wide group (designated "medium width") corresponded to waveforms predicted from the known conduction velocities, on the assumption of short-duration synchronization in the medulla. The medium-width group included examples with latencies definitely too short to be explained without presynaptic synchronization (e.g. Fig. 10.6c). In contrast, for the group of narrow peaks *all* the latencies were appropriate to a monosynaptic connection, these latencies being known for each bulbospinal neuron because each had been antidromically activated from the thoracic cord, and the narrow peaks were thus accepted as representing such a connection.

The half-width criterion here (1.1ms) is relatively wide and therefore would not be of much use *a priori* for individual peaks. Indeed, Davies et al. (1985a) still could not exclude the possibility that some disynaptic connections were present in addition to the monosynaptic ones, though the latencies indicate that this was unlikely to have been the main effect. Further, the contribution of a disynaptic connection even to the "tail" of the peaks was likely to be small: the half-widths of peaks generated in cross-correlograms between spindle afferents and motoneurons, believed to include a weak oligosynaptic component (Kirkwood and Sears 1982a) were on average longer (Kirkwood and Sears 1982b; Kirkwood et al. 1982b, Fig. 3).

Three general points can be made. Firstly, the interpretation of the results and the setting up of the criteria for acceptance of peaks as representing a monosynaptic connection were based on the whole population of results and on prior knowledge of some of the neuron properties. Secondly, *two* criteria, half-width *and* latency of the peaks, were used. Thirdly, the use of both of these was critically dependent on the presence of a long conduction path for the presynaptic fibres, which allowed temporal dispersion to significantly broaden peaks due to presynaptic synchronization and allowed relationships between conduction velocity and latency to appear and thus to make the inevitable scatter of latencies insignificant. Davies et al. (1985a) also showed that similar, reliable criteria

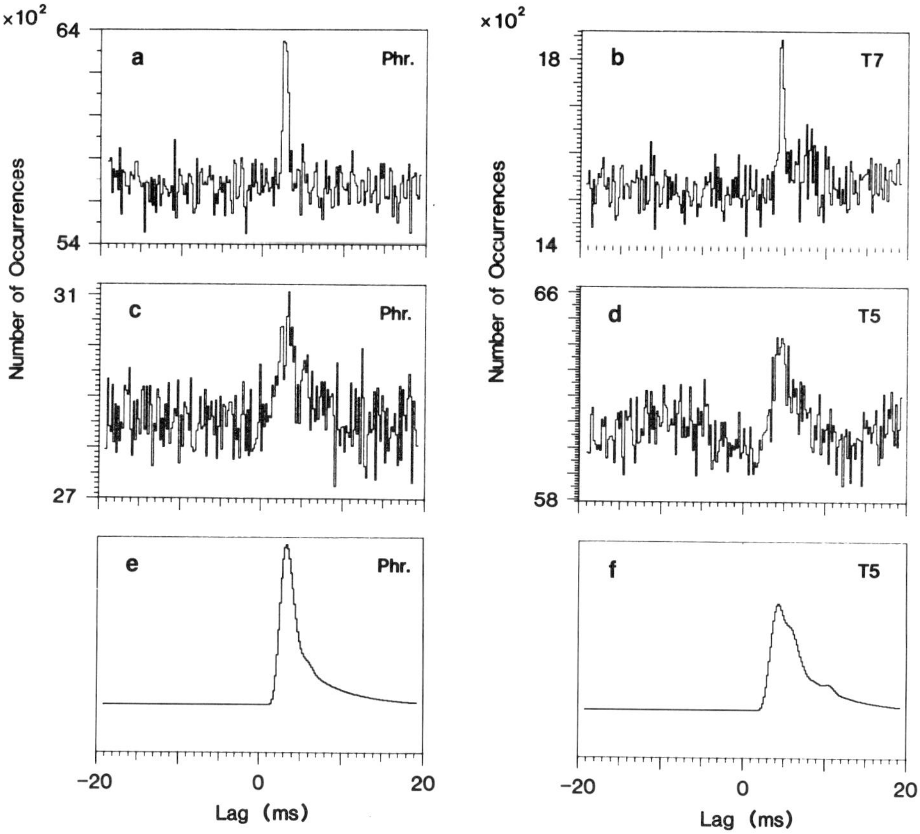

Fig. 10.6. a–d Typical examples of peaks in cross-correlograms constructed from the discharges of individual bulbospinal units (reference spikes) and motoneuron discharges. Each histogram comes from a different bulbospinal unit: **a** and **c** phrenic discharges; **b** and **d** external intercostal discharges from the indicated segments. **a** and **b** narrow peaks; **c** and **d** medium-width peaks. See text for definitions of types of peaks. Conduction velocities: **a** 100m/s; **b** 47.3 m/s; **c** 28.0 m/s; **d** 72.2 m/s. Numbers of reference spikes: **a** 52462; **b** 55215; **c** 40623; **d** 98675. **e,f** Theoretical time courses for medium-width peaks for phrenic or T5 intercostal discharges. See text for explanation. The slow conduction velocity of the unit in **c** means that the impulse in the bulbospinal axon did not reach the phrenic motor nucleus until a lag of 2.0ms, yet the latency of the peak is only about 1.5ms, which must also include synaptic delay and conduction to the periphery. Figure reproduced from the Journal of Physiology (Davies et al. 1985a). Bin width 0.192ms

for spike-triggered average PSPs were, if anything, harder to establish than for cross-correlation peaks.

Because of the ubiquitous nature of common inputs to neurons engaged in common tasks, presynaptic synchronization must be *expected* in cross-correlations or spike-triggered averaging studies of the connections of central neurones. A single criterion is unlikely to be of use in discriminating against such an effect and in many cases (without the long conduction distance described above) neither latency nor duration of a peak will provide a suitable criterion,

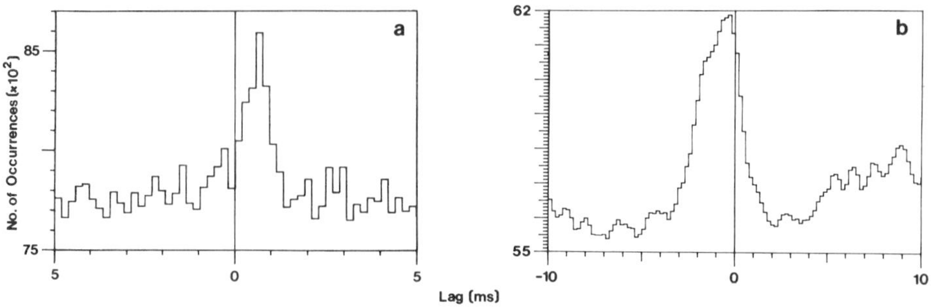

Fig. 10.7. **a** Cross-correlogram constructed from the discharges of a thoracic inspiratory interneuron in segment T6 with a contralateral descending axon (reference spikes) and of the alpha motoneurones of the contralateral external intercostal nerve of the same segment in an anaesthetized, paralyzed cat. Bin width 0.192ms. Previously unpublished data from Kirkwood et al. (1988). **b** Cross-correlogram constructed from the inspiratory discharges of alpha motoneurons in the external intercostal nerve of T4 (reference spikes) and in the internal intercostal nerve of the same segment of an anaesthetized, paralysed cat. Bin width 0.192ms. Histogram smoothed with a five-point triangular filter. Previously unpublished data from Kirkwood and Sears (1989)

except when duration is very restrictive or when latency is used to say that a common input effect must be present. The 1.1ms used by Davies et al. (1985a,b), though probably a good *descriptor* of cross-correlation peaks for monosynaptic connections in motoneurons, cannot be used as a *criterion* outside the particular circumstances of those experiments. As an example, Fig. 10.7a shows a cross-correlogram derived from the discharges of an inspiratory interneuron in the thoracic spinal cord and external intercostal motoneuron discharges of the same segment. The peak is narrow (half-width 0.8ms) but the latency is very short (0.2ms for the sharpest part of the peak or negative for the whole peak). The motoneuron conduction time to the recording site in the periphery was around 0.3 to 0.5ms, meaning that, even with some allowance for the possibility that the interneuron spike was recorded after delayed invasion of the soma following initiation at the initial segment, there was still insufficient time for a synaptic delay and the peak must have been largely generated by a common input effect. Thus if such a narrow peak *can* be generated by common inputs, then common inputs must be suspected as being the origin of *any* peak of this duration for this category of neurones whatever the latency, even when this may be appropriate for a monosynaptic connection.

10.5 Further Analysis of Motoneuron Synchronization

The insights gained from the above considerations have further implications for motoneuron synchronization. Firstly, the synchronization deduced for the bulbospinal neurons is likely to have some direct effects. However, if the direct connections from these neurons to the motoneurons are as weak as we believe, these effects will be correspondingly weak or rare (Davies et al. 1985b). The effects suspected should consist of a component of motoneuron synchronization whose time course is given by the convolution of a medium-width peak (Fig. 10.6) with itself (time reversed), i.e. a peak of width intermediate between short-term and broad peak. Such peaks are occasionally seen, though they were not recognized by Kirkwood et al. (1982a,b). An example which could be ascribed to such a mechanism is shown in Fig. 10.2g: a peak of relatively large k (1.24), a duration at the base of 15ms and a half-width of 5.5ms.

Moreover, recent experiments (Kirkwood and Sears 1989) using longer lengths of data and narrower bin widths, have revealed other effects of presynaptic synchronization concealed within the time course of short-term synchronization, which until now we had previously ascribed to the effects of branched-axon common inputs alone. In a rather similar way to the analysis of the bulbospinal connections, a critical step in the interpretation has been the timing of the peaks. The experiments were set up to compare inspiratory inputs to motoneurons innervating the dorsal areas of the external intercostal muscle (the region whose activity has been described above) and those innervating another area of the thorax, the parasternal, interchondral region, which is innervated by a different nerve, the internal intercostal nerve. Since this muscle is regarded by some (e.g. Duron, 1973) as more like the diaphragm in its activation, one possibility was that its motoneurons might receive a stronger monosynaptic input from the medulla than those innervating external intercostal muscle. This idea was supported by the observations of Duffin and Lipski (1987) who, measuring spike-triggered averaging of EPSPs, claimed that one group of neurons in the medulla gave relatively strong monosynaptic connections to inspiratory intercostal motoneurons. However, the intercostal motoneurones they recorded mostly had axons in the internal intercostal nerve and thus would have innervated the interchondral region. Thus the hypothesis was that the inputs to the two sets of motoneurones were largely separate, which could be tested by measuring the short-term synchronization between the inspiratory discharges of the external and internal intercostal nerves.

The main result, apparently contrary to the hypothesis, was that short-term synchronization between these two groups could clearly be demonstrated, even if it was weaker than between two branches of the external nerve in the same segment. However, the position of the peaks in the correlograms most often showed a time shift, the synchronized spikes in the internal nerve being on average about 1ms ahead of those in the external nerve (Fig. 10.7b). Because this was true for motoneurons *within* a segment, it was difficult to explain the time shift by differences in conduction distances. Instead, the alternative explanation was adopted that the common inputs which generated the synchronization included an extra synapse on the path to the motoneurons of the external nerve. The respiratory input was implicated in this asymmetric pathway because, when present, HFO showed a similar time shift and also because, when synchronization was measured between phrenic and intercostal motoneurons, this time shift was also seen superimposed on an expected additional time shift appropriate to bulbospinal or propriospinal fibres descending over 6 segments (C5 to T3).

The observations thus supported the hypothesis that the inspiratory drive was more direct to the parasternal motoneurons than to the external intercostal ones, but for the purposes of this article the most interesting conclusion was additionally that a disynaptic link could appear in a synchronization measurement involving the external intercostal motoneurons. Could this linkage therefore contribute to the normal short-term synchronization between two groups of these motoneurons? The answer seemed to be yes. As illustrated in Fig. 10.2h,i, the peaks in the correlograms constructed from the discharges of pairs of external nerves were always wider than those from internal nerves. One must therefore presume, by the logic set out in the earlier section of this chapter, that only peaks as narrow as those of the internal nerve pairs (half-widths never more than 2.1ms) can reliably be taken as representing simple branched-axon common inputs, and that a considerable degree of pre-synaptic synchronization is likely to be present in the rather wider peaks (typical half-width 3ms) of the external nerve pairs. Further, without any positive evidence to the contrary one must also assume that the same is true for motoneuron synchronization recorded elsewhere (e.g. in human studies) where the durations of the peaks may be similar.

Fortuitously, perhaps, the particular example of short-term synchronization used by Kirkwood and Sears (1978) for their comparisons with the ACE potential (the only one from Sears and Stagg (1976) measured with a sufficiently large number of counts per bin at a narrow bin width) was at the lower end of the half-width range for external intercostal nerve pairs (2.5ms). Thus the good correspondence between theory and measurement for the branched-axon mechanism of synchronization in Kirkwood and Sears (1978), using a realistic time course for the EPSPs, turned out to be appropriate for using the theory to deduce the primary correlation operator.

10.6 Conclusions

Study of the respiratory system has allowed insight to be gradually acquired into a variety of different types of common input circuits that are involved in synchronizing motoneurons. The measurements consisted of basically a single technique, cross-correlation between pairs of multi-unit discharges. The ability to identify the various synchronizing mechanisms that are involved in the widely convergent/divergent network of inputs to the motoneurons stems from the experimental flexibility allowed by the multisegmental arrangement of two or three individual muscles, and the controllable state-dependency of the respiratory and other inputs. Some general properties of the distribution of the last-order inputs to these motoneurons have been deduced, which should help us understand how the CNS controls a complex motor act involving simultaneous graded activation of motoneurons in many segments of the spinal cord.

In addition, the motoneuron synchronization measurements have led to an understanding of the basic unit of cross-correlation measurements, the primary correlation kernel, which has been confirmed by more direct correlations between the discharges of individual inputs and motoneurons and has allowed a start to be made in assessing the integrated effect of individual categories of inputs to the production of the motoneuron output in a particular behaviour. An emphasis has been placed on the need in such studies (or in spike-triggered averaging studies) for excluding the correlations derived from synchronization of inputs. The motoneuron synchronization studies suggest that such effects are likely to be very common and only in some circumstances may it be possible to discriminate against them.

References

Adams, L., Datta, A.K., Guz, A. (1989): Synchronization of motor unit firing during different respiratory and postural tasks in human sternocleidomastoid muscle. J. Physiol. **413**, 213–231

Berger, A.J., Davies, J.G., Sears, T.A. (1983): Time signatures of peripheral and central chemoreceptor drives to feline respiratory motoneurones. J. Physiol. **334**, 41–42P

Bryant, H.L., Marcos, A.R., Segundo, J.P. (1973): Correlation of neuronal spike discharges produced by monosynaptic connections and by common inputs. J. Neurophysiol. **36**, 205–255

Cohen, M.I. (1973): Synchronization of discharge, spontaneous and evoked, between inspiratory neurons. Acta Neurobiol. Exp. **33**, 189–218

Cohen, M.I. (1979): Neurogenesis of respiratory rhythm in the mammal. Physiol. Rev. **59**, 1105–1173

Cope, T.C., Fetz, E.E., Matsumura, M. (1987): Cross-correlation assessment of synaptic strength of single Ia fibre connections with triceps surae motoneurones in cats. J. Physiol. **390**, 161–188

Da Silva, K.M.C., Sayers, B.McA., Sears, T.A., Stagg, D.T. (1977): The changes in configuration of the rib cage and abdomen during breathing in the anaesthetized cat. J. Physiol. **266**, 499–521

Datta, A.K., Fleming, J.R., Stephens, J.A. (1985a): Effect of central nervous system lesions on the synchronization between motor unit discharge during voluntary contraction in human first dorsal interosseus muscle. J. Physiol. **365** 19P

Datta, A.K., Fleming, J.R., Stephens, J.A. (1985b): The effect of stroke on synchronization of human motor unit firing in large flexor and extensor muscles in the leg and a small hand muscle. J. Physiol. **369**, 29P

Davey, N.J., Ellaway, P.H. (1988): Control from the brainstem of synchrony of discharge between gamma motoneurones in the cat. Exp. Brain Res. **72**, 249–263

Davey, N.J., Ellaway, P.H., Friedland, C.L. (1986): Synchrony of motor unit discharge in humans with Parkinson's disease. J. Physiol. **377**, 30P

Davey, N.J., Ellaway, P.H., Friedland, C.L., Short, D.J. (1990): Motor Unit discharge characteristics and short term synchrony in paraplegic humans. J. Neurol. Neurosurg. Psychiatry **53**, 764–769

Davies, J.G., Forster, I.C. Sears, T.A. (1983): A modular interface for the computer acquisition of multichannel neural data. J. Physiol. **342**, 7–8P

Davies, J.G.McF., Kirkwood, P.A., Romaniuk, J.R. Sears, T.A. (1986): Effects of sagittal medullary section on high-frequency oscillation in rabbit phrenic neurogram. Respir. Physiol. **64**, 277–287

Davies, J.G.McF., Kirkwood, P.A., Sears, T.A. (1985a): The detection of monosynaptic connexions from inspiratory bulbospinal neurones to inspiratory motoneurones in the cat. J. Physiol. **368**, 33–62

Davies, J.G.McF., Kirkwood, P.A., Sears, T.A. (1985b): The distribution of monosynaptic connexions from inspiratory bulbospinal neurones to inspiratory motoneurones in the cat. J. Physiol. **368**, 63–87

Dietz, V., Bischofberger, E., Wita, C., Freund, H.J. (1976): Correlation between the discharges of two simultaneously recorded motor units and physiological tremor. Electroenceph. Clin. Neurophysiol. **40**, 97–105

Draper, M.H., Ladefoged, P., Whitteridge, D. (1960): Expiratory pressures and air flow during speech. Br. Med. J. **1**, 1837–1843

Duffin, J., Lipski, J. (1987): Monosynaptic excitation of thoracic motoneurones by inspiratory neurones of the nucleus tractus solitarius in the cat. J. Physiol. **390**, 415–431

Duron, B. (1973): Postural and ventilatory functions of intercostal muscles. Acta Neurobiol. Exp. **33**, 330–380

Farmer, S.F., Ingram, D.A., Stephens, J.A. (1990): Mirror movements studied in a patient with Klippel-Feil syndrome. J. Physiol. **428**, 467–484

Fetz, E.E., Gustafsson, B. (1983): Relation between shapes of post-synaptic potentials and changes in firing probability of cat motoneurones, J. Physiol. (London) **341**, 387–410

Gustafsson, B., McCrea, D. (1984): Influence of stretch-evoked synaptic potentials on firing probability of cat spinal motoneurones. J. Physiol. **347**, 431–451

Jack, J.J.B., Miller, S., Porter, R., Redman, S.J. (1971): The time course of minimal excitatory post-synaptic potentials evoked in spinal motoneurones by group Ia afferent fibres. J. Physiol. **215**, 353–380

Khatib, M., Hilaire, G., Monteau, R. (1989): Excitatory interactions between phrenic motoneurons: intracellular study in the cat. Exp. Brain Res. **74**, 131–138

Kirkwood, P.A. (1979): On the use and interpretation of cross-correlation measurements in the mammalian central nervous system. J. Neurosci. Meth. **1**, 107–132

Kirkwood, P.A., Munson, J.B., Sears, T.A., Westgaard, R.H. (1988): Respiratory interneurones in the thoracic spinal cord of the cat. J. Physiol. **395**, 161–192

Kirkwood, P.A., Sears, T.A. (1978): The synaptic connexions to intercostal motoneurones as revealed by the average common excitation potential. J. Physiol. **275**, 103–134

Kirkwood, P.A., Sears, T.A. (1982a): Excitatory post-synaptic potentials from single muscle spindle afferents in external intercostal motoneurones of the cat. J. Physiol. **322**, 287–314

Kirkwood, P.A., Sears, T.A. (1982b): The effects of single afferent impulses on the probability of firing of external intercostal motoneurones in the cat. J. Physiol. **322**, 315–336

Kirkwood, P.A., Sears, T.A. (1989): Dual descending pathways to inspiratory motoneurones in the anaesthetized cat. J. Physiol. **414**, 15P

Kirkwood, P.A., Sears, T.A., Stagg, D., Westgaard, R.H. (1982a): The spatial distribution of synchronization of intercostal motoneurones in the cat. J. Physiol. **327**, 137–155

Kirkwood, P.A., Sears, T.A., Stagg, D., Westgaard R.H. (1987): Intercostal muscles and purring in the cat: influence of afferent inputs. Brain Res. **405**, 187–191

Kirkwood, P.A., Sears, T.A., Tuck, D.L., Westgaard, R.H. (1982b): Variations in the time course of the synchronization of intercostal motoneurones in the cat. J. Physiol. **327**, 105–135

Kirkwood, P.A., Sears, T.A., Westgaard, R.H. (1981): Recurrent inhibition of intercostal motoneurones in the cat. J. Physiol. **319**, 111–130

Kirkwood, P.A., Sears, T.A., Westgaard, R.H. (1984): Restoration of function in external intercostal motoneurones of the cat following partial central deafferentation. J. Physiol. **350**, 225–251

Knox, C.K. (1974): Cross-correlation functions for a neuronal model. Biophys. J. **14**, 567–582

Knox, C.K., Poppele, R.E. (1977): Correlation analysis of stimulus-evoked changes in excitability of spontaneously firing neurons. J. Neurophysiol. **40**, 616–625

Lipski, J., Merrill, E.G. (1983): Inputs to intercostal motoneurones from ventrolateral medullary respiratory neurones in Nembutal-anaesthetized cats. J. Physiol. **339**, 25–26P

Lloyd, D.P.C. (1944): Functional organization of the spinal cord. Physiol. Rev. **24**, 1–17

Merrill, E.G., Lipski, J. (1987): Inputs to intercostal motoneurones from ventrolateral medullary respiratory neurons in the cat. J. Neurophysiol. **57**, 1837–1853

Nelson, P.G. (1966): Interaction between spinal motoneurons in the cat. J. Neurophysiol. **29**, 275–287

Sears, T.A. (1964): Efferent discharges in alpha and fusimotor fibres of intercostal nerves of the cat. J. Physiol. **174**, 295–315

Sears, T.A. (1977): The respiratory motoneurone and apneusis. Fed. Proc. **36**, 2411–2420

Sears, T.A., Stagg, D. (1976): Short-term synchronization of intercostal motoneurone activity. J. Physiol. **263**, 357–381

Stagg, D. (1973): Computer acquisition of multiunit nerve-spike signals. Med. Biol. Eng. **11**, 340–347

Takakusaki, K., Ohta, Y., Mori, S. (1989): Single medullary reticulospinal neurons exert post synaptic inhibitory effects via inhibitory interneurons upon alpha-motoneurons innervating cat hindlimb muscles. Exp. Brain Res. **74**, 11–23

Tuck, D.L. (1977): Investigation of intercostal neuronal intracellular processes and connectivity by signal analysis and computer stimulation. Ph.D. Thesis: University of London

Watt, D.G.D., Stauffer, E.K., Taylor, A., Reinking, R.M., Stuart, D.G. (1976): Analysis of muscle receptor connections by spike-triggered averaging. 1. Spindle primary and tendon organ afferents. J. Neurophysiol. **39**, 1375–1392

11. Correlated Activity of Neurons: A Neural Code for Higher Brain Functions?

Eilon Vaadia, Ehud Ahissar, Hagai Bergman, and Yizhar Lavner

Department of Physiology, Hadassah Medical School, The Hebrew University, Jerusalem, Israel

This chapter describes some response properties of single neurons and the interactions between them in both a sensory area (the auditory cortex) and an "association" area (the frontal cortex) of the monkey. The data described were obtained by simultaneous recording of several (6–10) neurons by six microelectrodes, in behaving animals.

It is shown that neurons in both areas can participate in prolonged processes and be involved in more than one process. Neurons in each recording site need not function in unison; the neighboring neurons show a variety of response patterns to identical stimuli and while some neurons are activated, others may remain unaffected.

Techniques to quantify the strength of interactions among neurons are discussed. Three different criteria to estimate the strength of effective connections between pairs of neurons are described. All the criteria show that the interactions between adjacent neurons are weak but can sometimes change upon sensory stimulation, selective attention or changes in behavioral state.

Such findings support the hypothesis that information is represented in the cortex by selective coactivation of sets (assemblies) of neurons, where each single nerve cell may be a member of several representation sets. Thus, each neuron may participate in more than one function and each small cortical area may contain members of several functional groups. The dynamic properties of neuronal interactions further suggest that neuronal assemblies are not rigid anatomical entities. Rather, neuronal groups can organize to serve specific tasks and then their members may reorganize to form new groups, serving new tasks.

11.1 Introduction

Most neurophysiologists and neuropsychologists have long agreed that the electrical activity of the neurons is involved in information processing and computations carried out by the brain. Yet there is very little understanding of the neural mechanisms underlying these processes, and no consensus has been reached

Springer Series in Synergetics
J. Krüger (Ed.)
Neuronal Cooperativity

about the nature of the neural codes. Therefore, it is essential to consider different alternatives, for each of them dictates not only a way of thinking about brain mechanisms, but also suggests experimental approaches toward electrophysiological studies of these mechanisms.

A common view suggests that the neural code lies within the spike activity of a single neuron. It is generally assumed that the most relevant parameter to be extracted from the spike train is the rate of spike occurrence. Barlow (1972) suggested that, "The frequency of neural impulses codes subjective certainty: a high impulse frequency in a given neuron corresponds to a high degree of confidence that the cause of the percept is present in the external world". Thus, according to this view, the neural code is the firing rate of a single neuron and the appropriate experimental tool is a microelectrode that can isolate and record the activity of a single neuron (single-unit activity). Indeed, single-unit studies in many cortical and sub-cortical areas have clearly demonstrated highly specific changes of firing rates that were related not only to sensory and/or motor processing, but also to cognitive and other higher brain functions such as perceptual analysis of sensory input, selective attention, short-term memory, etc. (Fuster 1980; Evarts et al. 1984; Perrett et al. 1982; Goldberg and Segraves 1987; Goldman-Rakic 1987; Niki and Watanabe 1976, 1979; Gottlieb et al. 1989).

A different view was first presented by Hebb. He suggested that the activity within groups of neurons ("*cell assemblies*") in a coherent manner, mediated by anatomical, plastic connections, is the basis of the neural code. The groups, he hypothesized, are formed by modifications of synaptic strength that obey the rule: "When an axon of cell A is near enough to excite cell B and repeatedly or persistently takes part in firing it, some growth process or metabolic change takes place such that A's efficacy, as one of the cells firing B, is increased." (Hebb 1949).

Several models and hypotheses followed Hebb's, ideas. (Braitenberg 1978; Edelman 1981; Abeles 1982; Hopfield 1982; Hopfield and Tank 1986; Palm 1982; Rumelhart et al. 1986; Amit 1987; Sompolinsky 1987, and others). Almost all the above models assume that "learning" is based on selective modifications of the connectivity among neurons and "computations" are carried out by selective activation of interconnected neurons. According to these hypotheses, the neural code lies in correlated activity within and between neuronal groups. Therefore, the experimental methodology must involve simultaneous recording of many neurons in parallel. Recording of single units, one at a time, is not an appropriate tool because it does not allow testing of intracortical interactions that are associated with internal representations by the hypothetical "*cell assemblies*".

Ideally, one should study the activity in a cell assembly by recording the activity of all the neurons in the region in which the cell assembly resides. With the current technology, one can hope to record spike activity of a few single neurons (up to six) through one microelectrode (Abeles and Goldstein 1977; McNaughton et al. 1983) and up to 30 neurons by using multiple electrodes (Kruger 1983). At best, therefore, we can tap only a few members of the hypothetical neuronal network. Nevertheless, it is appropriate to examine how the available physiological data support different concepts and models of neural codes, in the quest for an optimal research strategy. Indeed, in recent years there has been a growing interest in the simultaneous recording and analysis of spike trains.

The analysis of interactions among neurons consists primarily of cross-correlating spikes from pairs (sometimes triplets) of neurons and inspecting the results for possible signs of "functional" or "effective" connectivity (Moore et al. 1966; Perkel et al. 1967; Abeles 1983). Recent developments have expanded the scope of the studies; methods were developed in an effort to reveal cooperativity in groups of neurons (Gerstein and Aertsen 1985; Gerstein et al. 1985), detect spatio-temporal firing patterns of many neurons (Dayhoff and Gerstein 1983; Abeles and Gerstein 1988; Tam et al. 1988) and study in detail dynamic properties of the correlation between two neurons (Gerstein 1988; Aertsen et al.1989).

Application of these analysis methods to experimental data revealed that the modulations of firing rates of single neurons are just the tip of an iceberg of phenomena involved in cortical information processing. It was demonstrated that the correlated activity between pairs and among groups of neurons relates to the anatomical and functional organization of the cortex. It was also demonstrated that neuronal interactions may depend on sensory stimulation, arousal levels and behavioral state (Dickson and Gerstein 1974; Gassanov et al. 1980; Burns and Webb 1979; Frostig et al. 1983; Abeles et al. 1983; Levick et al. 1972; Ts'o et al. 1986; Eckhorn et al. 1988; Gray and Singer 1989; Kruger and Aiple 1988; Vaadia et al. 1989; and others).

Most of the above results could not possibly have been predicted from single neuron measurements. Nevertheless, studies of properties of single neurons are still very important; they provide a range of parameters for each element in the hypothetical "neuronal network" (maximal rate, response properties, statistical properties of the spike train etc.). They also provide description of the functional organization of the cerebral cortex. Thus, single-unit studies can show where and when certain processes take place. Furthermore, a research strategy that combines a study of both the individual properties of a neuron and its relation to other neurons allows for a confrontation of different hypotheses, using the same experimental data.

11.2 Functional Properties of Single Neurons

The brisk responses and well-defined receptive fields of many neurons in sensory cortical areas support the hypothesis that the role of these neurons is to extract specific features (the orientation of a line, the frequency of a tone, etc.) from the sensory information and to end their role in the process by transmitting the information to another, higher-order, cortical region for further processing. This view is well supported by a large number of studies that have described a hierar-

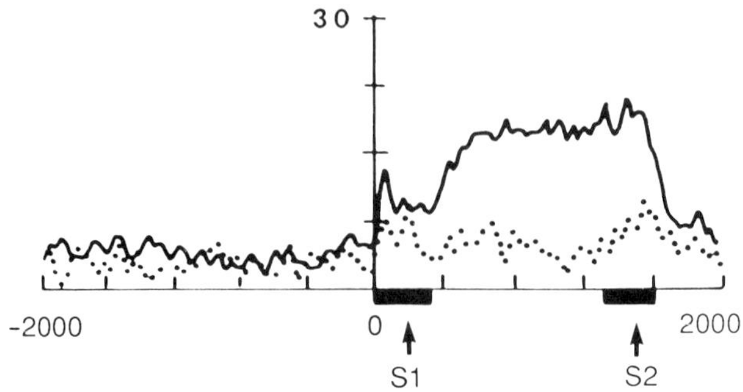

Fig. 11.1. Activity of a single neuron in the auditory cortex of the Rhesus monkey. The solid line shows a PST histogram collected during performance of the short-term memory task. The broken line shows the PST histogram collected when the stimuli were presented but thc monkey did not perform the task. The abscissa describes time from 2.0 s before the onset of the first stimulus (S1) until 0.45 s after the second stimulus (S2) ended. The ordinate describes the firing rate in spikes/s. (Adapted from Gottlieb et al. 1989)

chy of highly specific neurons at different levels of the visual cortical fields. The implication of this view is that the processing in each "station" is brief, and that each cortical region is associated with one specific function. Similarly, at the neuronal level, it is implicitly assumed that a single nerve cell is involved in a single function.

While this view can be justified for some sensory processes, it does not seem to hold in all cases. Under appropriate behavioral conditions, prolonged slow waves which mark distinct perceptual, cognitive and linguistic processes, lasting hundreds of milliseconds to several seconds, can be recorded in a single cortical area (Donchin 1984; Deecke et al. 1985). Furthermore, such long processing could also be found in the activity of single units recorded from sensory, association, premotor and motor cortices of behaving monkeys during some specified stages of behavior (Fuster 1980; Gottlieb et al. 1989; Wise and Mauritz 1985; Vaadia et al. 1989). An example from our studies is shown in Fig. 11.1. The Peri-stimulus-time (PST) historgrams in the figure illustrate the firing rate of a single unit in the auditory cortex of a monkey trained to perform an auditory short-term memory task: two tone bursts (S1 and S2), separated by a silent inter-stimuli interval of 1s, were presented in each trial. The monkey had to touch one key if the tones were identical and another key if they were different. The firing rate of the unit while the monkey was performing correctly is shown in a solid line. The broken line shows the activity of the same single unit when the water spout was removed and the monkey did not perform the task.

Though it is not possible to tell from such an experiment how the single unit shown in Fig. 11.1 was engaged in forming the short-term memory of pitch, it is clear that the neuron did not end its role when the stimulus was turned off. It took part in, or was affected by, a prolonged process that took place in the primary auditory cortex during the inter-stimulus interval, long after the information could have been transmitted to higher cortical levels. Approximately

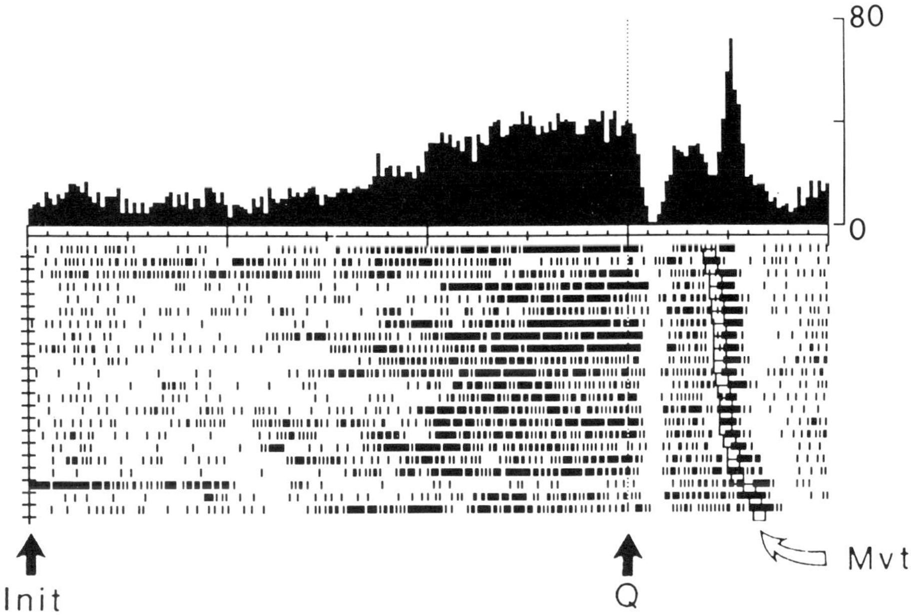

Fig. 11.2. Activity of a single neuron in the premotor cortex of the rhesus monkey. The raster display and PST histogram show 8 s of activity centered around the onset time of the spatial cue (Q). Trial initiation (INIT) is marked by a plus sign on each raster. The beginning of the limb movement (MVT) is marked by an empty square. (From the study by Vaadia et al. 1988)

60% of the recorded neurons in this study showed prolonged changes in their firing rates during the interval between the two tones (Gottlieb et al. 1989).

Furthermore, the activity of a single neuron may be rather complex. An example is shown in Fig. 11.2. The figure shows a raster display and PST histogram illustrating the activity of one neuron in the premotor cortex of a monkey (after Vaadia et al. 1988). The activity was recorded during performance of a stimulus-guided reaching task. The monkey initiated a trial ("INIT") by touching a key (located in front of him in a mid-sagital location) and waited for 6 s. Then, one of two keys (either on the left or right) was illuminated (the "cue", marked as "Q" in Fig. 11.2). The monkey had to maintain contact with the center key for an additional 500 ms and then touch the illuminated key.

Three components of activity can be seen in the PST histogram. The unit was first activated during the waiting period (from INIT to CUE, starting about 3s before cue onset). EMG recordings of superficial muscles indicated that the monkey did not attempt to make movements during the pre-cue interval. This and other considerations led Vaadia et al. to suggest that this activity was not related to preparation or execution of the movements, but rather to other aspects of task performance such as expectation or timing of events (Vaadia et al. 1988). The spike discharge of the same unit was suppressed in response to the directional-cue onset and the unit was then reactivated at the time of movement onset (marked by an empty square on each line of the raster display). This last component of activity could be easily matched with some of the EMG records and was also direction specific, suggesting that it was related to motor function. This type of complex activation does not support the notion that a single neuron is dedicated to processing of a single function.

In another study, we examined whether this view also holds for a primary sensory cortex. We trained a monkey to perform a "*habit-reversal*" tone discrimination task. The monkey had to discriminate between pure tones and noise bursts (S1 and S2) by moving a lever to the left or right respectively. Once the associations (S1 → Left and S2 → Right) were established, the contingencies were reversed to that of S1 → Right and S2 → Left. When this sensorimotor association was established, the contingencies were reversed again. By recording the activity of the same neurons under the two modes of behavior, we could distinguish between types of neuronal activity. Purely sensory activity would depend only on the stimulus regardless of the motor response that followed. Purely motor activity would depend only on the movement emitted, regardless of the stimulus which elicited the motor response. Activity related to sensorimotor association would depend on the stimulus-response pair (for details see Vaadia et al. 1982). Although almost all the recorded neurons in the auditory cortex had pure sensory responses, some also exhibited sensorimotor-associative activity. Thus, in the same trial, a single neuron could exhibit both pure sensory activity and associative activity (Vaadia et al. 1982).

Conclusions

(1) A single cortical neuron may be active for prolonged intervals that last hundreds of milliseconds to several seconds.
(2) A single neuron may participate in more than one process.

11.3 Firing Rates and Response Properties of Neighboring Neurons

A simplistic interpretation of the "columnar organization" structure of the cortex may lead one to suppose that all neurons in one column share similar inputs and therefore become active concomitantly when the appropriate conditions apply. Simultaneous extracellular recordings of spike activity from several neurons demonstrate that this interpretation does not hold even when primary sensory areas are considered.

In our studies, neuronal activity was recorded by glass-coated tungsten microelectrodes of 0.1 to 1 MΩ impedance. In the early experiments, a single electrode was inserted into the brain and several units were detected and isolated by a spike sorter (Abeles and Goldstein 1977). In later experiments, up to six microelectrodes were inserted into the cortex. The electrodes could be arranged in a circle or a row. In the circular arrangement, the distance between

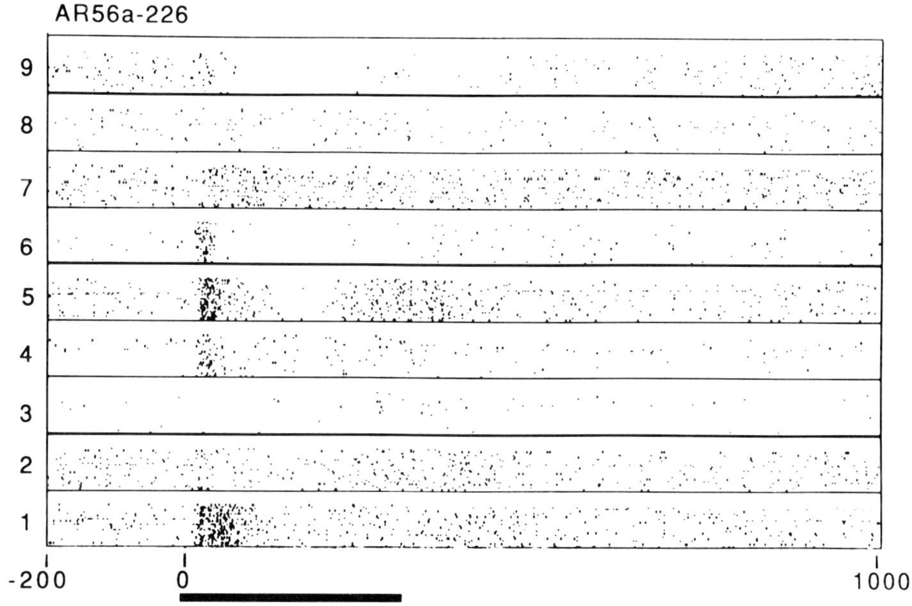

Fig. 11.3. Raster displays of the responses of 9 neurons to tone bursts in the primary auditory cortex of the rhesus monkey. The stimulus (a tone burst of 3.4 Khz at 68 dBspl) is marked by a black bar under the abscissa. The neurons were recorded by 5 microelectrodes. The first electrode recorded unit 1; the second, units 2 and 3; the third, units 4 and 5; the fourth, unit 6 and the fifth, units 7, 8 and 9

any two electrodes ranged from 300μ to 600μ. In the linear arrangement, the horizontal distance between adjacent electrodes was about 500μ. In all studies the monkeys sat in a sound-attenuating room and performed behavioral tasks. Spike activity was detected by spike sorters and dual-window discriminators. The shapes of the isolated spikes were displayed on an oscilloscope screen to allow for careful monitoring of the unit isolation quality.

In two sets of experiments, we have examined the relations of neighboring neurons in the primary auditory cortex and the granular frontal "association" areas. Figures 11.3 and 11.5 show the activity of adjacent neurons recorded simultaneously in each of these two cortical fields.

Figure 11.3 shows neuronal activity recorded by five microelectrodes arranged in a circle in the primary auditory cortex of an alert Rhesus monkey. Three electrodes recorded more than one isolated neuron each (units 2 and 3; units 4 and 5; and units 7, 8 and 9). Each of the nine raster-display bands depicts one second of the activity of a single neuron. Spikes are locked in time to the stimulus onset (at time = 0). The stimuli were 3.4 KHz tone bursts of 300ms duration, at an intensity of 68 dBspl. Figure 11.3 clearly demonstrates that neurons within the recording site shared some similar properties and were activated by the same stimulus (neurons 1, 4, 5, 6 and 7). However, it is also evident that the response patterns of the neighboring neurons differed, and that some of the neurons did not respond at all to the stimulus. While units 1 and 5 responded very strongly to the acoustic stimuli, the activity of units 2 and 8 was almost unaffected. Notice that the variability was seen even among neurons that were recorded by the same electrode (for example, units 7, 8 and 9). The distance separating these neurons is probably less than 100μ (Abeles et al. 1975).

The dogma of columnar organization of the auditory cortex implies that all neurons in the same column share the same best frequency. Although the results shown in Fig. 11.3 are not inconsistent with this dogma, they do not support the simplistic interpretation that neurons within a column are all activated in unison.

The study of neuronal activity in the "non-specific" association areas poses some difficulties. The functional organization is not very well known and the majority of neurons do not respond at all to sensory stimuli. It is however important to study these areas if we wish to examine mechanisms involved in higher brain functions. To examine properties of adjacent neurons related to higher functions, one should carefully select and match a recording area with an appropriate, controlled behavioral task.

Previous studies in primates have demonstrated that projections to the prefrontal cortex are segregated in the form of vertical columns (Goldman and Nauta 1977; Goldman-Rakic 1987), and that areas within the frontal cortex are related to higher brain functions (Butters and Pandya 1969; Niki and Watanabe 1976, 1979; Passingham 1975; Fuster 1980; Evarts et al. 1984; Petrides 1986; Watanabe 1986; Goldman-Rakic 1987; and others). It has also been found that many neurons in the vicinity of the arcuate sulcus responded to both auditory and visual stimuli, when the stimulus served to cue the direction of movements (Vaadia et al. 1986). The combined results of some of the above studies suggest that neuronal activities related to sensory analysis, sensorimotor association and preparation or execution of movements may be found in the same area. Thus, the frontal cortex seems to be a good case study of neuronal interactions in behaving animals. In order to study the activity of neighboring neurons in the area, we have designed a task that allows for dissociation between different functional components of the neuronal activity.

Rhesus monkeys were trained to perform a behavioral task with two paradigms. In the *localizing* paradigm, two spatial cues (S1 and S2 in Fig. 11.4) of different modalities (auditory and visual) were presented sequentially from two different locations. After 1–1.5 seconds, a "GO" signal was presented, informing the monkey by a color code which of the two stimuli was the relevant directional cue. The reinforced response was an arm movement in the direction of the relevant cue. In the second paradigm (*non-localizing*) the same sequence of stimuli (INIT, S1, S2, and GO) was presented in the same pattern as in the first task paradigm. However, the reinforced responses were movements to fixed targets, disregarding the directional cues and the GO signals.

Figure 11.5 shows the activity of nine neurons recorded during performance of the localizing paradigm. The raster display shows those trials in which the relevant spatial cues were auditory stimuli at 15° in the contra-lateral hemifield. Each of the nine upper strips shows a raster display for the activity of one unit. Behavioral events are displayed in the lowest strip (EV) and the vertical line

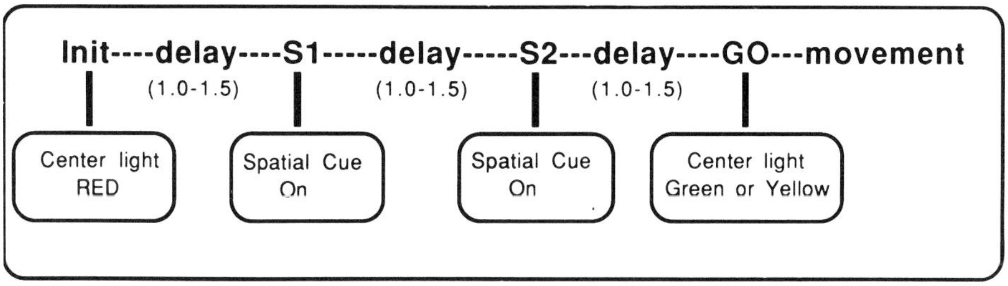

Fig. 11.4. The sequence of events from trial initiation to the behavioral response in the localizing task

marks the onset of the "GO" signal. The units were recorded by four microelectrodes. Units 6 and 7 were recorded by one microelectrode, and units 8, 9, and 10 by another. The figure illustrates that the firing rate of unit 6 and unit 8 were affected by the "GO" signal during performance of the task while other neurons, even those recorded in the same recording site by the same microelectrode, were not affected.

Note that the firing rate of the task-related neurons in Fig. 5 was increased after the "GO" signal when the monkey was performing the localizing task and responding by making limb movements to touch a key at 15° on the left. The same neurons were almost unaffected by the "GO" signal when the monkey made other movements, or when the same movements were made during performance of the non-localizing task (not shown in the figure). Thus, neurons 6 and 8 had a similar directional selectivity and behavioral dependency.

In many other recording sites we found that this pattern was likely to be repeated:

(1) task-related neighboring units were activated in similar conditions, but could also have separate or different response patterns and

(2) task-related neurons intermingled with many non-responsive neurons.

These findings held for both the auditory cortex and the frontal cortex with only quantitative differences. In the auditory cortex we usually found that a larger fraction of the simultaneously recorded neurons responded to at least one stimulus. In the prefrontal cortex, at most recording sites, the fraction of task-related neurons was smaller. Thus, based on these and other results, we hypothesize that the functional organization of each small cortical area (in both sensory and association fields) allows for the coexistence of functionally different neurons which may participate in several different processes.

Conclusion

Neighboring neurons may be functionally related and have common features. However, even when neurons were recorded by the same microelectrode, they were not all activated in unison. The findings suggest that different functional groups of neurons may reside within each small cortical area.

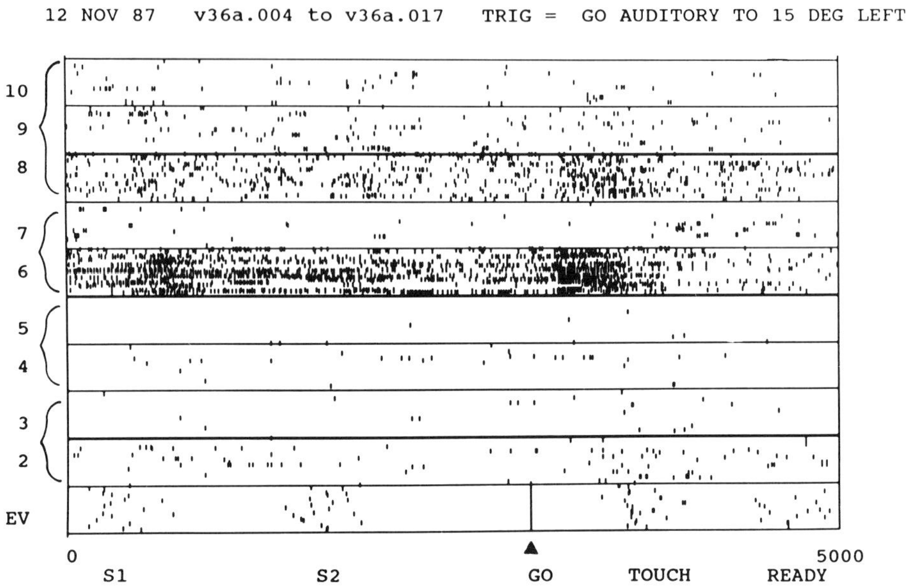

Fig. 11.5. Raster displays of the activity of 9 neurons in the frontal cortex of a monkey, during performance of the localizing task. The rasters show 5 seconds of activity centered around the onset time of the "GO" signal. The activity was collected from correct trials where the auditory spatial cues were presented in the left (contralateral) hemifield, at 15° from midline. Several events are marked in the lowest window (EV): the onset of the first spatial cue (cluster of bars, S1), the onset of the second spatial cue (S2), the time when the monkey hit the target ("touch") and the time when the monkey was ready for the next trial ("ready"). The units were recorded by 4 microelectrodes. The first recorded units 2 and 3; the second, units 4 and 5; the third, units 6 and 7 and the last, units 8, 9 and 10. Note the task-related activity of units 6 and 8, while their neighboring units were practically unaffected by behavioral events. (From Vaadia et al. 1989)

11.4 Cross-Correlation Analysis: Significance and Limitations

The broad definition of "*interacting neurons*" includes not only neurons making direct synaptic contact with one another, but also neurons sharing inputs from one or more common sources. A set of neurons forms a "functional group" if their spike trains show temporal relationships with some characteristic pattern. Raster displays as in Figs. 11.3 and 11.5 show the relationships of the neurons to a "trigger" event such as a sensory stimulus. It is difficult, however, to observe accurate temporal firing patterns among neurons in such displays.

The most direct way to examine inter-neuronal relationships is probably to record from all units intracellularly. This is not possible in the cortex, certainly not in behaving animals. We must examine the neuronal interactions indirectly, using statistical methods to analyze extracellular records of spike trains.

In the last twenty years, techniques have been developed to study neuronal connectivity by simultaneously recording spike trains. The most common technique is known as the *cross-correlation* technique (Moore et al, 1966). When two spike trains are considered as two point processes, the cross-correlogram (cross-correlation histogram) describes the firing rate (or firing probability) of one neuron as a function of elapsed time after a spike occurred in another neuron. The correlation $R_{12}(j)$ between two units 1 and 2, at a time bin j is given by

$$R_{12}(j) = \frac{1}{N_t \Delta t} \sum_{i=1}^{N_t} S1(i)S2(i+j), \tag{11.1}$$

where $S1(i)$ and $S2(i+j)$ are the spike count at bin i and bin $i+j$ respectively, N_t is the number of spikes of unit 1 (used as "triggers" for the averaging) and Δt is the time bin width. When $j\Delta t$ extends from $-T$ to $+T$, the cross-correlogram for positive $j\Delta t$ values ($0 \leq j\Delta t \leq T$) expresses the average firing rate of unit 2 at time $j\Delta t$, given that unit 1 fired a spike at time 0. Similarly, the cross correlogram at $-T \leq j\Delta t \leq 0$ gives the firing rate of unit 1, after unit 2 fired a spike.

The shape of the cross-correlogram provide indications for the "effective connectivity" between neurons. Typical shapes of cross-correlograms feature peaks or troughs near the origin. When the peaks or troughs extend to both sides of the origin ("double-sided"), they are usually interpreted as reflecting a shared input to the two neurons. Narrow peaks or troughs, restricted to one side of the cross-correlogram, are usually interpreted as reflecting a direct synaptic interaction.

These interpretations were justified by simulation studies (Perkel et al. 1967; Moore et al. 1970) and by cross-correlation of the spike trains of neuron pairs that were also studied by intracellular recording (Moore et al. 1970; Lindsey and Gerstein 1979; Kirkwood and Sears 1982; Fetz and Gustafsson 1983).

Although cross-correlation analysis is the most direct and valuable tool to study neuronal connectivity when intracellular recordings are impractical or impossible, there are number of difficulties one must cope with:

A. The sensitivity of cross-correlation analysis is determined by two factors: (1) the impulse discharge rate of the two neurons and (2) the duration of recording. As these two factors increase, the sensitivity increases and weaker neuronal interactions can be detected. Most cortical neurons fire spontaneously at low rates (0 to 20 spikes/s). Therefore, a long time is required for collection of the hundreds and sometimes thousands of spikes needed to observe statistically significant phenomena. The best solution, whenever possible, is to design experiments that allow for a long enough recording time, preferably to compute the cross-correlogram on-line and continue data collection for as long as necessary to reduce the statistical fluctuations. In addition, the use of statistical analysis is indispensable in determining the significance of deviations from the values expected in the absence of interaction.

B. The validity of cross-correlation analysis depends on the stationarity of the spike trains. Unfortunately, non-stationary spike trains are quite common. When the source of non-stationarity can be defined, as is the case when we study the units' activity during presentation of sensory stimuli, it is possible to use correction procedures such as the "*shift predictor*" (Perkel et al. 1967). This correction is carried out by first estimating the direct stimulus effects on each of the two spike trains and then subtracting these effects from the cross-correlogram. The shift-predictor correction is highly valuable when the temporal relation of neuronal activity to the external trigger event is reproducible. However, statistical fluctuation in the "difference-correlogram" may be bigger and therefore the recording time must be further increased. When the non-stationarity is not well defined, it is very difficult to interpret the cross-correlogram. In all cases, one should test each of the two spike trains for stationarity.

C. The results of the cross-correlation analysis may reflect not only interactions between the two neurons, but also the firing pattern of the individual neurons. The most common phenomenon is the tendency of some neurons to fire in bursts. The effect of bursts on the cross-correlogram shape may lead to an incorrect interpretation. When two neurons are synaptically connected, and the pre-synaptic neuron fires in bursts, the peak of the cross-correlogram is no longer confined to one side of the origin but resides around it. This may bring about the mistaken interpretation that the two neurons share a common input.

   Assuming that the synapse behaves as a linear system, it may be predicted that deconvolution of the cross-correlogram with the autocorrelogram of the presynaptic neuron can suppress these effects. We have tested the efficiency of this procedure and found that the method was not very effective. Another simple procedure that usually gives good results consists of filtering out the bursts from the spike train by ignoring all but the first spike in a burst. This procedure may change the quantitative estimate of the synaptic efficacy, but it serves as an efficient tool for qualitative interpretation of cross-correlograms. The method is useful for most cortical "bursting" cells where the number of spikes in a burst is small.

D. Another assumption of the analysis is that the strength of the interaction between the two neurons does not change during data collection. This is not necessarily true (as shown later in this chapter). The strength of interaction may be modified by external events such as sensory stimuli, or by the activity of other units that were not recorded. One important goal of cross-correlation analysis, especially for data collected from behaving animals, is to detect modifications of neuronal interactions. One way to expose such modifications is to parse the data into sections, and to construct the cross-correlograms from data in selected sections. Provided that the statistical properties of individual spike trains are maintained, changes in cross-correlogram shapes may reflect modifications of the strength and nature of the interaction between neurons. For example, let us consider the neuron illustrated in Fig. 11.2. A reasonable way to study its interaction with other neurons would be to parse the data to sections containing spikes that were collected between INIT to CUE onset and to construct the cross-correlogram for these sections, then to parse the data to sections containing the activity around the onset of movements, construct another cross-correlogram and compare the results. The major difficulties involved are firstly, that it is practically impossible to determine precisely how to parse the data, and secondly, that the number of spikes used to construct each cross-correlogram is reduced.

Another way to detect modifications of the effective connectivity on a short time scale was recently developed by Aertsen et al. (1989): the *Joint PST Histogram*, which can be used instead of cross-correlograms. This method and its application is described in Sect. 11.7.

Conclusion

Cross-correlation analysis can provide valuable qualitative and quantitative infomtation about the functional properties of neuronal connectivity. The analysis must be used with caution, accompanied by statistical procedures to avoid misinterpretations.

11.5 Estimating the Efficacy of Neuronal Interactions

Several methods to estimate "synaptic efficacy" from cross-correlograms were previously described (for example: Levick et al. 1972; Toyama et al. 1981; Aertsen and Gerstein 1985; Aertsen et al. 1989; Abeles 1982, 1991). Three methods that we examined and used are described in this section.

Let us consider the schematic cross-correlogram between two "neurons", U1 and U2, as shown in Fig. 11.6. The hill to the right of the origin indicates that the firing rate of U2 was increased for 3ms after U1 fired a spike. One can intuitively see that the size of the hill represents the efficacy of the "synapse" between the two neurons.

One way to estimate the efficacy of the connection is to examine how well the synapse transmits each spike of U1 to U2. When each pre-synaptic spike

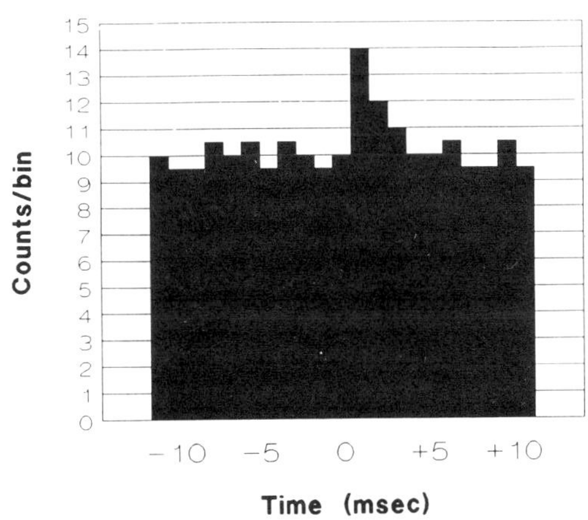

Fig. 11.6. A schematic cross-correlogram R_{12}. U1 fires at 1 spike/s, U2 fires at 10 spikes/s. $N1_t = 1000$ spikes, $N2_t = 10\,000$ spikes

adds one spike to the post-synaptic spike train, the gain of the synapse is 1. Levick et al. used the term "effectiveness" to describe this estimate (Levick et al. 1972). We use here the term "asynchronous gain" (Abeles 1982) to emphasize that this measure is valid when pre-synaptic spikes are not correlated with other excitatory inputs. There are also no strict constraints on the accurate delay between the pre-synaptic and the post-synaptic spikes; the estimate takes into account the total spike count above the expected spikes count of the post-synaptic neuron for $0 < t \leq T$. When the cross-correlogram displays the total spike count in each bin, the "asynchronous gain" (ASG) is given by

$$ASG = (N2_s - N2_e)/N1_t, \tag{11.2}$$

where $N2_s$ is the total spike count under the peak of the histogram, $N2_e$ is the expected spike count during a time interval T_s, of the same length, when U1 does not fire or does not affect U2 and $N1_t$ is the number of spikes of U1 used as triggers to construct the cross-correlogram.

Anatomical and physiological evidence indicates that the number of synapses converging on each cortical neuron is extremely large. A neuron is usually activated by summation of many EPSPs (excitatory post-synaptic potentials) evoked by spikes of many neurons. Therefore, another approach to estimate the effectiveness of the excitatory connection is to assess the average contribution of a spike discharge of the pre-synaptic unit to the spike train of the post-synaptic unit (Levick et al. 1972). Following the rationale above, we use the term "asynchronous contribution" (ASC) for this measure. Constructing the cross correlogram R_{21} instead of R_{12}, we get the ASC by

$$ASC = (N1_s - N1_e)/N2_t, \tag{11.3}$$

where $N1_s$ and $N1_e$ are defined as $N2_s$ and $N2_e$, and $N2_t$ is the number of U2 spikes used as triggers. Equation (11.3) represents the number of spikes of the pre-synaptic unit that preceded each spike of the post-synaptic unit during the interval Ts before U2 fired a single spike.

From (11.1) it can be shown that $N1_s = N2_s$ and that $N1_e = N2_e$. Therefore, we define $N_s = N1_s = N2_s$ and $N_e = N1_e = N2_e$. Substituting in (11.2,3) we get

$$ASG = (N_s - N_e)/N1_t, \tag{11.4}$$
$$ASC = (N_s - N_e)/N2_t. \tag{11.5}$$

Equations (11.4) and (11.5) thus differ only in the denominator: while the ASG is normalized by the firing rate of U1, the ASC is normalized by the firing rate of U2.

To demonstrate the difference between the two estimates, let us consider again the cross correlogram of Fig. 11.6. The expected count during an interval of 3 ms is 30 spikes. The total count under the peak is 37 spikes, the number of U1 spikes used as triggers to construct R_{12} is 1000 and the number of U2 spikes used as triggers to construct R_{21} (not shown) is 10 000. Therefore

$$ASG = Ns\text{-}N_e/N1_t = (37\text{-}30)/1000 = 0.007, \tag{11.6}$$
$$ASC = Ns\text{-}N_e/N2_t = (37\text{-}30)/10\,000 = 0.0007. \tag{11.7}$$

The ASG estimates the connection as being 10 times stronger than the ASC, indicating that the pre-synaptic unit (U1) contributes a very small fraction of the spikes of the post-synaptic unit (U2). The different firing rates of U1 and U2 are the source of these different estimates.

We may conclude, therefore, that the two estimates can be used to describe different aspects of the excitatory connection, namely, the *gain* of the synapse (ASG) and the *contribution* of the synapse to the post-synaptic spike train (ASC).

The time scale of Fig. 11.6 does not affect the values of the computed ASC and ASG. Thus, accurate temporal considerations are neglected. It can be shown, however, that the relationship between EPSP magnitude and firing rate is non-linear. Therefore, cortical neurons are very sensitive to synchronous excitation (Abeles 1981). If EPSPs were evoked by several excitatory inputs at the same time, the probability of the post-synaptic neuron reaching threshold may increase by several fold. If several excitatory inputs were able to excite the post-synaptic neuron in "synchronous mode", the apparent efficacy of each of the synapses would increase. Therefore, a measure that estimates the "synchronous gain" is highly relevant to the study of neuronal interactions in cortical networks. From the shape of the peaks in the cross-correlograms, we can estimate the number of excitatory synapses evoking EPSPs of the same size and shape *in synchrony* needed to bring the membrane potential to threshold at a precise instant of time. Thus, the "synchronous gain" (SG) is high in comparison to the ASG when the cross-correlogram peak is high and narrow. A method to estimate the synchronous gain was developed and described in detail by Abeles (1982, 1991).

The method is based on an approximation of the effect of the EPSP on the membrane potential from measurements of the neurons' firing rates. The approximation does not hold in all cases, but it is probably valid for most of the cortical neurons whose firing rate is low, and whose membrane potential is affected by many relatively small excitatory and inhibitory sources. The membrane potential can be treated in this case as a random variable with a Gaussian distribution. Figure 11.7 (modified from Abeles 1982, p21) illustrates the relationship between the membrane potential fluctuations and the firing rate of the cell. Random fluctuations are shown in Fig. 11.7A, with a Gaussian curve (Fig. 11.7B) that represents the probability density function of the membrane potential. The membrane potential fluctuates around some average voltage V. When the membrane potential exceeds the threshold T, a spike is generated. The probability of this event is the area under the Gaussian above the threshold T. The firing rate is proportional to this probability. This relationship is given by the equation

$$\lambda(t) = \mathrm{K}P\{\mathrm{V}(t) > \mathrm{T}\} = \mathrm{K}\frac{1}{\sqrt{2\pi}} \int_{\mathrm{T}/\sigma}^{\infty} e^{-y^2/2} dy, \tag{11.8}$$

where $\lambda(t)$ is the instantaneous firing rate at time t, and y is the difference between the membrane potential at time t and the average membrane potential in

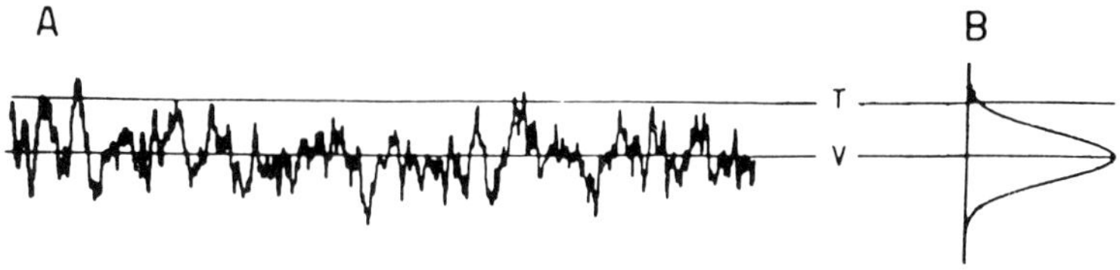

Fig. 11.7. **A:** A schematic representation of membrane potential fluctuations around a mean voltage (V); **B:** The Gaussian curve, turned on its side, represents the probability density of the membrane potential to reach threshold T. (Adapted with permission from Abeles 1982)

standard deviation units. The integral on the right gives the probability of the membrane potential to be above T, and K is a constant used to translate from probability to rate. K depends on the frequency of the membrane potential fluctuations and the refractory period of the cell. If spikes can occur every 1 ms then $K = 1000$. The variable controlling the firing rate is the ratio of the threshold to the standard deviation of the membrane potential fluctuations. This ratio is simply the distance of the threshold from the mean in standard deviation units. When the "threshold distance" ($TD = T/\sigma$) is 2.58 standard deviations, the probability of the membrane potential being above threshold is 0.005 and the firing rate is 5 spikes/s (for $K = 1000$). An EPSP may bring the membrane potential closer to the threshold by shifting the Gaussian to the right (upwards in Fig. 11.7): TD is decreased and the probability of firing is increased. The ratio between the change of distance from threshold and the expected distance from threshold (in absence of the EPSP) is defined as the "synchronous gain" (SG) of the "synapse"

$$SG = [TD(\lambda_e) - TD(\lambda_s)]/TD(\lambda_e). \tag{11.9}$$

Note that TD is a function of the firing rate λ. Its value can be computed by numeric evaluation of (11.8) for any given λ. Figure 11.8 shows the threshold distance in standard deviation units as a function of the firing rate.

The membrane potential of neurons that fire at low rates lies far from threshold. The threshold is therefore on the decaying part of the Gaussian, where the area under the Gaussian changes rapidly in relation to the changes of TD. It can be shown that the effects of several EPSPs on the firing rate of such neurons strongly depends on the synchronization of the EPSPs. To illustrate this property, let us examine again the cross correlogram of Fig. 11.6. We can use (11.9) and the graph of Fig. 11.8 to evaluate the synchronous gain of the synapse between U1 and U2 from the cross correlogram. The average firing rate of U1 and U2 are 1 and 10 spikes/s respectively. From Fig. 11.8, we get the expected TD for a "spontaneous" firing rate of the post-synaptic neuron [$TD(\lambda_e) = TD(10) = 2.32$]. Then, from the firing rate at the peak of the cross correlogram, we compute the TD value at the peak [$TD(\lambda_s)$]. The instantaneous firing rate *at the peak* of the cross correlogram is

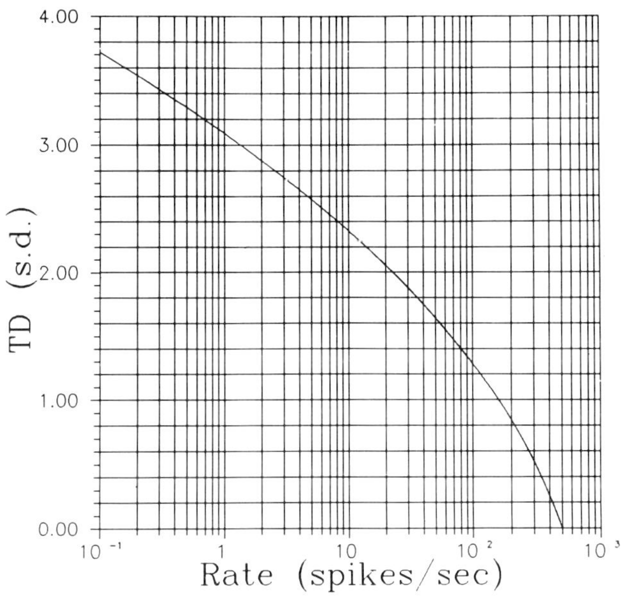

Fig. 11.8. Threshold distance TD (in standard deviation units) as a function of firing rate. The graph was obtained by numerical evaluation of (11.8) (K = 1000)

$$N_s/N1_t\Delta t = 14/(1000 \cdot 1 \cdot 10^{-3}) = 14 \text{ spikes/s},$$

and the SG is

$$SG = [TD(10) - TD(14)]/TD(10) = (2.32 - 2.19)/2.32 \approx 0.056.$$

Thus about 17 presynaptic spikes synchronized with 1 ms are needed to bring the membrane potential to threshold (at threshold, the probability that U2 fires is 0.5). If time constraints are less strict the SG decreases, meaning that more EPSPs are needed in order to bring the membrane to threshold. If we count all spikes that neuron U2 fires within a 3 ms time window after neuron U1, we get a total count of 37 spikes at the peak, but the firing rate is

$$N_s/N1_t\Delta t = 37/(1000 \cdot 3 \cdot 10^{-3}) = 12.3 \text{ spikes/s}.$$

The magnitude of the SG is much smaller:

$$SG = [TD(10) - TD(12.3)]/TD(10) = (2.32 - 2.24)/2.32 = 0.034.$$

Therefore about 33 synapses of the same efficacy should be coactivated within a 3 ms interval to bring the membrane potential to threshold.

Note that if one records one pair of neurons, one may get very different estimates for ASG, ASC and SG depending on the number of pre-synaptic synchronized inputs that converge on the post-synaptic neuron. If a neuron such as U1 is activated in synchrony with other cells (which we do not record), the apparent gain of the connection between U1 and U2 is higher. This property predicts that cross correlograms between two units may change if sampled under different physiological or behavioral conditions. For example, the two neurons 6 and 8, described in Fig. 11.4 above, were activated near the onset of the "GO" signal. If the two neurons were indeed coactivated *in synchrony*, their effective connectivity with a third (post-synaptic) neuron would have been increased for a while and when the degree of synchronization decreased, so would the "synaptic efficacy". This prediction can be further evaluated by cross correlation or Joint PST analysis.

Conclusions

The "synaptic efficacy", as estimated from cross-correlation analysis, is not defined in a unique way. The three estimates described can be used to study different aspects of direct "synaptic" interactions. When pre-synaptic neurons are activated in synchrony, their effect on a mutual post-synaptic unit may increase by several fold without modification of the individual synapse.

11.6 Cross-Correlation Analysis: Experimental Results

Interactions between pairs of neurons were examined by cross-correlation analysis; data were obtained from the experiments described in Sect 11.3. The unit activity was recorded during performance of a behavioral task and during "non-performance" periods when no stimuli were delivered.

Figure 11.9 illustrates types of cross correlograms obtained from pairs of neurons in the auditory cortex and frontal cortex during "non-performance" periods. Fig. 11.9A displays uncorrelated activity: the firing rate of cell 9 did not change significantly after the firing of cell 2 (from x-axis origin to the right), and the firing rate of cell 2 did not change after the firing of cell 9 (from the origin to the left). About 30% of the neuron pairs recorded from the same electrode showed such uncorrelated activity.

In some cases a narrow peak confined to one side of the zero delay time could be found (Fig. 11.9B). The interpretation that these correlograms reflect a direct excitatory connection between the two neurons is supported by the fact that most of the "one-sided" peaks were found in the cross correlogram when the two neurons were recorded by the same microelectrode, and were never found when the two neurons were more than 600μ apart. Signs of direct inhibitory connections ("one-sided" trough, as in Fig. 11.11B) were rarely found. The small number of "one-sided" troughs is not surprising, considering the low sensitivity of the cross-correlation technique to inhibitory connections (Aertsen and Gerstein 1985). In the sample of neuronal pairs recorded by the same microelectrode, we found signs for such "direct connections" (excitatory and inhibitory) in about 20% of the cross-correlograms.

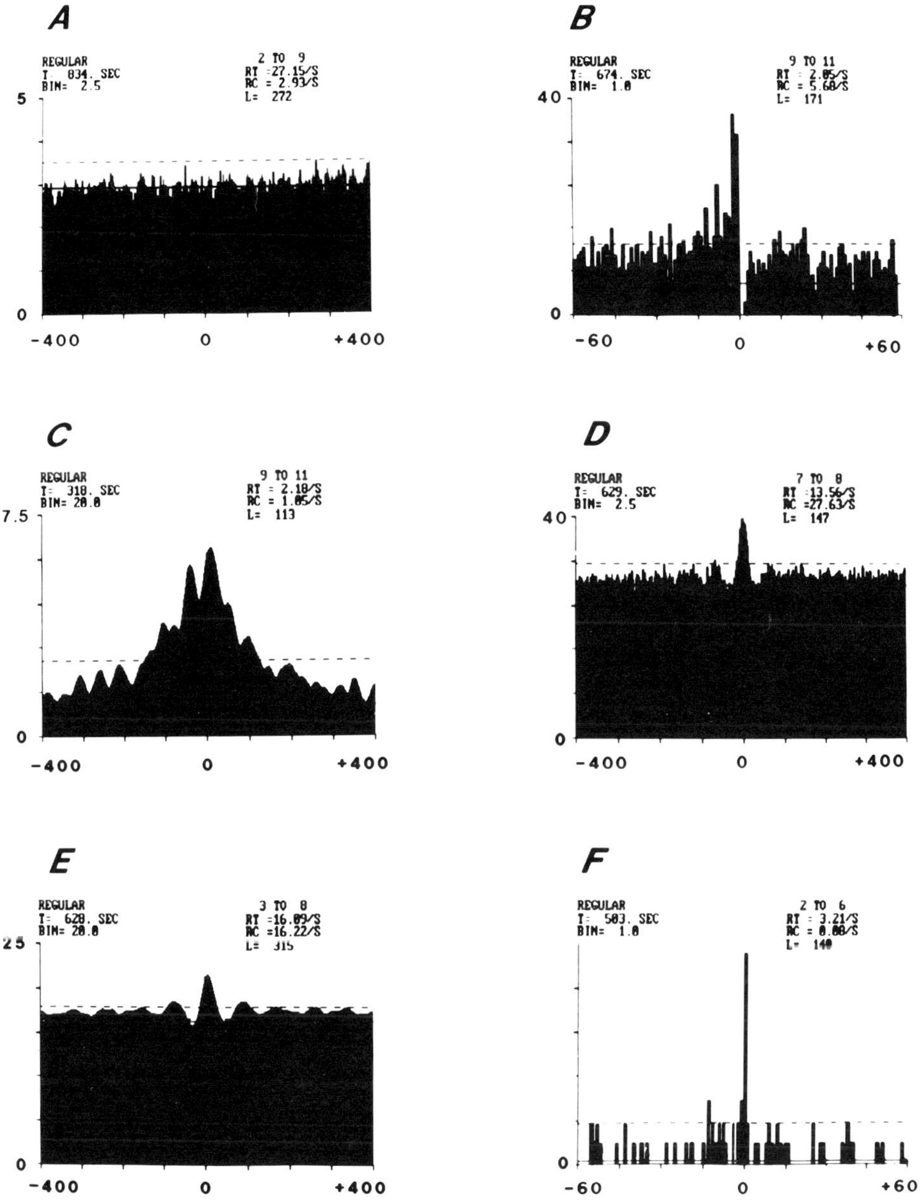

Fig. 11.9. Cross-correlograms between pairs of cortical neurons. The abscissa describes the time (ms) around the firing of one neuron and the ordinate describes the firing rate of the other neuron in spikes/s. The labels on each correlogram give the total data collection time (T), the average firing rate of the two neurons (RT and RC), the bin size (BIN) and the number of data sections (L). For each correlation, a band of 99% confidence limits for the equivalent, independent Poisson processes is shown by broken lines.
A Uncorrelated activity; **B** "One-sided" narrow peak (note the time scale: ±60 ms.); **C** "Double-sided" wide peak; **D** "Double-sided" narrow peak; **E** "Double-sided" peak with slight signs of periodicity; **F** "Ultra-narrow" peak at time 0 (time scale: ±60 ms.)

The strength of the interactions was in many cases weaker than illustrated in Fig. 11.9B. Abeles reported that the average asynchronous gain in his experiments was 0.06 (with maximal value of about 0.25) and the average synchronous gain was in the same range. Anatomical and theoretical considerations suggest that the average synaptic efficacy is much lower (Abeles 1982, 1991), but it is difficult to detect weak connections by cross-correlograms. The fact that we found a low efficacy of direct connections between neurons indicates that few pre-synaptic spikes are transmitted by a single connection. On average, more than 94% of pre-synaptic spikes are not translated to spikes in the post-synaptic neuron. The majority of pre-synaptic spikes have only sub-threshold effects, unless some external or internal events change the activity in the cortex so that additional neurons fire in synchrony and bring the post-synaptic neuron to threshold. Therefore, cooperation among neurons may be an important factor in modulation of synaptic efficacy.

Cross-correlograms reflecting indirect interactions are shown in Fig. 11.9C–E. Figure 11.9C shows correlated activity represented by a "wide double-sided" peak, while Fig. 11.9D shows a cross-correlogram with a "narrow double-sided" peak. Cross-correlograms of these patterns could emerge when the two neurons shared common excitatory or inhibitory inputs. "Double-sided" peaks in cross-correlograms could also emerge when the two neurons excited each other (although this interpretation is highly unlikely for the wider peaks). The cross-correlograms of about 30% of the pairs were of this type. In most cases the peak gradually decayed. In a few cases, however, there were weak signs of periodicity in the cross-correlograms, as demonstrated in Fig. 11.9E. The periodicity was in all cases at a frequency of less than 14 Hz. The area under the "double-sided" peak was, on average, 0.06 spikes. This area represents the correlation coefficient which indicates to what extent the two neurons tend to discharge in synchrony. The low value usually measured suggests that most impulse discharges were not correlated. Most of the spikes fired by two neurons that shared a common excitatory input originated from EPSPs of other excitatory inputs, which drove each of the neurons independently. This finding confirms the results of our previous studies, where triple cross-correlograms were used to demonstrate that a single neuron may share common input with an adjacent neuron from one source, while another input drives the same neuron in unison with another neuron (Abeles et al. 1983; Frostig et al. 1983; Vaadia and Abeles 1987).

The width of the peak around the origin could vary from several milliseconds (as with the "narrow peak" of Fig. 11.9D) to very wide peaks of more than 1000 ms. In a preliminary analysis of 41 neuron pairs, we found an uneven distribution of peak widths in the cross-correlograms: narrow peaks around 34 ± 10 ms (26 out of 41 cases) and wide peaks around 500 ± 175 ms (15 out of 41 cases). In some cases, both wide and narrow peaks could be observed in the same cross-correlogram. The data also showed that the probability of finding a narrow peak in the cross-correlogram decreased when the anatomical distance between the two neurons increased. The decrease in the probability of observing a wide peak was much less pronounced. The width of the peak reflects the degree of synchronization of the two neurons, and thereby the time scale of the process that affects the neurons. Whether these findings imply that "narrow"

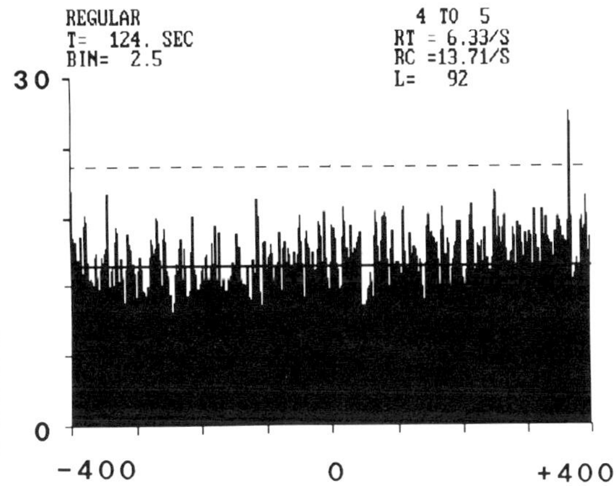

Fig. 11.10. An "ultra-narrow" peak at a long time delay. The cross-correlogram shows random, statistically insignificant fluctuation at all times except ror the peak which exceeds the confidence line at +366 ms

and "wide" peaks reflect different mechanisms of interactions is an open question. Nevertheless, it is reasonable to hypothesize that the narrow peaks reflect local direct coactivation of relatively small groups of neurons, while the wide peaks reflect a general coactivation of larger groups of neurons in a relatively wide-spread cortical area.

Several previous studies reported that narrow peaks of 1 ms or less were also found in cross-correlograms (Toyama et al. 1981; Allum et al. 1982; Murphy et al. 1985, Lurito et al. 1988). An example of an "ultra-narrow" peak is shown in Fig. 11.9F. This correlogram indicates that some spikes of the two units were highly synchronized. This type of cross-correlogram was infrequent in our studies. A possible explanation may lie in the technological difficulty of detecting two spikes that occur at the same time using a single electrode.

In addition to peaks at or near the origin, we discovered narrow peaks at long time delays. Figure 11.10 depicts a neurons pair with no observable interaction at short time delays. However, a statistically significant "ultra-narrow" peak can be observed in two bins of the cross correlogram (at delays of 366 and 367 ms). Such a peak indicates that neuron 5 tends to fire a spike 366–367 ms after neuron 4. Such a temporal firing pattern cannot be explained by a simple mechanism of neuronal interactions. The pattern was repeated 45 times during 124 seconds of stable and stationary activity. The same pattern was repeated in the data obtained during an additional section of 143 seconds. The peak at bins of 366–367 ms delay in the cross correlogram could emerge neither from few bursts of spikes nor from changes in the firing rate of the individual neurons. Statistical tests indiciate that it is unlikely that such a coincidence could happen by chance alone (the probability of finding 45 repetitions of any such a pattern in the two spike trains during 124 seconds of recording is less than 1%). Such cross correlograms were found in both the auditory cortex and in the frontal cortex. In the latter, we examined cross-correlograms of 440 neuron-pairs and found statistically significant "special patterns" in 39 pairs while the expected number, if these patterns were to occur by chance, would be 4.4. We can therefore conclude that these cross correlograms represent two spike trains in which most of the spikes are not correlated, but from time to time a certain temporal pattern appears.

We suggest that the ultra-narrow peaks at the origin are one example of a general case. Although spike trains of different neurons are usually loosely correlated, occasionally the spikes of two neurons are highly synchronized so that the two neurons can fire spikes at accurate time delays. At present, one can only speculate about the mechanisms that underlie such interactions and the possible functions they serve. It is unlikely that two neurons would fire in accurate temporal patterns without a synchronization mechanism. If information is transmitted in asynchronous mode, every EPSP increases the probability of firing for a period of time (the duration of the EPSP) and creates a jitter in the actual firing times of the post-synaptic neurons. If spikes were transmitted in sequence through several synapses in asynchronous mode, we would never find narrow peaks at long delays because of accumulating jitter in each synapse. In synchronous mode, however, there is very little jitter in each synaptic transmission; in this mode a group of neurons fires spikes at the same time (as with the two neurons depicted in Fig. 11.9F). As a result, post-synaptic neurons are most likely to fire a spike at an accurate time – at the peak of synchronous volley of EPSPs. If the post-synaptic group fires in synchrony, then (and only then) the transmission continues to the next group. Thus, the synaptic jitter is kept very small even after many synapses. This mode of activity was called "synfire activity" by Abeles (1982, 1991).

Conclusion

The interactions between neuron pairs are usually weak. Neurons that interact also receive other independent inputs. We speculate that the three types of "double-sided" peak-width in cross correlograms reflect different mechanisms of neuronal interactions: the "ultra-narrow" peaks reflect the activity in "synfire chains" the "narrow" peaks may reflect local coactivation of neurons and the "wide" peaks reflect a more general asynchronous mechanism affecting large cortical areas. The existence of "ultra-narrow" peaks at different time delays suggests that groups of neurons can fire in synchrony and transmit information with high temporal accuracy. We also hypothesize that cooperativity among neurons and the degree of synchronization can control the synaptic efficacy. This hypothesis predicts that the effectiveness of interactions among neurons may depend on environmental or internal factors that change the degree of synchronization.

11.7 Dynamics of Neuronal Interactions

To look for evidence that supports or refutes the hypothesis that neuronal interactions may change dynamically and depend on external events or behavioral state, we have parsed the data to sections around specified behavioral events and

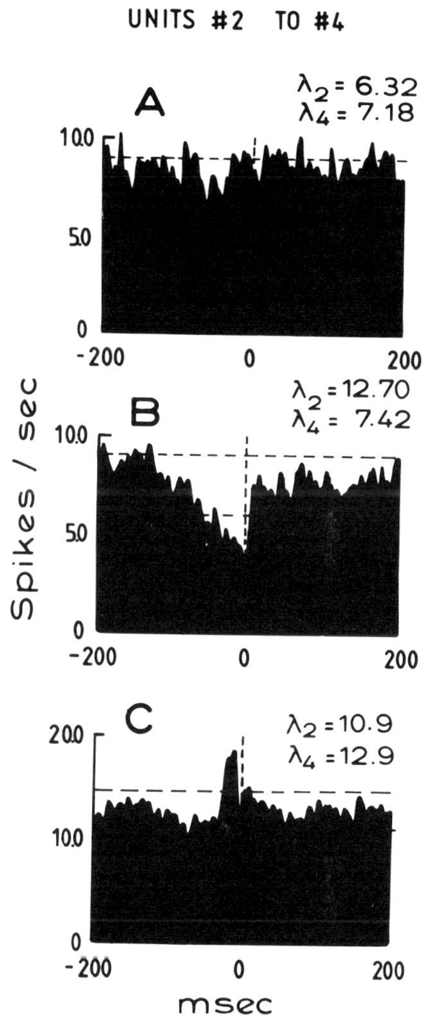

Fig. 11.11. Three cross-correlograms of one pair of neurons recorded during performance of the habit-reversal task. **A** "Flat" cross correlogram with "random" fluctuations. This correlogram was computed from the inter-trial intervals; **B** "one-sided" trough near the origin. This correlogram was computed from the activity of the same pair of neurons in sections of data between stimulus presentation and the initiation of movements, when the monkey responded to S1 (a tone stimulus) by shifting a lever to the right; **C** "one-sided" peak. This correlogram was computed from the activity of the same pair of neurons when the monkey responded to S1 by shifting the lever to the left. (The narrow trough at t = 0 in the three correlograms reflects the inability of the spike sorter to detect two spikes that occur simultaneously at the same electrode)

computed a separate cross-correlogram for each behavioral condition. In all cases the "shift predictor" was computed also to test for direct stimulus effects. In many cases we found that there was little or no change in the interactions. In some cases, however, statistically significant modifications were observed. Examples of behavioral modifications of the interaction between pairs of neighboring neurons are demonstrated in Fig. 11.11 13.

The data for Fig. 11.11 were obtained by simultaneous recording of neuronal activity by a single electrode in the auditory cortex of a monkey during performance of the "habit-reversal" task. The three cross correlograms of Fig. 11.11 were computed from the firing of one pair of neurons recorded under three behavioral conditions. The practically flat curve in Fig. 11.11A were computed from the firing of two neurons during inter-trial intervals. The cross-correlogram in Fig. 11.11B, with a trough near the origin, was computed from firing of two neurons during the periods between the tone (S1) onset and the initiation of the behavioral response, when the sensorimotor association "S1 → Right" was reinforced (the monkey responded correctly to S1 by shifting a lever to the right). The cross-correlogram of Fig. 11.11C, with the "one-sided" peak, was computed as for Fig. 11.11B, from firing of two neurons in the period between the tone onset and initiation of the behavioral response, but the reinforced sensorimotor association was "S1 → left". The trough and peak in Fig. 11.11B,C are weak but statistically significant at 1%. The ASG of the excitatory connection (Fig. 11.11C) is about 0.14 post-synaptic spikes per pre-synaptic spike. The synchronous gain was lower (SG = 0.08). As can be seen from the cross-correlograms and by comparing the ASG to the SG, the duration of these direct interactions is rather long. In both cases, therefore, the ASG is greater than the SG. The peak of Fig. 11.11C and trough of Fig. 11.11B were not due to direct effects of stimuli on the two neurons: the respective "shift-predictor" correlograms were flat.

Another example of a modified interaction is shown in Fig. 11.12. This pair of neurons was recorded in the frontal cortex of a monkey. The data for the histogram in Fig. 11.12A were collected continuously during 6 minutes of ongoing ("spontaneous") activity in the "non-performing" condition. The data in Fig. 11.12B were collected from sections of the inter-trial intervals (3s each) during performance of the localizing task. Direct excitatory interaction can be detected in neuronal activity when the monkey was not performing the task, while the spike trains recorded during the inter-trial intervals show no correlation.

To conclude, the cross-correlation analysis of behavioral data indicates that, on average, the interactions between neurons are weak but may be modified. Thus, the prediction that the effective connectivity may depend on behavioral conditions is met by the data.

The cross-correlation analysis does not allow examination of the time course of the changes. We have proposed above that the effective connectivity between two neurons may be modified on a short time scale, due to modulation of discharge synchronization among neurons. In a preliminary analysis, we examined the dynamic properties of interactions between pairs of neurons using *Joint PST histograms* (JPSTH). The JPSTH analysis is particularly useful for examining possible effects of stimuli or behavioral events on connectivity. Details of the algorithm can be found in Aertsen et al. (1989), and a brief description follows.

The JPSTH is constructed by drawing a joint scatter diagram of firings of the two neurons as a function of the time elapsed since the stimulus. Bins falling along (near) the diagonal correspond to near-coincidental events in the two spike trains. Bin counts for the diagonal region represent the time-locked average for near coincidences in the same sense that the PST histogram represents the time-locked average of the single-unit spike discharge. It can be shown that summation along each stripe of diagonal bins in the JPSTH, normalized for the number of bins in each sum, gives the cross-correlogram of the two spike trains. As in ordinary cross-correlograms, the JPSTH includes the effects of the stimuli on the individual firing rates of the two neurons. Aertsen et al. (1989) have

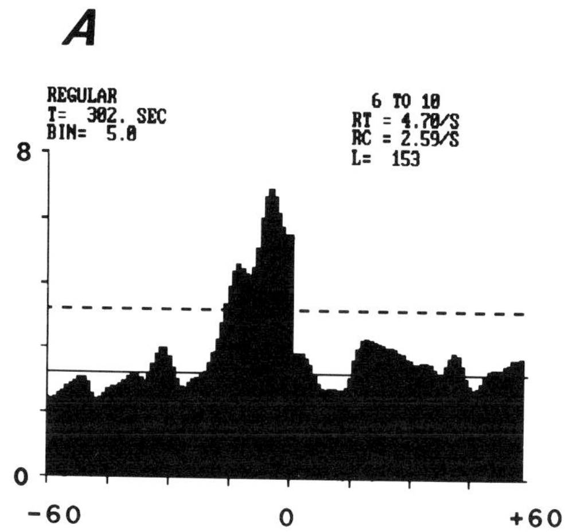

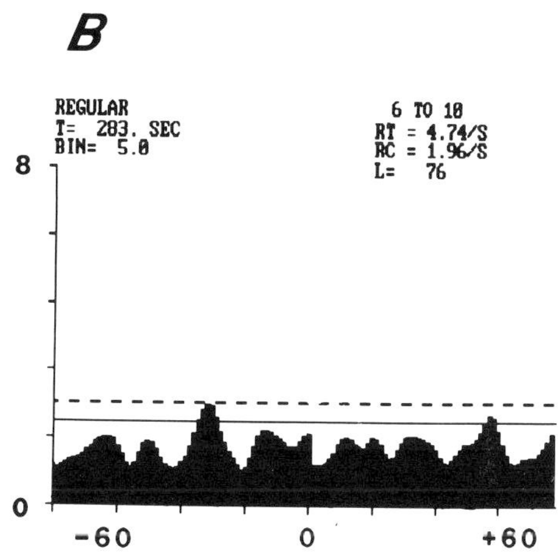

Fig. 11.12. Two cross-correlograms of one pair of neurons recorded in the prefrontal cortex. The cross-correlogram in **A** was computed from sections of "spontaneous activity". The cross-correlogram in **B** was computed from sections or "inter-trial intervals" during performance of the localizing task

shown that an optimal correction for this factor is to subtract the bin-by-bin product of the two PST histograms, and then to divide by the bin-by-bin product of the standard deviations of the two PST histograms. This correction procedure filters out the direct stimulus effects. Any residual peaks in the JPSTH exposed after the correction procedure correspond to the effective connectivity between the neurons. Therefore, it is possible to examine if and how these peaks change as a function of time after presentation of the stimulus.

An example of a JPSTH, computed for data obtained from single units recorded during performance of the "localizing" task, is shown in Fig. 11.13. The JPSTH displays the interaction between one pair of neurons recorded in the prefrontal cortex. Direct stimulus effects were filtered out by the correction procedure described above. The histograms show the activity within a time window of 1000 ms around the onset of the "GO" signal. In the scatter-diagram matrix, a

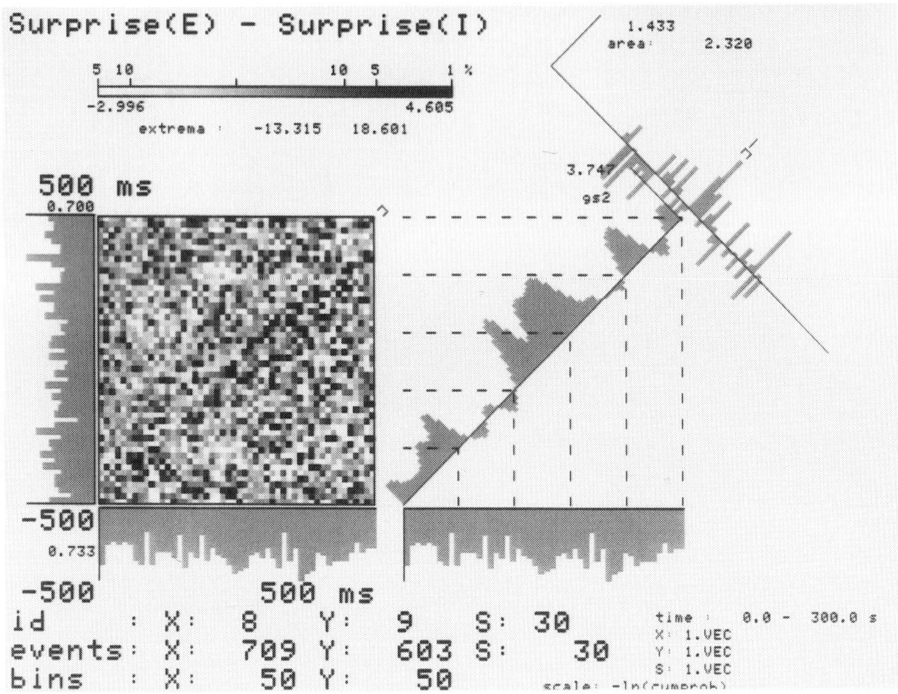

Fig. 11.13. Dynamic properties of the interaction between two neurons in the frontal cortex during performance of the localizing task. The Joint PST histogram shows 1s of activity around the onset of the "GO" signal (at time = 0). The matrix on the left shows the amount of correlated firing (direct stimulus effects are filtered out). The gray level scale of the matrix is shown at the top. The histograms on the horizontal and vertical axes show the individual PST histograms of the two neurons. The counts within a narrow band along the diagonal of the correlation matrix are shown as a diagonal histogram on the right, running from bottom left to upper right (note the increase in correlated activity near the onset of the "GO" signal). The cross correlogram is shown at the top right

gray level code is used to show the amount of correlated firing of the two cells at different time combinations (the higher the value, the darker the gray). Along the horizontal and vertical axes, the individual PST histograms are shown with the "GO" signal (at time = 0). It is evident that the firing rate of the neuron displayed on the vertical axis was not correlated at all with the "GO" signal onset. The other neuron (the PSTH shown upside down on the horizontal axis) had a very weak response to the "GO" signal. The cross-correlogram (top right corner, running from upper left to lower right) is noisy, but includes a "double-sided" peak around the origin. The PST-coincidence histogram (running diagonally from lower left to upper right) clearly shows that within the period near the onset of the "GO" signal, the number of correlated firings of the two units increased for about 250 ms. This result suggests that the interaction between two neurons changed within a fraction of a second, for a duration of 250 ms in relation to a behavioral event (in this case, the onset of the "GO" signal). Had we studied each of the two neurons separately, we would have assumed that their activity was not related to behavioral events.

The JPST histograms therefore indicate that while the firing rate of the individual single neuron was almost unaffected by task-related events, the two neurons could still be involved in, or affected by, task-related processes. This result supports the notion that the degree of correlation of neuronal discharge with other neurons is an expression of a neural encoding mechanism. We are tempted to speculate that the two neurons illustrated in Fig. 11.13 are part of a group of neurons which cooperate for about 250 ms, around the onset of the "GO" signal, when the monkey perceives the signal and then selects and prepares the appropriate arm movement. As soon as the execution of the movement starts, the neurons "depart" from the group and become available for other tasks.

Conclusion

Neuronal interactions between neuron pairs can be modified in relation to behavioral state. The modification may be linked to external events and may occur on very short time scale, within a fraction of a second.

11.8 Concluding Discussion

Our studies indicate that single units may participate in prolonged processing and be involved in more than one process. Adjacent neurons share common properties, but at the same time can also maintain individual, independent activity. On average, any interactions existing between adjacent neurons are weak, even when the neurons are coactivated in relation to an external event. All these properties are consistent with the idea that neurons within a small cortical area can be activated in the processing of several different functions, and thereby each small area contains several potential functional groups of neurons. Neurons that exhibit similar evoked activity in a given condition are probably members of one such group.

When members of such a group which converge to common post-synaptic neurons are coactivated, they can transmit information reliably even if the synapses are weak. On the other hand, other neurons which are synaptically connected to only a few of this neuronal group may remain unaffected. Thus, several groups of neurons may reside in each small cortical area without "disturbing" each other. These arguments are further supported by the finding that some pairs of neurons show highly synchronized activity even at long time delays, suggesting that optimal cooperation among neurons is achieved if they interact in synchronous mode. Thus, information can be represented and transmitted by synchronous activation of selected neuronal groups, as proposed by Abeles's "synfire chain" hypothesis (Abeles 1982, 1991).

It is hard to reconcile the data presented here with the hypothesis that the independent firing rate of single neurons is the sole basis for information processing and encoding. The weak synaptic interactions among neurons implies that even when a single neuron is discharging at a high rate, it is very difficult to reliably convey this information to other cells without the cooperation of other neurons. Further, coactivation of specific neurons in correlation with behavioral events can be found, even when each of the neurons does not change its average firing rate. Thus, the independent firing rate of the neuron is not the most significant factor, neither for encoding nor for transmitting information. The "elasticity", on a second-by-second basis, of the correlations between cells suggests that "neuronal groups" are dynamic entities, defined not only by anatomical connections, but also by the correlated electrical activity among their members. This property allows for one and the same neuron to participate in several functional sets. Taking into consideration that every neuron in the cortex makes contacts with thousands of other cells, it becomes evident that when such a neuron is coactivated with one set of cells, only part of its connections may become effective. When the same cell is coactivated with another set, a different part of its connections becomes effective. There is no need for each of the synaptic connections to be particularly strong because they become effective through synchronous activity with other neurons.

Acknowledgments. We thank Dr. Moshe Abeles and Mr. Israel Nelken for their comments on an earlier version of the manuscript. We also thank Dr. Ad Aertsen for the preparation of the JPST histograms in the context of a joint project supported by a grant from the G.I.F., the German-Israeli Foundation for Scientific Research and Development. The studies were supported in part by The Fund for Basic Research administered by the Israel Academy of Sciences and Humanities and by the Endowment Fund for Basic Research in the Life Sciences: Charles A. Revson Foundation.

References

Abeles, M. (1981): The role of the cortical neuron: Integrator or coincidence detector. Isr. J. Med. Sci. **18**, 83–92

Abeles, M. (1982): Local cortical circuits: an electrophysiological study. Studies of brain function. Vol. 7 (Springer, Berlin, Heidelberg)

Abeles, M. (1983): The quantification and graphic display of correlations among three spike trains. IEEE Trans. Biomed. Eng. BME-**30**, 236–239

Abeles, M. (1991) Corticonics: Neural circuits of the Cerebral Cortex. Cambridge University Press, Cambridge

Abeles, M., Goldstein, M.H. (1977) Multiple spike train analysis. Proc. IEEE **65**, 762–773

Abeles, M., Gerstein, G.L. (1988): Detecting spatiotemporal firing patterns among simultaneously recorded single neurons. J. Neurophysiol. **60**, 909–923

Abeles, M., Assaf, J., Gottlieb, Y., Hodis, H., Vaadia, E. (1975) Speculation on a neural substrate for immediate memory. Sensory Physiology and Behavior **15**, 117–126

Abeles, M., de Ribaupierre, F., de Ribaupierre, E. (1983): Detection of single unit responses which are loosely time-locked to a stimulus. IEEE Trans. Syst. Man. Cyber. SMC-**13**, 683–691

Aertsen, A.M.H.J., Gerstein, G.L. (1985): Evaluation of neuronal connectivity: sensitivity of cross-correlation. Brain Res. **340**, 341–354
Aertsen, A.M.H.J., Gerstein, G.L., Habib, M.K., Palm, G. (with the collaboration of P. Gochin and J. Kruger) (1989) Dynamics of neuronal firing correlation: modulation of "effective connectivity". J. Neurophysiol. **61**, 900–917
Allum, J.H.J., Hepp-Reymond, M.C., Gysin, R. (1982): Cross-correlation analysis of interneuronal connectivity in the motor cortex of the monkey. Brain Res. **231**, 325–334
Amit, D.J. (1987) The properties of models of simple neural networks. Heidelberg Colloquium on Glassy Dynamics. van Hemmen J.L. and Morgenstern I. (eds.), Springer, Heidelberg
Barlow, H.B. (1972): Single units and sensation: A neuron doctrine for percepteral psychology. Perception **1**, 371–394
Braitenberg, V. (1978): Cell assemblies in the cerebral cortex. Theoretical Approaches to Complex Systems. Lecture Notes in Biomathematics. Ed. by Heim R, Palm G (Springer, Berlin, Heidelberg) Vol. **21**, 171–188
Burns, D.B., Webb, A.C. (1979): The correlation between discharge times of neighboring neurons in isolated cerebral cortex. Proc. Roy. Soc. London B **203**, 347–360
Butters, N., Pandya, D. (1969): Retention of delayed alternation: Effect of selective lesions of sulcus principalis. Science **165**, 1271–1273
Dayhoff, J.E., Gerstein, G.L. (1983): Favoured patterns in spike trains. II. Application. J. Neurophysiol. **49**, 1349–1363
Deecke, L., Kornhuber, H.H., Lang, W., Lang, M., Schreiber, H. (1985): Timing function of the frontal cortex in sequential motor cortex in motor and learning tasks. Human Neurobiol. **4**, 143–154
Dickson, J.W., Gerstein, G.L. (1974): Interactions between neurons in auditory cortex of the cat. J. Neurophysiol. **37**, 1239–1261
Donchin, E. (ed.) (1984): Cognitive Psychophysiology. The Carmal Conferences. Vol. 1 (Erlbaum, New Jersey)
Eckhorn, R., Bauer, R., Jordan, W., Brosch, M., Kruse, W., Munk, M., Reitboeck, H.J. (1988): Coherent oscillations: A mechanism of feature linking in the visual cortex? Biol. Cybern. **60(2)**, 121–30
Edelman, G.M. (1981): Group selection as the basis for higher brain function. The organization of the cerebral cortex. Proceedings of a neuroscience research program colloquium. Ed. by Schmitt et al. (MIT Press, Cambridge) pp. 536–563
Evarts, E.V., Shinoda, Y., Wise, S.P. (1984): *Neurophysiological approach to higher brain function.* (Wiley, New York)
Fetz, E.E., Gustafson, B. (1983): Relation between shapes of post-synaptic potentials and changes in firing probability of cat motoneuron. J. Physiol. (Lond.) **341**, 387–410
Frostig, R., Gottlieb, Y., Vaadia, E., Abeles, M. (1983): The effects of stimuli on the activity and functional connectivity of local neuronal groups in the cat auditory cortex. Brain Res. **272**, 211–221
Fuster, J.M. (1980): The Prefrontal Cortex. Raven Press, New York Gassanov, U.G., Galashina, A.G., Bogdanov, A.V. (1980): A study of neuron systems activity in learning. In *Neural Mechanisms of Goal-Directed Behavior and Learning* ed. by Thompson, R.F. et al. (Academic, New York) pp. 341–352
Gazanov, U.G., Galashina, A.G., Bogdanov, A.V. (1980) A study of neuron systems activity in learning. Neural Mechanisms of Goal-Directed Behavior and Learning 341–352
Gerstein, G.L. (1988): Information flow and state in cortical neural networks: Interpreting multineuron experiments. In *Organization of Neural Networks*, Ed. by von Seelen W. et al. (VCH, Weinheim) pp. 53–85
Gerstein, G.L., Aertsen, A.M.H.J. (1985): Representation of cooperative firing activity among simultaneously recorded neurons. J. Neurophysiol. **54**, 1513–1526
Gerstein, G.L., Perkel, D.H, Dayhoff, J.E. (1985): Cooperative firing activity in simultaneously recorded populations of neurons: detection and measurement. J. Neuroscience. **5**, 881–889
Goldberg, M.E., Segraves, M.A. (1987): Visuospatial and motor attention in the monkey. Neuropsychologia **25**, 107–118
Goldman, P.S., Nauta, W.J.H. (1977): Columnar distribution of cortico-cortical fibers in the frontal association, limbic and motor cortex of the developing Rhesus monkey. Brain Res. **122**, 393–413
Goldman-Rakic, P.S. (1987): Circuitry of primate prefrontal cortex and the regulation of behavior by representational knowledge. Mountcastle VB, Plum F (eds) The Nervous System, Vol. V, Handbook of Physiology, Am. Physiol. Soc., Bethesda, MD, pp. 373–417
Gottlieb, Y., Vaadia, E., Abeles, M. (1989): Single unit activity in the auditory cortex of a monkey performing a short term memory to tones task. Exp. Brain Res. **74**, 139–148

Gray, C.M., Singer, W. (1989) Stimulus-specific neuronal oscillations in orientation columns of cat visual cortex. Proc. Natl. Acad. Sci. USA, **86**, 1698–1702

Hebb, D.O. (1949): *The organization of behavior: a neuropsychological theory* (Wiley, New York)

Hopfield, J.J. (1982): Neuronal networks and physical systems with emergent collective computational abilities. Proc. Natl. Acad. Sci. USA, **79**, 2554–2556

Hopfield, J.J., Tank, D.W. (1986): Computing with neural circuits: A model. Science **233**, 625–633

Kirkwood, P.A., Sears, T.A. (1982): The effect of single afferent impulses on the probability of firing of external intercostal motoneurons in the cat. J. Physiol. **322**, 315–336

Kruger, J. (1983): Simultaneous individual recordings from many cerebral neurons: Techniques and results. Rev. Physiol. Biochem. Pharmacol. **98**, 177–233

Kruger, J., Aiple, F. (1988): Multimicroelectrode investigation of monkey striate cortex: spike trains correlations in infragranular layers. J. Neurophysiol. **60**, 798–828

Levick, W.R., Cleland, B.G., Dubin, M.W. (1972): Lateral geniculate neurons of cat: Retinal inputs and physiology. Investigative Ophthalmology **1**, 302–310

Lindsey, B.G., Gerstein, G.L. (1979): Interactions among an ensemble of chordotonal organ receptors and motor neurons of the crayfish claw. J. Neurophysiol. **42**, 383–399

Lurito, J.T., Schwartz, A.B., Petrides, M., Kettner, R.E., Georgopoulos, A.P. (1988): Cross-correlations between motor cortical cells simultaneously recorded during reaching task in the monkey. Soc. Neurosc. Abst. **14**, 142.5

McNaughton, B.L., O'Keefe, J., Barnes, C.A. (1983): The stereotrode: A new technique for simultaneous isolation of several single units in the central nervous system from multiple unit records. J. Neurosci. Methods **8**, 391–397

Moore, G.P., Perkel, D.H., Segundo, J.P. (1966): Statistical analysis and functional interpretation of neuronal spike data. Ann. Rev. Physiol. **28**, 493–522

Moore, G.P., Segundo, J.P., Perkel, D.H., Levitan, H. (1970): Statistical signs of synaptic interactions in neurons. Biophysical J. **10**, 876–900

Murphy, J.T., Kwan, H.C., Wong, Y.C. (1985): Cross-correlation studies in primate motor cortex: Synaptic interaction and shared input. Can J Neurol. Sci. **12**, 11–23

Niki, H., Watanabe, M. (1976): Pre-frontal unit activity and delayed response: Relation to cue location versus direction of response. Brain Res. **105**, 79–88

Niki, H., Watanabe, M. (1979): Prefrontal and cingulate unit activity during timing behavior in the monkey. Brain Res. **171**, 213–224

Palm, G. (1982): Neural assemblies, an alternative approach to artificial intelligence. Studies of Brain Function, Vol. 7 (Springer, Berlin, Heidelberg)

Passingham, R.E. (1975): Delayed matching after selective prefrontal lesions in monkeys (macaca mulatta). Brain Res. **92**, 89–102

Perkel, D.H., Gerstein, G.L., Moore, G.P., (1967): Neuronal spike trains and stochastic point processes: II. Simultaneous spike trains. Biophys. J. **7**, 419–440

Perrett, D.I., Rolls, E.T., Caan, W. (1982): Visual neurons responsive to faces in monkey temporal cortex. Exp. Brain Res. **47**, 329–342

Petrides, M. (1986): The effect of periarcuate lesions in the monkey on performance of symmetrically and asymmetrically reinforced and auditory go, no-go tasks. J. Neurosci. **6**, 2054–2063

Rumelhart, D.E., Hinton, G.E., Williams, R.J. (1986): Learning internal ed. representation by error propagation. In Parallel Distributed Processing, ed. by Rumelhart DE et al. (MIT Press, Cambridge)

Sompolinsky, H. (1987): The theory of neural networks: the Hebb rule and beyond. Heidelberg Colloquium on Glassy Dynamics, ed. by van Hemmen JL and Morgenstern, I. (Springer, Berlin, Heidelberg.)

Tam, D.C., Ebner, T.J., Knox, C.K. (1988): Conditional cross-interval correlation analyses with applications to simultaneously recorded cerebellar Purkinje neurons. J. Neurosci. Methods **23**, 23–33

Toyama, K., Kimura, M., Tanaka, K. (1981): Crosscorrelation analysis of interneuronal connectivity in cat visual cortex. J. Neurophysiol. **46**, 191–201

Ts'o, D., Gilbert, C.D., Wiesel, T. (1986): Relationships between horizontal architecture in cat striate cortex as revealed by cross-correlation analysis. J. Neurosci. **6**, 1160–1170

Vaadia, E., Abeles, M. (1987): Temporal firing patterns of single units, pairs and triplets of units in the auditory cortex. J. Isr. Med. Sci. **23**, 75–83

Vaadia, E., Gottlieb, H., Abeles, M. (1982): Single unit activity related to sensorimotor association in auditory cortex of a monkey. J. Neurophysiol. **48**, 201–213

Vaadia, E., Benson, D.A., Goldstein, M.H., Hienz, R.D. (1986): Unit study of monkey frontal cortex: Active localization of auditory and visual stimuli. J. Neurophysiol. **56**, 934–952

Vaadia, E., Kurata, K., Wise, S.P. (1988): Neuronal activity preceding directional and nondirectional cues in the premotor cortex of Rhesus monkeys. Somatosensory Res. **6**, 207–230

Vaadia, E., Bergman, H., Abeles, M. (1989): Neuronal activities related to higher brain functions – Theoretical and experimental implications. IEEE Trans. Biomed. Eng. **36**, 25–35

Watanabe, M. (1986): Prefrontal unit activity during delayed conditional Go/NoGo discrimination in Monkey I. Relation to stimulus. Brain Res. **382**, 1–14

Wise, S.P., Mauritz, K.H. (1985): Set related neuronal activity in the premotor cortex of rhesus monkeys: effects of changes in motor set. Proc. R. Soc. Lond. B **223**, 331–354

12. Systems Neuronal Activity in Learning

U.G. Gassanov

Institute of Higher Nervous Activity and Neurophysiology, USSR Academy of Sciences, Moscow, USSR

It is reasonable to assume that learning as a physiological problem has played a provocative role in studying the brain integrative function. Efforts are being made to find out the mode and construction of learning induced inter- and intracentral integrity. Hebb (1957) developed perhaps the first comprehensive systems analysis of brain function and learning. His notion of assembly activity of neurons is the most popular idea in neuro-physiology. For many years functional neuronal integrations have been based on the anatomical knowledge of what is connected to what. At present such anatomical knowledge is not sufficient to provide an unambiguous interpretation of experimental findings on a cellular level. Thus, columnar organization of cortical cells supposes a possible structural organization of cortical activity, but it is not valid in the explanation of interneuronal functional relation within a given column.

The idea of the leading role of network activity in brain adaptive responses is so commonplace today that it does not need to be discussed. The point is to find a physiological approach to the examination of this problem. One such approach concerns single cell activities. For many years the functional network organization was believed to be constructed from the data obtained on single unit activity. Logically, it is assumed that each neuron are in the working brain reflects network influences that are sufficient to judge systems neuronal activity. This approach has some disadvantages. Firstly, it is difficult, probably even impossible, to distinguish between intrinsic and network contributions to the function of a cell. Secondly, this approach does not provide information about the participation of unresponding neurons to network organization. Thirdly, there is a limitation in describing the spatial organization of network activity.

A few scientists are convinced that the data obtained with single-unit recording cannot be used to understand network organization (Szenthogothai 1967; Kostyuk 1973). The above basic assumption is essential and appears to be justified. A large number of accumulated data provides only a suitable model of the neural network, and does not describe real network organization with dynamic and spatial features.

There exists an opinion that slow potentials can be considered as evidence of the assembly activity of cells (John 1972, 1980). Indeed, field potentials are known to be related to a large population of cells and reflect their functional

Springer Series in Synergetics
J. Krüger (Ed.)
Neuronal Cooperativity

state. It is important to realize, however, that the main advantage of slow potential lies in the reflection of cellular synchronous activity. Besides, this method suffers from shortcomings when a student tries to obtain information about the functional connections between the cells.

As Abeles (1982) states, single neuron and slow potential recording, despite their great role in organization and function of the cortex, have not enabled us to examine the spatio-temporal organization of cells in any detail.

12.1 Cortical Network Activity

The development of the theory of cell assemblies evolved together with the successful experimental method of statistical description of neural networks. In recent years a powerful and flexible method based on computer analysis of multiunit activity has been developed to study interneuronal connections (Gerstein 1970; Bryant et al.1973). The principle consists of a cross correlation analysis of two trains of action potentials, triggered with the action potentials of one train and averaging over the other with respect to time. By reversing the process and triggering with the action potentials of the second train, a correlation might be found in the opposite direction. Prominent in counts in cross-correlograms are interpreted as evidence of the dependence of the two neurons.

This statistical method introduces a procedure which can be used to find interneuronal connections. Amassian et al. (1959) were probably the first to use this method to obtain functional links between cortical cells by means of manual calculation. Later, Gerstein (1970) developed special techniques and computer programming for studying statistical interneuronal correlations. At present, there are different technical approaches to the problem of statistical relations within a large population of neurons (Gerstein et al. 1985, Vaadia and Abeles 1987).

It should be noted that the statistical analysis of interspike cross correlations does not provide a complete determination of interneural connections. Physiological interpretations of computed histograms are not as clear as one would wish. There is a wide-spread opinion that structures in cross-correlograms can be attributed to both direct interneuronal connections and the manifestation of shared input activity. The latter suggestion is not in favour of systems (interconnected) neural activity. According to accepted notions, cross-correlograms with a small bin width (0.5–2.0 ms) and narrow peaks confined to one side of the zero delay time are related to an excitatory synaptic connection between two cells (Kirkwood 1979; Dickson, and Gerstein 1974). Cross correlations with a narrow symmetrical peak around the zero position or a wide bilateral peak are interpreted in terms of shared input, that is, correlograms resulted from the influence of a third source. From such considerations the neural network organization seems rather difficult to analyze by the cross correlation method. Considerable research activity is directed towards elucidating the physiological origin of cross-correlograms.

Experiments with simulation did not clarify the problem; they supported the generally accepted view of two different types of correlograms.

Today, the study of cortical neuron correlations stimulates a new approach to the understanding of cross-correlograms. The accumulated data indicate that interpretations of correlograms require a consideration of the morphology of interneuronal connections.

It is well known that synaptic connections are significantly stronger between closely spaced neurons and decrease with interneuronal distance. The morphological principle is confirmed by cross correlation analysis. Any type of correlation is stronger and more frequent between neighhouring neurons recorded by one electrode. At the distances of 1 or 2 mm correlograms usually indicate independency (Abeles 1982; Gassanov, 1987; Gassanov et al. 1985).

There is one further morphological principle put forward by Lorento de No (1949): the principle of interneuronal connections. This implies that bidirectional interneuronal relations are a fundamental form of functional connections. Finally the wide peaks in the histograms can be explained in terms of distributed synaptic contacts with neighbouring neurons. Some correlograms obtained in the study of neurons in different cortical regions, showing a single narrow unilateral peak, seem to support the interpretation that structures in cross-correlograms are related to neuronal contact density.

In any case, the morphophysiological approach to the interpretations of cross-correlograms appears to be a fruitful one, all the more since the so-called third sources have not yet been identified by anybody.

The other complicated point of statistical analysis of neuronal networks is the necessity of multicellular recordings. Temporal relations between only two cells can be interpreted in terms of a network to which they belong. It is clear that advances in network studies depend on the technique of multiunit recording. Different techniques of multispike and multielectrode recording of neuronal activity are reviewed by Krüger (1983). Among them, multiunit recording is noteworthy in two respects. Firstly, recorded by one electrode, multiunit activity enables us to investigate the interactions of adjacent neurons in a small neural cluster. Secondly, multiunit activity reveals distinctive functional features of neurons, depending on the spike amplitude. This was shown in experiments with pharmacological and cross correlation testing (Verzeano et al. 1970; Gassanov 1986; Merzhanova 1985), and with antidromic stimulation (Weber et al. 1983). However, there are different speculations as to spike segregation in multiunit recordings. Technical difficulties arise concerning discriminative procedures. The general method consists of window discrimination with parallel examination of the spike configuration on an oscilloscope. A good deal of work has been carried out to try to improve the process of spike segregation, using a computer. This question still needs to be solved.

One issue is the definition of amplitude and temporal variabilities of spikes generated by one neuron. Extracellular recording contributes to the complementary modification of spike form.

Despite the shortcomings of the cross correlation method and multiunit discrimination, they have provided a new insight into the properties of neural network activity. Knowledge about cell assemblies is of value due to its bearing upon the neurophysiology of learning.

12.2 Two Views on the Function of Cortical Neurons in Learning

Throughout the discussion so far, network organization, assembly activity and neural aggregation have been concerned with systems of interconnected or interacting neurons. One may firmly state that any kind of behaviour occurs due to the organization of a particular neuronal system. The decisive role of such systems in learning can hardly be overestimated. Various considerations show, however, that the role of neural systems in learning is not unequivocal.

The most promising approaches toward discrimination of the neural circuitry underlying associative learning are those of Weinberger and Diamond (1986). They postulate two functions of the learning brain: one is responsible for sensory-motor reactions the other for cognitive processes. These two higher nervous functions are mutually dependent, although they can exist separately in the form of memory or image.

Livanov (1972) was the first to work out an experimental systems approach to the neurophysiological study of conditioning. He discovered a new form of electrical activity of cortical cells, and called it spatial synchronization. In different regions of the cortex, rhythmic activity usually has a low coherence of EEG frequencies. In the course of conditioning, coherent activity appears among different cortical regions and is gradually confined to the cortical projection region of associated stimuli. Livanov has argued that synchronous cortical activity promoted the best impulse conduction.

Abeles (1982) advances the synfire activity idea. According to his view, convergent-divergent connections among neural sets create the possibility of synchronous activity of cells as a consequence of set formation. This assumption is based on the results of cross correlation analyses of multiunit activity.

The studies have two important points. One might be described as a systems approach to the problem, i.e. spatial-temporal analysis of the processes. The other concerns the organization of neural circuitry controlling the reflex pathway.

Some scientists argue that our approach to neural mechanisms of learning must involve the problem of cell specialization. Weinberger and Diamond (1986) and Abeles (1982) describe experimental evidence of specific sensory and associative functions of one and the same sensory neuron. An associative function was revealed in pyramidal cells of the motor cortex (Tanji and Evarts, 1976). In connection with cellular polyfunctional properties, it is necessary to mention Konorsky's view about gnostic neurons in the cortex (1967). He suggested that in the course of learning some cortical cells acquired a new gnostic function and some played a transitory role.

In the light of recent concepts, the integrative processes seem to be the key problem in the programming of behaviour. Pavlov's (1951) structural dynamic complex, Hebb's (1957) cell assembly, Anokhin's (1968) afferent synthesis, Asratyan's (1970) local conditioned state and John's (1980) hyperneurons all represent different theoretical approaches to the problem of interneuronal functional organization underlying associative learning in vertebrates.

The main outcome of the theoretical and experimental work lends support to the idea of a double system of neural mechanisms of learning: one for the communication along the neuronal chains and the other for recognition and fixation of biological significance of the stimuli. Synaptic processes play an important role in organization of any kind of neuronal circuits, but the specificity of these circuits is believed to depend on the spatial distribution of interneuronal connections. Perhaps the most fundamental problem of contemporary neurophysiology of learning is that of neuronal organization and localization of the engram containing the behaviour program. According to this view, it should be accepted that the responses to meaningful stimuli are the result of activation of a neuronal network bearing cognitive functions. In that case, the question aims at searching for the location and functional characteristics of the neuronal bases of the engram.

As mentioned above, there is strong experimental evidence that associative properties are inherent in the cortical sensory neurons (Weinberger and Diamond, 1986; Abeles 1982). From this viewpoint it seems reasonable to investigate neural set activity in cortical sensory projections of the stimulus.

The next problem concerns the type of cortical neuronal activity. There is a growing body of data presenting evoked and spontaneous cellular activity in the course of learning. The neurophysiological literature shows that the principal concepts of learning are based on findings of impulse evoked responses. Meanwhile, spontaneous impulse activity is considered to be related to the excitation of local cortical circuits (Beritashvili, 1966). Anticipatory activity in monkey motor cortex neurons for several seconds prior to muscular contraction illustrates the sensitivity of ongoing firing to learning (Tanji and Evarts 1976). There is sufficient evidence to believe that the so-called spontaneous activity might be the subject of promising research into the engram of learning.

12.3 Two Types of Cortical Neuronal Network in Learning

The present experiments on cats were undertaken to determine the extent to which the aforementioned ideas are correct (Gassanov et al. 1980; Galashina and Bogdanov 1978; Bogdanov and Galashina 1987). Multiunit activity was recorded from the auditory and motor cortex by means of a chronically implanted band of nichrom electrodes. Seven electrodes of 50 μm diameter were attached in parallel and soldered to a socket which was fastened by protocril to the skull. The behaviour was studied in a small chamber on freely moving animals.

After recovery from the operation, two or three electrodes were active during 1–3 weeks and in some cases for a longer period. Multiunit activity was

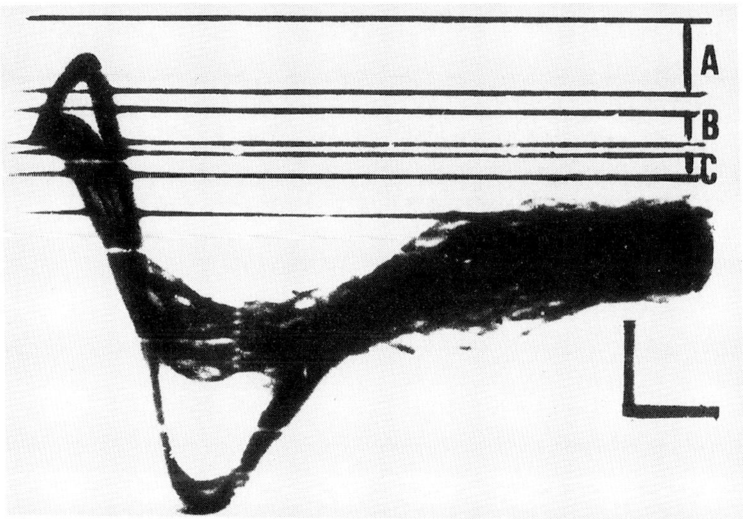

Fig. 12.1. Selection of spikes from multiunit recordings. Amplitude discrimination of large (gate A) and medium (gate B) amplitude. Gate C is for spikes with small amplitude. Beneath gate C there is a line indicating the level of noise elimination. Calibration: 50 μv; 0.2 ms

recorded on a four-channel tape recorder. Each multiunit recording was subjected to a discrimination of the first (negative) phases of spikes to obtain three impulse trains. The amplitude digcrimination was corrected by oscilloscopic examination of selected spikes (Fig. 12.1). Cross correlograms were computed both within one multiunit recording and between neuronal pairs from different multiunit recordings. Therefore, it was possible to analyze interneuronal relations among closely spaced, neighbouring, and remote, space-separated neurons. A cross correlation analysis made in awake cats displayed sharp, one bin-wide unilateral peaks in histograms with 0.5–2.00 ms bin width (Fig. 12.2). Such cross correlograms were found in 10–15% of adjacent neuron pairs. The same type of cross correlogram was seen in 2–5% of remote neuron pairs. No wide bilateral peak was observed in the histograms with small bin width. It was noteworthy that in most cases interneuronal relations were defined by spike amplitudes of neuron pairs. Usually, the functional connections were directed from the neurons having higher spike amplitude towards those with the smaller spikes. Such directed connections were observed in the auditory and in the motor cortex. Special functional testing of neurons with different spike amplitudes revealed their individual properties in relation to antidromic and afferent stimulation (Gassanov 1986; Weber et al. 1983). The findings are thought to be interpretable in terms of morphological differentiation of the neurons contributing to the multiunit recording. It seems reasonable to suppose that in the multiunit activity the relative amplitude of spikes can be assumed to be an indication of cell size, whereas the absolute amplitude is defined by the distance to the

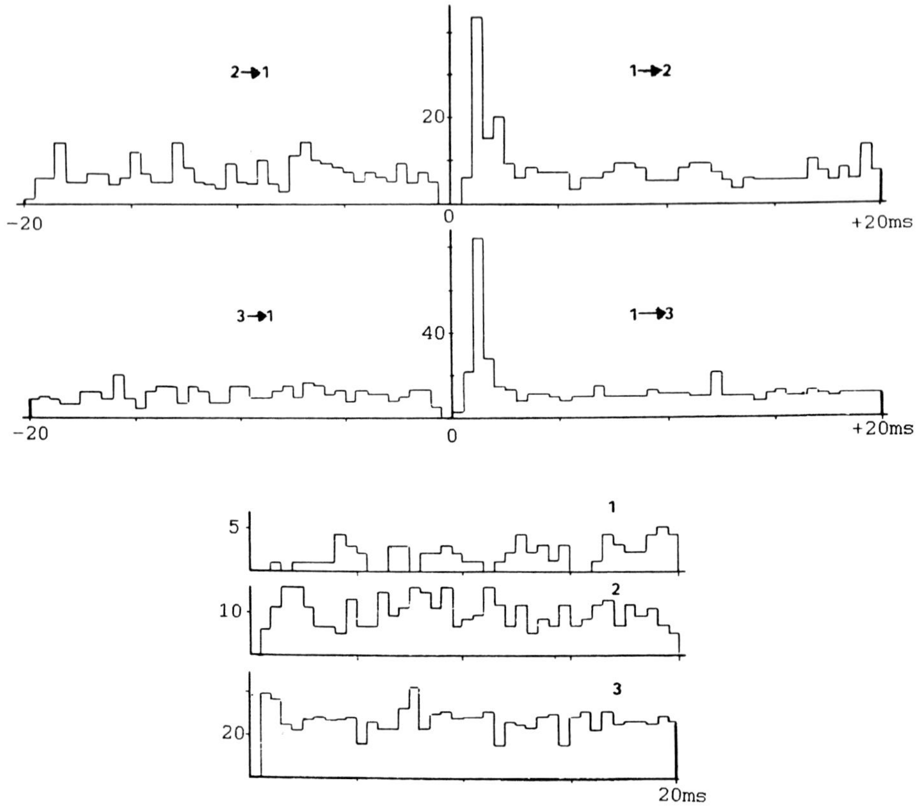

Fig. 12.2. Examples of cross-correlograms (top two histograms) and auto-correlograms (bottom three histograms; for neurons 1, 2 and 3). The ordinate indicates the number of spikes, the abscissa ±20 ms

electrode. Verzeano's study of multiunit activity with pharmacological testing seems to support this supposition (Verzeano et al. 1970). These experiments favour the view that, in the case of multiunit registration, one might deal with large (high amplitude), medium (medium amplitude) and small (low amplitude) neurons.

Anatomical parallels can be found in relation to multunit activity as a whole. In the course of electrode penetration into the cortex during the operation, multiple unit activity was encountered as frequently as single neuron activity. Furthermore, if the electrode came into contact with a neural aggregation, all the neurons were involved in a damage reaction (Gassanov 1986).

These observations, together with the above-mentioned intensive neural interconnections in a microset, allow us to think that multiunit activity can be attributed to an anatomical cluster organisation of cortical neurons. The existence of such neuronal clusters in the cortex is described in anatomical papers

(Peters and Walsh 1972; Batujev and Babmindra 1977). Anatomo-physiological parallels are important for the subsequent analysis of cross-correlograms because they render it possible to describe the cortical network not only in space but also according to connectivity. To explore the systems neural mechanisms of conditioning in a series of experiments, conditioning was achieved by pairing clicks with electrical shock, applied to the forelimb (Gassanov et al. 1985). A train of clicks with a frequency of 10Hz for 5s was followed by electrical shocks with the same frequency for 1s. Conditioned responses were assessed by electromyographic reactions of the stimulated forelimb. Multiunit activity was recorded from the cortical auditory and motor projections of the click and the forelimb, respectively. Thirty second fragments of spontaneous multiunit activity (prior to stimulus presentation) in trained and untrained cats were statistically processed.

Figure 12.3 illustrates the systems activity in microsets (under one electrode) and among the microsets of cortical neurons during different behavioural tasks. Positive conditioned motor reflexes would appear when the background activity was characterized by a high level of interneuronal correlations in the auditory cortex both within single microsets and between different. In the motor cortex, there were only intermicroset connections that showed enhancement. In single microsets, no change in cross correlation occurred. During inhibitory conditioning (e.g. extinction), the number of interacting neurons in the auditory cortex microsets was as large as in the case of positive conditioning. Among microsets, however, their number significantly decreased in the auditory and motor cortex. The data allow us to draw two conclusions: 1) in the cortical projection of the conditioned signal, the network organization of neurons is more complicated than in the motor cortex which performs the effector function, and 2) microsets in the auditory cortex play an active role in realising positive (motor-inducing) and negative (motor-suppressing) conditioning.

The experiment based on this study was planned in such a way as to enable us to investigate the neuronal sets in the course of prolonged conditioning with subsequent substitution of a negative reflex by a positive one. The most effective approach to such conditioning is thought to be timing behaviour, in which the animal is trained to space its response on a differential-reinforcement schedule. In our experiments a reinforcement (a piece of meat) was delivered on a fixed two-minute interval and a cat had to push a movable window to grasp the food (Galashina and Bogdanov 1978; Bogdanov and Galashina 1987; Gassanov 1987).

Usually a session started with time-independent motor responses and evolved to a distinct temporal discrimination behaviour. It should be noted that motor manifestations of timing conditioning were seen in the second minute of the intervals. Some of them coincided with the end of the interval.

After acquisition of the timing behaviour, when the waiting period was limited to the first minute and active motor responses were displayed in the second minute, four cats were subjected to operations where an electrode band was implanted in the motor cortical projection of the working forelimb. The motor responses were usually accompanied by a neuronal manifestation of tem-

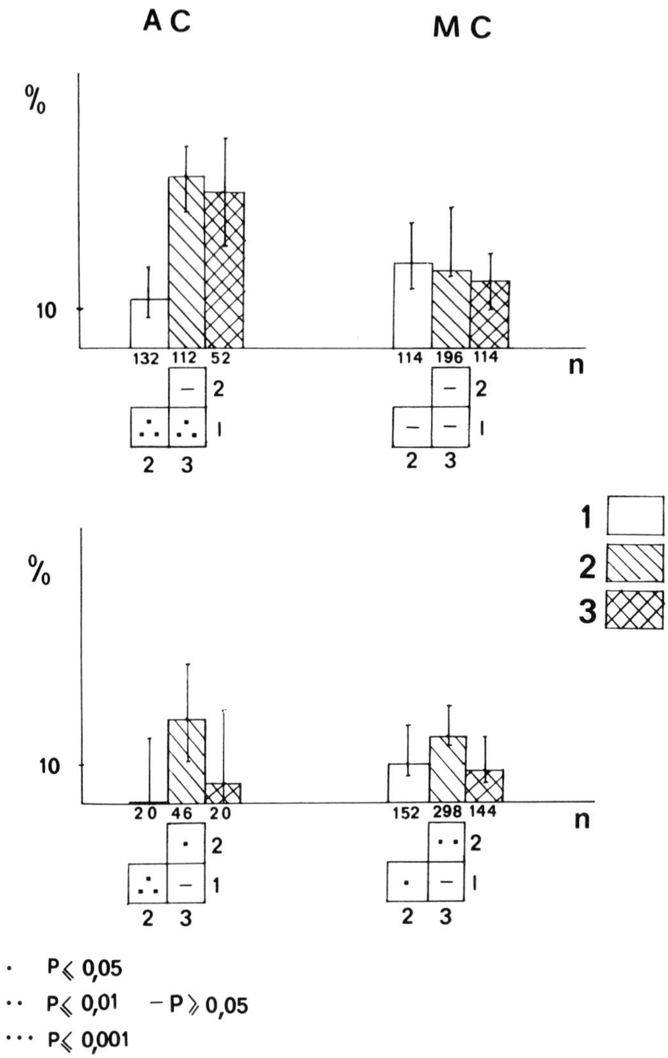

Fig. 12.3. Interneuronal-dependent connections in auditory (AC) and motor (MC) cortical areas, 1: in naive cats (white). 2: after acquisition of the stable CR (oblique hatching) and 3: after extinction (cross hatching). The number of all cross-interval histograms (CIHs) for each state was taken as 100%. Top: interneuronal-dependent relations within microsets. Bottom: interneuronal-dependent relations between neurons located in different microsets (distance: 70–300 μm). n: total number of CIHs; levels of significance are given according to the ϕ method

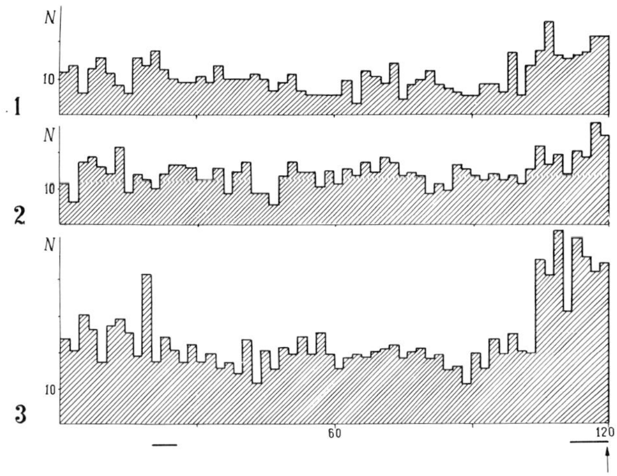

Fig. 12.4. Variability curves with excitation accompanying the movement. 1, 2, 3: neurons according to decreasing spike amplitude. N: number of impulses per two seconds. The bottom horizontal lines indicate pushing of the window. The arrow indicates delivering the meat; 60, 120: time in seconds

poral discrimination shown by an increase of impulse frequency. It was noteworthy that small spike neurons showed stronger motor activations (Fig. 12.4).

Multiunit activity was recorded in the course of a session and then one minute segments were chosen for cross-correlation analysis. The correlograms were divided into three groups: firstly, the control group before learning, in which one minute segments of the record did not depend on time sequences; secondly, the experimental group with motor reactions in time and during the first minute (from the last food reinforcement up to the second half of the interval) and second minute (the second part of the interval up to the next food reinforcement); thirdly, the experimental group with incorrect behavior in time (with the same one-minute recording periods as in the previous group).

The resulting correlograms obtained in the trials with motor temporal discrimination are summarized in Fig. 12.5. The number of correlated neuron pairs within microsets and among microsets is clearly different in the first and second minutes of the fixed intervals. They were significantly more notable in the second halves of the intervals than in the first halves. An exception was the intermicroset activity in the first minute. It is reasonable to believe that such time-dependent neuronal organizations reflect different behavioural states: a waiting or inhibitory state and a preparatory or excitatory state. A more appropriate definition of the latter state would be a state of focused attention. A comparison of systems neuronal activity in associative and temporal conditioning reveals definite features of similarity: more intensive functional cooperativity of adjacent neurons in the cortical regions corresponding to 1) the conditioned stimulus during classical defensive conditioning and 2) preliminary motor activation during alimentary instrumental conditioning. As for the connections among the remote neurons in the same cortical areas, they were activated as a rule

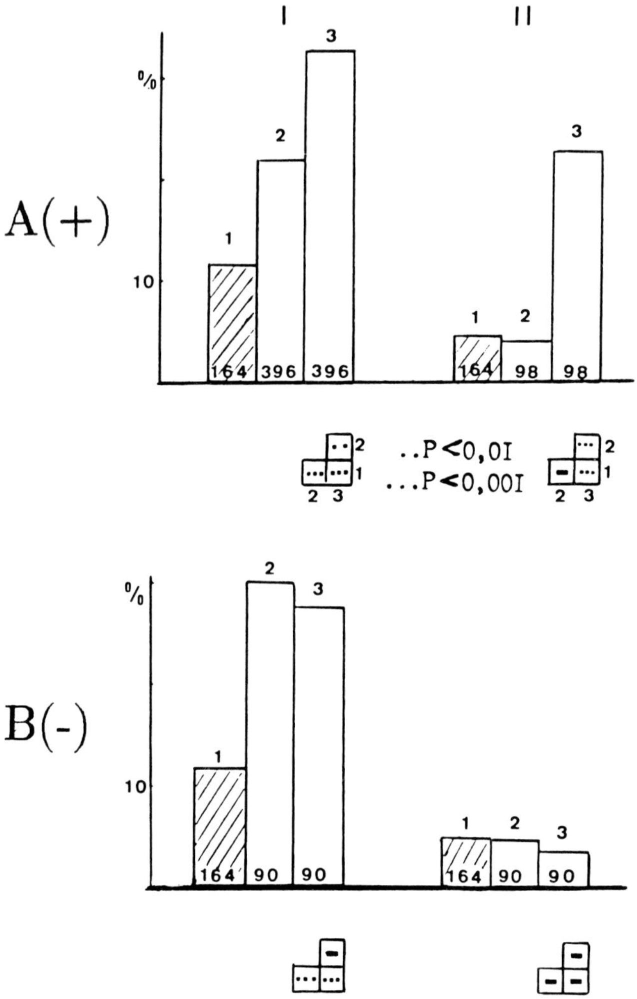

Fig. 12.5. Cortical interneuronal correlations in timing behaviour. A (+): in the waiting state; B (−): during incorrect motor responses; I: neighbouring neurons; II: remote neurons; 1: control data; 2: the first minute; 3: the second minute. The number inside the column is the number of cross-interval histograms. The ordinate indicates the percentages of identified connections. Levels of significance are given according to the ϕ method

during positive conditioning, while the neighbouring neurons interacted during both positive and negative (inhibitory) conditioning (Figs. 12.3 and 12.5A).

It is of interest that in experiments with temporal discrimination a complementary control presented evidence for a state-dependent origin of the observed network organization. The control data were concerned with the intervals characterised by irregular motor actions for two minutes. As a result of correlation analyses, it appeared that a number of interacting neuron pairs

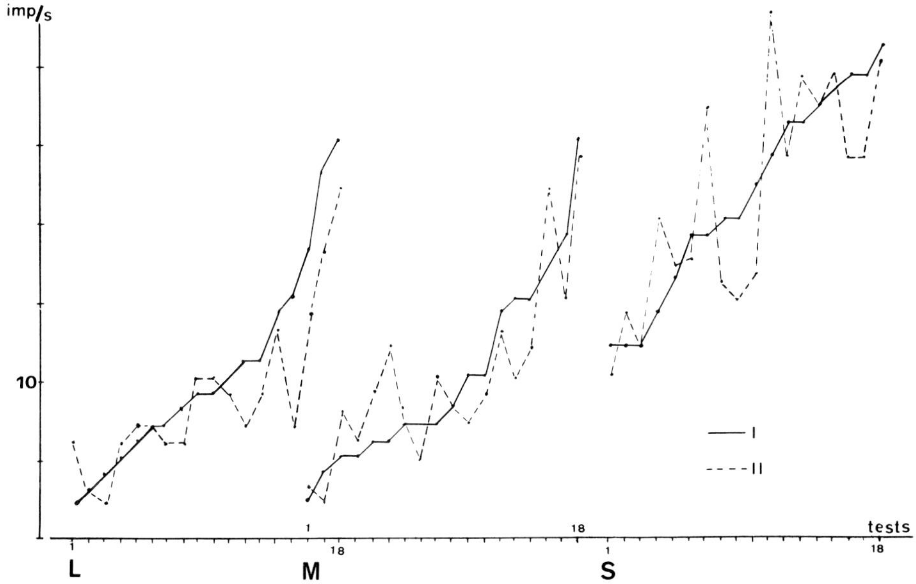

Fig. 12.6. Mean impulse frequencies in the first (solid lines) and second (dashed lines) minutes. Ordinate: impulses/s; abscissa: tests; L: neuron 1 (large spike); M: neuron 2 (medium spike); S: neuron 3 (small spike). The tests are ordered on the abscissa according to increasing frequencies in the first minute. These data underwent a cross-correlation analysis (Fig. 12.5)

attained a high level within microsets and did not change among the remote neurons (Fig. 12.5B).

There was one more distinction between neuronal organization in intervals with time-dependent and time-independent movements, namely that of equal level of microset systems activity during the latter type of behaviour. These observations confirm the assumption that the functional state manifesting itself in the form of systems activity is not related directly to motor actions.

Another characteristic feature of network function might be seen in the activity of units participating in a set organization. Figure 12.6 demonstrates mean frequencies of units in intervals with distinct temporal discrimination, taken for correlation analysis (see Fig. 12.5). Apparently there were no significant differences in the mean frequencies of selected neurons during the first and second minutes. Only neurons with large spikes (L) had shifts of the mean frequency in some intervals.

The results indicate that single and systems activity can be attributed to different functional properties of the cortical neurons. All the results together lead to the idea of two types of local cortical network: microset and inter-microset, each having different functional designations in learning, and being related to the behaviour of animals.

12.4 Specificity of Cortical Networks

So far, the analyses have been concerned with the total number of interacting neurons without detailed consideration of available connections among the different neuron pairs. As was pointed out above, the selection of spike trains was carried out according to the amplitude of spikes, which rendered it possible to distinguish three neuronal classes: large, medium and small neurons. Despite the insufficient evidence of single-unit-discrimination purity, the data obtained allowed us to state that in multiunit activity spike amplitude has a feazing on certain functional properties. It was mentioned that the activity of small neurons in the motor cortex was more significantly related to forelimb movements than the activity of large and medium neurons (Fig. 12.4). Clear differences between amplitude-selected neurons were displayed in cross-correlograms. More than half of the histograms with dependent relations had definitly directed connections, namely from the larger to the smaller neuron (Fig. 12.2).

To obtain a distribution of interneuronal connections within three neurons of a microset population, all cross-correlograms were considered with respect to the spike sizes of the neuron pairs. In Fig. 12.7, the dominant connections are shown for the motor cortex. It is seen that during the waiting period of temporal discrimination, the number of connections increases in comparison to that of the inexperienced animals. This enhancement consists mainly in the appearance of additional connections with different directions, and it depends on the behaviour stage. The most stable connections are those which are directed from large to small neurons. In addition, one can see the network specialization of single neurons. Thus, large neurons have mainly outward connections to neurons recorded by another electrode, whereas small neurons have inward connections. The data obtained give evidence that there are state-dependent cortical microsystem organizations with specific distribution of interneuronal connections (Gassanov 1980).

Convincing support of the notion of network specificity was provided by a cross-correlation analysis of neurons in the visual and motor cortex in cats trained to lick conditioning (Gassanov et al. 1985; Merzhanova 1985). Two cats were taught a special form of classical conditioning, i.e. transswitching. The same conditioned stimulus (CS), 2Hz frequency electrostimulation of the optical pathways, was reinforced after 3 s either by milk or by a 5% solution of $CaCl_2$ in one session. Both reinforcements in equal 2 ml volumes were delivered into the mouth through an indwelling tube. The two situations were differentiated by means of background sounds presented continuously and used as transswitchers. A background tone of 400 Hz was presented in the alimentary situation when electrical stimulation of the optic pathways was reinforced by milk. A background tone of 800 Hz sounded in the defensive situation when the same CS was paired with $CaCl_2$. Thus, one and the same CS in a session could be connected either to licking (alimentary reaction) or to spitting out (defensive reaction). The acquired conditioned reflexes (CRs), recorded by a mechanical system, had different shapes of mechanograms. Figure 12.8 presents the distribution of domi-

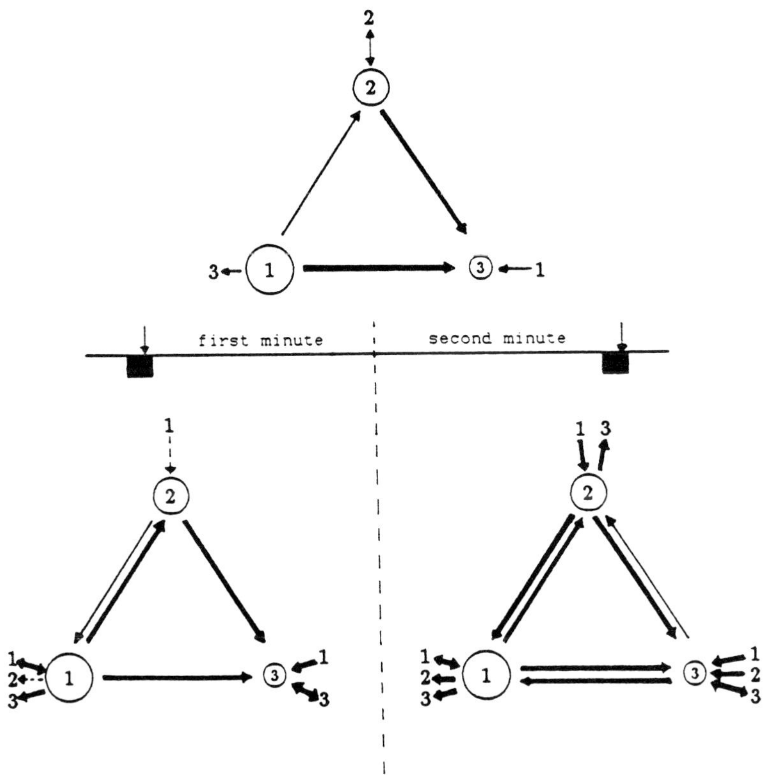

Fig. 12.7. Distribution of functional connections among motor cortex neurons before (top) and after (bottom) acquisition of timing behavior. Left: first minute; right: second minute. Long (short) arrows indicate connections with neurons recorded by the same (a different) electrode, respectively. Numbers at the short arrows indicate spike sizes of partner neurons. Thick arrows point to cross correlations appearing in 40% or more of all the cross-correlograms in the given direction, thin arrows: in less than 40%; 1,2,3: neuron ordered according to decreasing spike amplitude

nant connections in the projections of the conditioned stimulus and the effector (the tongue).

It is noteworthy that in the visual cortex background, the interneuronal functional organization is strictly related to the ongoing experimental situation (Fig. 12.9 A,C). In other words, the network activity in the visual microsets was defined by biological features of the conditioned signal. At the same time, in the motor cortex microsets, the configuration does not distinguish defensive and alimentary situations. In the case of disorders of conditioning, the network arrangements are changed in unpredictable ways (Fig. 12.8B,D).

It should be noted that subtle learning-induced differences in neuronal correlations were described by Abeles (1982) in monkey auditory cortex. He pointed out that the network reorganization was related to the side of the associated hand. The data were obtained from adjacent neurons recorded by one electrode.

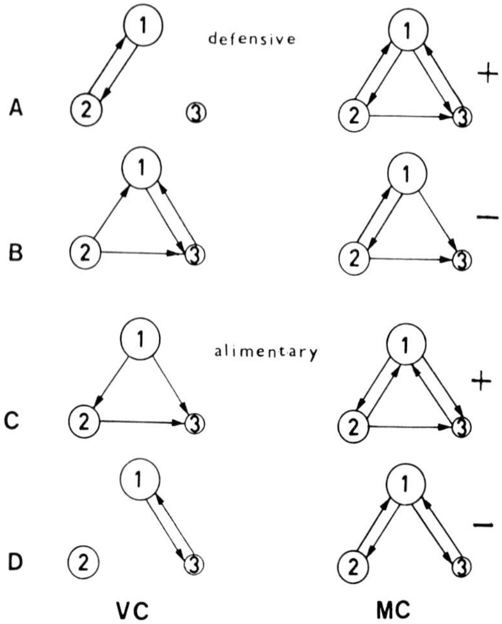

Fig. 12.8. Patterns of interneuronal connection in visual (VC) and motor (MC) cortical areas after acquisition of conditioned transswitching of defensive (A,B) and alimentary (C,D) CRs. A and C: conditioned reflexes (CRs) were performed; B and D: CRs were absent. The arrows point to the functional connections appearing in 50% or more of all cross-correlograms in the given direction; 360 histograms for each direction were analyzed

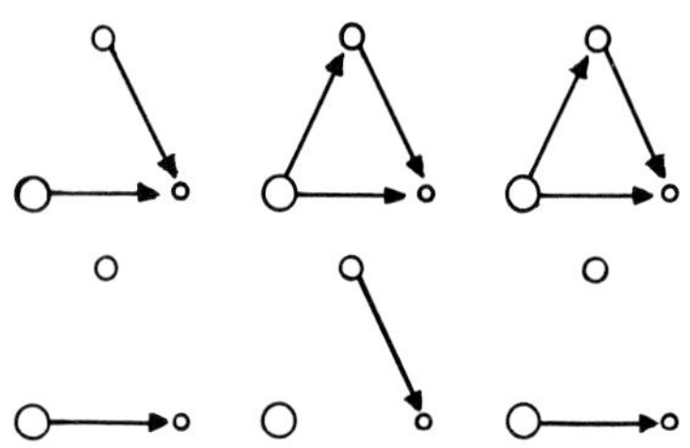

Fig. 12.9. Two simultaneously recorded microsets (top and bottom) of motor cortex neurons before (left diagrams) and after two subsequent trials of conditioning (middle and right diagrams). It can be seen that the upper microset develops stable connections while the connections remain variable in the lower set

All this evidence leads to the conclusion that one of the most important properties of cortical networks consists of the formation of specific neuronal interconnections.

It would be erroneous to think that the specific distribution of interneuronal connections can be attributed to most microsets under study. On the contrary, each microset usually shows a single and rather changeable connection.

Simultaneous recording of two or more microsets enabled us to demonstrate parallel existence of different microsystems. One example of such multiplicity of the cortical networks in the course of one session of classical conditioning is presented in Fig. 12.9. Three subsequent trials show that in the motor cortex, one microset of neurons displays a stable frame of connections while the other one shows extremely variable configurations of interneuronal connections. It is important to note that such a combination of stable and unstable cortical networks seems to be inherent in any brain activity. One can add that there are microsets which do not undergo any changes during learning. Thus, on the cortical microset level it is possible to state that they reveal a multiplicity both in the internal organization and in the general composition of different networks.

12.5 Conclusion

It has become clear that associative learning involves many neuronal mechanisms beyond that of signal–response circuitry. Some of them could be the neuronal basis of cognitive functions of the brain. There is reason to believe that sensory receptive areas of the cortex might be responsible for mediating association. The above-mentioned data suggest that the systems approach to the problem makes it accessible to experimental exploration. It is necessary to note that properties of cortical networks are shown not to be predictable by single neuron activity. This was revealed by cross-correlation analysis of neuron pairs. This method has proven to be a powerful instrument in describing the network organization of neurons.

In order to gain some insight into the cortical connectivity by quantitative analysis of the probability of connections, we decided to investigate the spontaneous activity of cortical neurons in projections of meaningful signals. This view was based on the assumption that state-dependent neuronal activity appeared to reflect specific memory traces, related to the specific learning procedure. The correctness of this view is supported by such facts as the discrepancy between spontaneous and evoked impulses during conditioning.

The experimental findings have led to conclusions supported by anatomical data about the structural organization of cortical cells. A microset of cortical cells can be considered as being separate from the neighbouring cell assembly with relatively strong interneuronal connectivity. Functional connections between different microsets are rather weak. Learning processes increase interneuronal connections in microsets unrelated to excitatory or inhibitory types of behaviour. At the same time, intermicroset connections are strengthened only by the excitatory type of behaviour. All these forms of the neuronal systems activity are observed in ongoing spontaneous impulses and are not involved directly in acquired motor actions. On the basis of the evidence obtained, it is suggested that learning results in the organization of a state-dependent cortical network where single microsets of neurons are distinguished by the complexity and patterns of their interneuronal connections. This allows us to point out the specificity of different behaviour-induced functional states within a local cortical region.

It is apparent that the arbitrarily selected three-neuron groups are far from being an authentic neuronal assembly. Nevertheless, data on cross-correlation analyses are principally concerned with specific features of neuronal connectivity which are not detectable in the study of single neurons. Striking differences in microset and intermicroset connectivity with respect to training call our attention to the functional role of intermicroset connections in motor programming. Undoubtedly, the microsystems organization is an elementary network with narrowly restricted functions and there should be a more complex network consisting of interacting microsets of the cortical neurons.

References

Abeles, M (1982): *Local cortical circuits* (Springer, Berlin, Heidelberg) p. 95

Amassian, V.E., Macy J., Hardless J., Waller H.J. (1959): Simultaneous recording of the activities of several individual cortical neurons. Trans. N.Y. Acad. Sci., II, **21**, 395–410

Anokhin, P.K. (1968): *Biologia i neurofisiologia uslovnogo refleksa* (Medicina, Moscow) p. 547 (in Russian)

Asratyan, E.A (1970): Ocherki po fisiologii uslovnykh refleksov. Nauka, Moscow, p. 359, (in Russian)

Batujev, A.S., Babmindra, V.P. (1977): Cortical neuronal cooperativity. Zhurnal visshej nervnoj dejatelnosti, **37**, 715–720, (in Russian)

Beritashvili, I.S. (1966): *Memory of verterbates: characteristics and origin* (Tbilisi, Mezniereba) p. 137 (in Russian)

Bogdanov, A.V., Galashina, A.G. (1987): Reorganization of functional connections of neighbouring neurons of cat's motor cortex at elaboration of conditioned food-producing reflex to time. Zhurnal visshej nervnoj dejatelnosti, **37**, 1165–1167, (in Russian)

Bryant, H.L., Marcos, A.R., Segundo, J. (1973): Correlations of neuronal spike discharges produced by monosynaptic connection and common inputs. J. Neurophysiol. **36**, 205–225

Dickson, J.W., Gerstein, G.L. (1984): Interaction between neurons in auditory cortex of the cat. J. Neurophysiol. **37**, 1239–1261

Galashina, A.G., Bogdanov, A.V. (1978): Analysis of cat's motor cortex neuron activity at food-producing reflex to time. Zhurnal visshej nervnoj dejatelnosti **37**, 766–768 (in Russian)

Gassanov, U.G. (1980): Specific and unspecific neuronal mechanisms of cellular operant conditioning. Acta Neurobiol. Exp. **40**, 137–148

Gassanov, U.G. (1986): Amplitude principle of structural and functional classification of cortical neurons. Zhurnal visshej nervnoj dejatelnosti **37**, 3–11 (in Russian)

Gassanov, U.G. (1987): Gnostic microsets of cortical neurons. Zhurnal visshej nervnoj dejatelnosti **37**, 634–641 (in Russian)

Gasaanov, U.G., Galashina, A.G., Bogdanov, A.V. (1980): A study of neuron system activity in learning. In: B. Lomov, R. Thompson and V. Shvyrkov (Eds.) *Neural mechanisms of goal-directed behaviour and learning* (Academic, New York) pp. 341–352

Gassanov, U.G., Merzanova, G.Ch., Galashina, A.G. (1985): Interneuronal relations within and between cortical areas during conditioning in cats. Behav. Brain Research **15**, 137–146

Gerstein, G.L. (1970): Functional association of neurons: detection and interpretation. In: The Neurosciences: Second Study Program ed. by F.O. Schmitt (Rockefeller University Press, New York) pp. 648–661

Gerstein, G.L., Perkel, D.H., Dayhoff, J.E.(1985): Cooperative firing activity in simultaneously recorded populations of neurons: detection and measurement. J. Neurosci. **5**, 881–889

Hebb, D.O. (1957): *The organization of behaviour. A neurophysiological theory* (Wiley, New York) p. 335

John, E.R. (1980): Switchboard versus statistical theories of learning and memory. Science **177**, 850–864

John, E.R, (1980): A neurophysiological model of purpose behavior. In Neural mechanisms of goal-directed behaviour and learning. Ed. by R.F. Thompson, L.H. Hicks, V.B. Shvyrkov (Academic, New York) 93–118

Kirkwood, P.A. (1979): On the use and interpretation of cross-correlation measurements in the mammalian central nervous system. J. Neurosci. Meth. **1**, 107–129

Kostyuk, P.G. (1973): Osnovnye mekhanismy obiedinenija neironov v nervnom centre. In: P.K. Anokhin (Ed.). *Principy sistemnoi organizacii funkcii*. (Nauka, Moscow) p. 115–124 (in Russian)

Konorsky, J. (1967): *Integrative activity of the brain*. (University of Chicago Press. Chicago) p. 410

Krüger, J. (1983): Simultaneous individual recordings from many cerebral neurons: techniques and results. Rev. Physiol. Biochem. Pharmacol. **98**, 177–233

Lorento de No, R. (1949): Cerebral cortex: architecture, intracortical connections, motor projection. In: *Physiology of the nervous system*, 3rd ed. ed by Fulton J.F. (Oxford Univ. Press, New York) p. 288–312

Livanov, M.N. (1972): Spatial organization of brain processes. Moscow. Nauka, p. 182. (in Russian)

Merzhanova, G.Ch. (1985): Activity of cortical three-neuronal microsystems in cats at conditioned switch-over. Zhurnal visshej nervnoj dejatelnosti **35**, 435–441 (in Russian)

Pavlov, I.P. (1951): Polnoje sobranije sochinenij, Moscow-Leningrad, Izdatelstvo AN SSSR, Vol. 3, book 2, p. 438 (in Russian)

Peters, A., Walsh, T.M. (1972): A study of the organization of apical dendrites in the somatic sensory cortex of the rat. J. Comp. Neurol. **144**, 253–261

Szentagothai, J. (1967): The anatomy of complex integrative units in the nervous system. In: K. Lissak (Ed.), Results in neuroanatomy, neurochemistry, neuropharmacology and neurophysiology. Akademiai Kiado, vol. 1. Budapest, pp. 9–45

Tanji, F., Evarts, E.V. (1976): Anticipatory activity of motor cortex neurons in relation to direction of an intended movement. J. Neurophysiol. **39**, 1062–1068

Vaadia, E., Abeles, M. (1987): Temporal firing patterns of single unit pairs and triplets of units in the auditory cortex. Israel J. Med. Sci. **23**, 75–83

Verzeano, M., Dill,R., Navarra, G., Vallecalle, E. (1970): The action of metrazol on spontaneous and evoked activity. Physiol. Behav. **5**, 1099–1109

Weber, N.V., Rapoport, S.Sh., Sylkis, I.G. (1983): On correlation between amplitude of action potentials and velocity of axon conductivity of sensorimotor cortex neurons. Zhurnal visshej nervnoj dejatelnosti **33**, 1114–1120 (in Russian)

Weinberger, N.M., Diamond, D.M. (1986): Physiological plasticity in auditory cortex: rapid indication by learning. Prog. Neurobiol. **2**, 200–220

Subject Index

40 Hz 177
absorption 33, 42
acetylcholine 172
acoustic system 82
active points 77
α-wave 113
anti-co-firing 62
apparent RF size 91
assemblies see neuronal assemblies
asynchronous contribution 262, 268
asynchronous gain 262
auditory cortex 73, 249, 257, 269, 271
average common excitation (ACE) potential 233
average cortical neuroanatomy 89

background stimulus 79, 95
behavioral task 252 ff., 266
binocular correspondence 169
brain-slices 33

cell assemblies see neuronal assemblies
center-surround 81
charge-coupled device (CCD) 43 ff.
checkerboard model 155
chromophore 33
circuit selection 168
coding properties 202
coherence 165 ff., 184, 202
coherent features 166
coincidence detection 170
coincidence of oscillation 75, 76, 97
colour selectivity 113
columnar organization 254, 256
common excitation, common input 8, 54, 63, 138, 147, 243, 259 ff., 287
complex cell 5
connectivity, structural 53
condensation 57
conditioning 287
conduction velocity 112
context-dependent 56
contour planes 82
contribution 16, 146
correlation profiles 199, 200
correlational assemblies 53
cortical architecture 69 ff., 99
cortical columns 53
cortical layers 89
cortical network 281
  specificity of cortical networks 292
cortical slice 24
cortico-geniculate transmission 11
covariability 115–119, 120–123
covariance computer 189
co-firing 62
cross-correlation, cross correlogram 5 ff., 52, 76, 105–115, 120–123, 135, 179, 199, 206, 211, 251, 258, 259, 264 ff., 271
cumulative PSTH 96
cytochrome 33, 42
cytochrome oxidase blobs 148

damped aperiodic oscillation 74 ff., 97 ff.
delay scale 112
dendritic field 68, 90, 91, 99
deoxyglucose 30
development 169, 175
direct connections 54
direction of motion 93, 98
direction selectivity directionality 20, 95, 113, 144
disynaptic connections 242, 246
double color opponent 151
double line model 9
double-barrel microelectrode 7
dura 47, 48
dynamic cooperativity 52
dynamic reorganization 56
dynamic RF properties 82 ff., 99, 100

early vision 165
effective connectivity 52, 249, 251, 259–261, 272

elementary events 98, 100
end stopping 68, 83, 95
environment 105
excitatory postsynaptic potential (EPSP) 23, 225, 228, 262 ff., 268, 270
experience 169
exuberant connections 176
e.m.g., human 227, 228

feature detector 68, 99
feature linking 184, 186
feedback 80, 97
feeding network 187, 201
firing rate modulation 62
fluorescence 31, 33, 42
frequency discrimination 89
frontal cortex 249, 256, 269, 272
functional connectivity 53
functional group 99, 249

gating systems 171
geniculo-cortical transmission 9
gravitational clustering 55
gravity representation 55

habit-reversal 254, 271
Hebb rules 170
hebbian assemblies 53
Hebb-synapse 106
hemoglobin 33, 42, 47
high frequency oscillation 71 ff.
higher brain functions 250, 256
horizontal connections 134
horseradish peroxidase (HRP) 135
hypercomplexity 68, 83, 95, 98

idiosyncratic temporal structure 101
inhibition 56, 69, 91, 92
inhibitory postsynaptic potential (IPSP) 21
interareal interaction 96, 201
interareal synchronization 207
intercostal motoneurons 244, 225–227
  inputs from bulbospinal neurons 240, 241
  pattern of inputs 238–240
  spatial distribution of synchronization 238–240
interneurons 54
interpeak interval 74, 88
interspike interval 88
intracolumnar interaction 18, 167, 175
intracolumnar transmission 12
intrinsic dynamic properties 100
isoluminant color contrast 152
IT 70, 96, 97

Joint Peri Stimulus Time Histogram (JPSTH) 54, 61 ff., 261, 266, 272, 273

latency 112
lateral geniculate nucleus (LGN) 71, 74, 81, 96 ff.
light-scattering 33, 42
linking features 186, 212
linking field 185, 217
linking network 185, 201, 211
local assembly 217
local field potential (LFP) 40, 191
local mass signals 190
local oscillator 212
long-term potentiation 173
low frequency oscillation 69, 71 ff., 97
LP Pulvinar 70

membrane potential 263, 265
microset 287
model neurons 185, 212
modified type II 151
modulatory systems 171
monosynaptic connection 241, 242
monosynaptic input 138
motoneuron synchronization 225, 228, 244
  broad peak synchronization 229
  high frequency oscillation (HFO) 229, 245
  presynaptic synchronization 228, 232, 245, 246
  purring 231
  respiratory periodicity 229
  short-term synchronization 231, 245
motoneurons 225
MT 70
multiple microelectrode 6, 106, 184, 250
multiple unit activity (MUA) 192
multi-channel photodiode array 24
multi-neuron recordings 6, 56, 106, 184, 250
multi-unit spike trains 52
mutual inhibition 63

narrow peak 263, 268
net correlation 7
neural code 250, 251
neural coordination 56
neuron pools 53
neuronal assemblies 40, 50, 52 ff., 65, 76, 99, 166, 177, 181, 202, 210, 214, 249 ff.
NMDA 66, 172
nonlinearity 70
non-spiny stellate cell 69
noradrenaline 172
normalization procedures 52

now print signals 173
nuclear magnetic resonance (MRI) 30

object integration 187
ocular dominance 76, 109, 110
ocular dominance column 12, 47
off subarea 5
olfactory bulb 34
on subarea 5
optical flow 79, 98
optical recording 24
optic-tectum 34
orientation column 43
orientation selectivity 20, 68, 83, 84, 95, 98, 126, 130
orientation tuning 87, 195, 197
oscillation 74, 96, 113–115, 177
oscillation spindles 198
oscillator units 211
oscillatory episodes 98
oxymetry 42

parallel processing 70, 96
pattern recognition 186
phase locking 210
photodetector 31
piecewise separability 79 ff.
PMLS 70, 73, 75, 92, 96, 97
pool activity 66
positron emission tomography (PET) 30
prefrontal cortex 256, 273
presynaptic synchronization 242–244
presynaptic inhibition 66
primary correlation operator 232–238
processing submodality 77, 92
projection 59
pseudo-correlation 7
PST coincidence histogram 54
PST histogram 54
psychological time scale 88, 100
purring 231
pyramidal cell 69

rapid modulations 57
rapidly changing synapses 66
reaction time 88
receptive field (RF) 68 ff., 70, 90 ff., 98 ff., 109
  spatiotemporal RF 80
receptive field size 82, 83, 87
respiratory movements 225
response plane 64, 77
response selectivity 20
response synchronization 178
retinal ganglion cell 81, 109
retinotopy 91, 109
RH 155 24

segmentation 165
selective attention 250
self-organization 175, 207
sensorimotor association 254, 256, 272
separability 77
serial excitation 8
serial inhibition 8
shared excitation see common excitation
shared input see common input
shared modulation 54
shift predictor 7, 55, 62, 108, 109, 259, 271
short term memory 250, 252
signal content 130
signal importance 130
simple cell 5
slow-wave field potentials 189
small field stimulation 101
somatosensory areas 92
somatosensory cortex 34, 39, 48, 73, 76
spatial filter 68, 79, 80, 100
spatiotemporal pattern 30, 33, 123, 126, 130
spectro-temporal 82
spike size 113
spike-triggered averaging 21, 41
steady state 75, 79, 98, 100
stimulus coding 64
stimulus discrimination 95, 100
stimulus dynamics 100
stimulus recognition 123–129
stimulus shape 93, 100
stimulus speed 93, 98
stimulus-induced oscillation 191, 192, 202
stimulus-induced synchronization 187, 197, 204, 210, 216, 218
stimulus-locked response 193
strength index 142
striate cortex see visual cortex
subfield stimulation 77
superposition problem 167
surround antagonism 161
surround suppression 152
synaptic efficacy 260, 261, 266, 268
synaptic latency 138
synaptic plasticity 170
synchronization 101, 177, 184
synchronized activity 214
synchronized oscillations 198, 210, 212, 217
synchronous gain 263, 264, 268
synfire 270, 276

temporal averaging 79, 82, 99
temporal discrimination behaviour 287

temporal encoding 88, 98, 100, 130, 185 ff., 212
temporal filter 79, 100
threshold distance 265
time windows 63
topographic representation 92
transcolumnar interaction 18
transient grouping 198
transswitching 292
tuning curve 64, 86, 93
two-channel recording 6

variability 115–119, 120–123
visual experience 176
visual latencies 89
visual cortex 6 ff., 33–35, 40, 43, 45, 47, 53 ff., 70 ff., 96 ff., 106 ff., 133 ff., 166 ff., 185 ff., 281 ff.
visual segmentation 188

wavelet activity 194
weak correlations 190
wetware 89
white noise 82
w-type 70, 81

x-y-w-input 89

y-type 70